Lecture Notes in Mathematics

Edited by A. Dold, Heidelberg and B. Eckmann, Zürich

369

Victoria Symposium on Nonstandard Analysis

University of Victoria 1972

Edited by
Albert Hurd, University of Victoria, Victoria/Canada
Peter Loeb, University of Illinois, Urbana, Il/USA

Springer-Verlag
Berlin · Heidelberg · New York 1974

AMS Subject Classifications (1970): 02H25, 26-02, 26A98

ISBN 3-540-06656-X Springer-Verlag Berlin · Heidelberg · New York
ISBN 0-387-06656-X Springer-Verlag New York · Heidelberg · Berlin

Offsetdruck: Julius Beltz, Hemsbach/Bergstr.

FOREWORD

This volume is a record of the Symposium on Nonstandard Analysis held at the University of Victoria in Victoria, Canada during the period May 8-11, 1972. The symposium followed in spirit the previous symposia in nonstandard analysis held at the California Institute of Technology (1968), and Oberwolfach, Germany (1970). The principal invited speakers were Abraham Robinson, W.A.J. Luxemburg, H. Jerome Keisler, and Elias Zakon. It was our intention, hopefully almost completely realized, to have everyone working in nonstandard analysis in North America attend the conference.

Most of the papers included here are based on lectures presented at the symposium, but in several instances we have included papers submitted in addition to those read at the conference. The following is a list of the papers which were presented at the conference.

Andrew Alder	Model theoretic ideas in topology
Michael Behrens	Analytic and meromorphic functions in the open unit disk
Steven F. Bellenot	A nonstandard theory of topological vector spaces
Harry Gonshor	Enlargements of Boolean algebras
C. Ward Henson	The nonstandard hulls of a uniform space
Joram Hirschfeld	Existentially complete models for arithemtic
Albert E. Hurd	Nonstandard dynamical systems
H. Jerome Keisler	Freshman calculus and measureable cardinals
Peter J. Keleman	Applications of nonstandard analysis to Quantum Mechanics and Quantum Field Theory
Anders Kock	Elementary Topoi Nonstandard extensions and the theory of Topoi
Lawrence D. Kugler	Weak almost periodicity
Peter A. Loeb	Some results in nonstandard measure theory
W.A.J. Luxemburg	On a theorem of Helly and a theorem about liftings
L. J. Moore	The nonstandard theory of topological vector spaces

Louis Narens — Homeomorphisms of generalized metric spaces

Rohit Rarikh — Conditional probability can be defined for all pairs of sets of reals

Robert Phillips — Omitting types in arithmetic

David Pincus — The power of the Hahn-Banach theorem

Abraham Robinson — Nonstandard exchange economies

Keith D. Stroyan — Infinitesimal relations on the space of bounded holomorphic functions

Janes K. Thurber — Applications of fractional powers of delta functions

Frank Wattenberg — Two different topologies with the same monads

Elias Zakon — A new variant of nonstandard analysis

Metrization and completeness of the hyperreals

In cases where the work presented at the conference is being published elsewhere the abstract included in the table of contents is the only record of the symposium lecture.

Also participating in the symposium were Gary L. Bender (U. of Colorado), D. S. Carter (Oregon State U.), David Cozart (Duke U.), Martin Davis (New York U.), Don Easton (Brandon U.), Bill Glassmire (Oregon State U.), Melvin Hausner (New York U.), James P. Jones (U. of Calgary), Steve Kloster (Simon Fraser U.), Susan Lenker (U. of Montana), George McRae (U. of Montana), Stephan Sperling (U.C.L.A.), Arthur L. Stone (Simon Fraser U.), Doug Super (Simon Fraser U.).

Almost all of the papers included here were typed at the institutions of the respective authors, but we have proofread the manuscripts and have had them refereed.

ACKNOWLEDGMENTS

The symposium was co-organized by one of us (A.E.H.) together with Professor C. Robert Miers. It would not have been possible without the enthusiastic support of Professor Phoebe Noble and the generous financial aid provided by the University of Victoria. That the conference flowed as smoothly as it did is a tribute to the staff of the Department of Mathematics and especially to Mrs. Ruth McRae. Helping us with the refereeing were David Berg, Earl Berkson, John Gray, Lester Helms, Carl Jockusch, Thomas McLaughlin, Lee Rubel, Gaisi Takeuti. To everyone mentioned above our sincere thanks.

Albert E. Hurd
Peter Loeb

October 1, 1973

CONTENTS AND ABSTRACTS

Date following abstract is date of receipt.

Every topological Boolean algebra is isomorphic to a topological subalgebra of the ordered space on a non-standard model of the rationals. So in particular a countable topological boolean algebra is isomorphic to a topological subalgebra of the ordered space on an η_1 set. An atomic countable T_3 dense-in-itself algebra is isomorphic to a topological subalgebra of the space of rationals. (January 10, 1973)

Nonstandard techniques are used to investigate the maximal ideal space $\mathfrak{M}(D)$ of the Banach algebra $H^\infty(D)$ of bounded analytic functions on a planar domain D, and, more especially, to study the analytic structure in $\mathfrak{M}(D) - D$. A relatively complete discussion is presented for the unit disk, and a few results for infinitely connected domains are discussed. (August 1, 1973)

Several classical boundary value theorems including the Gross principal value theorem are proved using nonstandard methods. (August 20, 1973)

An inverse function theorem is proved under a local assumption which is weaker than differentiability in a neighborhood. (August 20, 1973)

A simple short nonstandard proof of the uniqueness of finite dimensional Hausdorff topological vector spaces. (October 17, 1972)

This paper applies the nonstandard partition method of measure theory to the problem of integrating unbounded functions in a linear fashion, i.e., without truncation. As an application, one obtains a projection T of the extension $^*L_1(X,\mathfrak{M},\mu)$ of an arbitrary L_1 space onto a * finite dimensional subset of $^*L_1(X,\mathfrak{M},\mu)$ so that if $h \in L_1(X,\mathfrak{M},\mu)$ then $||T^*h-^*h|| \simeq 0$. (March 5, 1973)

In this paper we characterize certain topological properties whose definitions depend on a particular infinite cardinal by using ultrapowers over sets of that cardinality. In particular we obtain a nonstandard characterization of α-Baire spaces in terms of certain α-indexed ultrapowers. As a corollary, Baire spaces have a nonstandard characterization valid in any non-trivial countably-indexed ultrapower. This characterization may be used to give nonstandard proofs of results which depend on the Baire category theorem. This provides, at least in part, a solution to problem number 9 asked by Abraham Robinson in his retiring presidential address delivered to the Association for Symbolic Logic, January 1973 in Dallas, Texas. The paper also uses countably-indexed ultrapowers to examine certain countable equivalence conditions between topological spaces. (April 20, 1973)

An exchange economy consists of a set of traders each of whom is characterized by an initial endowment and a preference relation. In addition, one usually assumes that the set of traders is finite. But in order to model perfectly competitive markets, i.e., markets where each trader's economic influence is negligible, we assume that the economy has ω traders, where ω is an infinite integer, and the average endowment of each trader is infinitesimal.

In these nonstandard exchange economies, we examine the relationship between outcomes obtained through bargaining, called core allocations, and the allocations arising out of the competitive price system. We show that Edgeworth's conjecture, that every core allocation is a competitive allocation, is true in nonstandard exchange economies. As a consequence of this theorem we also show that core allocations in large finite standard economies are approximately competitive allocations.

References: Brown, D. J. and A. Robinson, "Nonstandard Exchange Economies", Econometrica (to appear).

———, "A Limit Theorem on the Cores of Large Standard Exchange Economies," Proc.Nat.Acad.Sc., U.S.A., Vol.69, No. 5, 1258-1260.

Harry Gonshor

In this paper we show that various types of completions in different senses may be obtained as subquotients of enlargements. Among the examples considered are the following: The rationals as a "completion" of the integers, the Stone-Cech compactification of a completely regular space, the second conjugate space of a Banach space, rings of quotients of rings of continuous functions, the projective cover of a compact Hausdorff space, and the completions of a Boolean algebra. The last example is studied in detail. (March 22, 1973)

C. Ward Henson and L. C. Moore, Jr.

Let E,F be vector spaces over R or C paired by a nonsingular bilinear form $\langle \ldots,\ldots \rangle$ and let θ be a locally convex topology on E which is admissible for the pairing. Given an enlargement $^*\mathfrak{M}$ of a set theoretical structure $\mathfrak{M}$ which contains E and F, let $(\hat{E},\hat{\theta})$ be the associated nonstandard hull of (E,θ). Also let $\hat{F}$ be the space of linear functionals on $\hat{E}$ which are represented by those points q in $^*F$ for which $\langle p,q \rangle$ is finite for all θ-finite p in $^*E$. Then $\hat{F}$ is contained in the dual space $\hat{E}'$ of $(\hat{E},\hat{\theta})$ and $\hat{\theta}$ is an admissible topology relative to the pairing between $\hat{E}$ and $\hat{F}$.

The principal result is that $(\hat{E},\hat{\theta})$ is semi-reflexive if and only if $\hat{F}$ is $\beta(\hat{E}',\hat{E})$-dense in $\hat{E}'$. (This extends a result for normed spaces proved earlier by the authors.) Moreover, a geometric condition on (E,θ) is given which is equivalent to the semi-reflexivity of $(\hat{E},\hat{\theta})$ (and therefore this property of $(\hat{E},\hat{\theta})$ does not depend on the particular enlargement $^*\mathfrak{M}$). The main technical tool is the following result which seems of interest itself.

<u>Retraction Theorem</u>: If $^*\mathfrak{M}$ is κ-saturated and S is a subspace of $\hat{E}$ which has Hamel dimension less than κ, then for each $\phi \in \hat{E}'$ there exists $\psi \in \hat{F}$ such that $\phi(x) = \psi(x)$ for all $x \in S$. (March 8, 1973)

<u>C. Ward Henson and L. C. Moore, Jr.</u>

Invariance of the nonstandard hulls of a uniform space

Let $(X,\mathfrak{U})$ be a uniform space and let $^*\mathfrak{M}$ be an enlargement of some set-theoretical structure $\mathfrak{M}$ which contain X. It is shown that if $p \in {}^*X$ is not $\mathfrak{U}$-pre-nearstandard and if $^*\mathfrak{M}$ is κ-saturated, then the filter monad $\mu(\mathrm{Fil}(p)) = \cap\{{}^*Y \mid p \in {}^*Y\}$ intersects at least κ distinct $\mathfrak{U}$-monads $\mu_{\mathfrak{U}}(q)$. As a consequence, either (i) the nonstandard hulls of $(X,\mathfrak{U})$ (constructed using the $\mathfrak{U}$-finite elements of $^*X$) are all equal to the completion of $(X,\mathfrak{U})$ and are therefore independent of $^*\mathfrak{M}$ or (ii) the nonstandard hulls of $(X,\mathfrak{U})$ can be made arbitrarily large by varying the choice of $^*\mathfrak{M}$. Similar results hold for the classes of nonstandard hulls defined by Luxemburg. In case (i) holds, we say $(X,\mathfrak{U})$ has <u>invariant nonstandard hulls</u> (extending the terminology introduced by the authors for locally convex topological vector spaces).

A metric space (X,d) has invariant nonstandard hulls if and only if every subset of X on which every uniformly continuous function is bounded must be totally bounded. This equivalence is not true, however, for arbitrary uniform spaces. It is also shown that if $(X,\mathcal{T})$ is a separable, topologically complete (metrizable) space, then $\mathcal{T}$ can be defined by a metric d such that (X,d) is complete and has invariant nonstandard hulls.
(March 8, 1973)

S. Feferman, D. Scott, and S. Tennenbaum proved that every non-trivial homomorphic image of the semi-ring R of recursive functions fails to be a model of arithemtic. The aim of this paper is to show that every countable model of full arithmetic can be embedded in such a homormorphic image. To prove this we modify a theorem by H. Friedman to obtain a sufficient condition for a model of arithmetic to be embeddable in a model of a fragment of arithmetic. We then introduce recursive ultrapowers - homomorphic images of R which are models of that fragment. Finally, given a model, we show how to construct a recursive ultrapower which satisfies the condition of Friedman's theorem. (April 3, 1973.)

If $D = (X,\pi,R)$ is a dynamical system then in an enlargement $*D = (*X,*\pi,*R)$ we see that the flow can be followed for infinite time. This leads to simple characterizations of such standard notions as limit set, stability, prolongations, etc. as well as some notions in the enlargement which have no standard counterpart. At the same time X has been blown up to $*X$ which is a sort of compactification. This leads to effective ways of studying flows on non-compact phase spaces.

Using these ideas we can give necessary and sufficient conditions for topological equivalence of compact metric dynamical systems (Amer. J. Math. 93 (1971), 742-52). Results on almost periodicity have also been obtained.

A nonstandard construction is given for the Bohr compactification of a class of uniformly closed, translation invariant subalgebras of the almost periodic functions on a group G which includes the almost periodic functions themselves. The compactification is obtained by putting an appropriate "weak" topology on the quotient group obtained from *G by factoring out the subgroup of common near periods of all the functions in the subalgebra. A crucial lemma shows that intersections of finitely many sets of translation elements are relatively dense. (June 20, 1973)

H. Jerome Keisler

Let F be an ordered field. F is said to be monotone complete if every bounded increasing function $h: F \to F$ has a limit $\lim_{x\to\infty} h(x)$ in F . The cofinality of F is the least cardinal of an unbounded subset of F . The field R of real numbers is the unique monotone complete field of cofinality ω. In nonstandard analysis one uses a proper elementary extension *R of R . *R can never be complete; in this paper we show that *R can be taken to be monotone complete. (Jan. 16, 1973)

Peter J. Kelemen

Applications of nonstandard analysis to quantum physics are considered. Quantum field theory, the theory of systems with countably infinite degrees of freedom is constructed as an external intermediary between quantum mechanics of finite degrees of freedom and its nonstandard extension, hyper-quantum mechanics of star-finite degrees of freedom. The possibility that hyper-quantum mechanics, perhaps modified with external assumptions will provide the foundations for a theory of elementary particles is also discussed. (January 11, 1973)

Anders Kock and Chr. Juul Mikkelsen

Topos-theoretic factorization of nonstandard extensions 122

An organization in terms of topos theory and closed category theory of the change in the higher order logic of nonstandard arguments taking place by means of the word 'internal'. Formally, we prove a factorization theorem for first-order logic preserving functors between toposes. Such a functor can be factored into two such functors, where the first preserves higher order logic and the second preserves elements. (January 23, 1973)

Peter A. Loeb

A nonstandard representation of Borel measures and σ-finite measures 144

It is shown that any extension of the real line contains a linearly ordered *finite collection of compact sets on which every finite Borel measure is essentially concentrated. A nonstandard proof of Lusin's theorem is given, and a method of transforming σ-finite measures into counting measures is established using partitions of the measure space. (March 5, 1973)

Peter A. Loeb

A note on continuity for Robinson's predistributions 153

It is shown that a predistribution which is standard at each standard point has a continuous standard part. (March 5, 1973)

W.A.J. Luxemburg

On a theorem of Helly and a theorem about liftings

Helly's theorem in the theory of normed spaces concerning the existence of a solution of a system of linear equations is essentially a statement of the type that a certain binary relation is concurrent. This fact can be used to show that the second dual space of a given normed linear space can be imbedded in the nonstandard hull of the given normed space. From this fact the various theorems concerning

reflexivity of normed spaces follow. In particular, a very simple and direct nonstandard proof of the theorem that every uniformly convex Banach space is reflexive, exists.

The existence of liftings is usually obtained from the existence of a lower density in the sense of Caratheodory. Extending a result of B. Eifrig (Ein nicht-standard Beweis für die Existenz eines starken Liftings, Archiv der Math. 23 (1972), 425-427) it is shown by using nonstandard methods that given an algebra of sets and a Boolean equivalence relation which admits a lower density, then it admits already a lifting.

A generalized metric space X is a space whose topology is "metrizable" by a "distance" function into an ordered abelian group $(G,<)$. $(G,<)$ is said to be of type Ω if and only if Ω is the smallest cardinality of a subset Y of positive elements of G such that for each positive element of g of G there is a $y \in Y$ such that $y \leq g$. X is said to be of type Ω if its topology is metrizable by an ordered abelian group $(G,<)$ of type Ω . It is shown that if X is of type Ω and the cardinality of X is at most Ω , then there is a nonstandard model of the rational numbers Q with the order topology such that X is homeomorphic to a subset of Q . Furthermore, if the generalized continuum hypothesis is assumed, it is shown that for each regular cardinal Ω there is a nonstandard model of the rational numbers Q of cardinality Ω with the order topology such that each generalized metric space X of type Ω that has cardinality at most Ω is homeomorphic to a subset of Q . (April 19, 1973)

It is shown that if X and Y are generalized metric spaces of type Ω , X and Y have cardinality Ω , and X and Y have no isolated points, then X and Y are homeomorphic. (April 19, 1973)

Rohit Parikh & Milton Parnes

We show how to give a nonstandard construction for conditional probability functions on the real line which satisfy strong regularity conditions (these are not unique). Applications are given to define the notion of uniformly distributed set, and to the theory of determinate pairs of sets A,B where the conditional probability does not depend on the particular probability function chosen.(April 16, 1973)

R. G. Phillips

In this paper we show that each model M of Peano's Axioms has a proper elementary extension, called a conservative extension, which defines no new relations on M. We show that conservative extensions have a canonical additive group which is minimal in structure. Finally we compare conservative extensions with Gaifman's minimal extensions. (February 10, 1973)

David Pincus

The Hahn-Banach theorem is proved to be strictly weaker than the prime ideal theorem for Boolean algebras. A "standard" proof of this for set theory with atoms appears in Bull. A.M.S. 78 (1972), 766-770. A proof for full Zermelo Fraenkel set theory uses nonstandard analysis. There are also some results on statements which are weaker than the Hahn-Banach theorem. (April 20, 1973)

Abraham Robinson

Enlarged sheaves are considered, particularly for sheaves of holomorphic functions of several complex variables. A monadic section is the restriction of a section to a monad in its domain. Let g_{ji}, $i = 1,\dots,m$, $j = 1,\dots,k$ be standard monadic sections at a standard point z, and let g'_{ji} be the corresponding germs at z. Let f_j, $j = 1,\dots,m$ be internal monadic sections with germs f'_j at z.

Theorem. If $(f'_1 \ldots, f'_m)$ belongs to the module generated by $(g'_{j1}, \ldots, g'_{jm})$, $j = 1, \ldots, k$ then there exist monadic sections q_j at z, $j = 1, \ldots, k$ such that $(f_1, \ldots, f_m) = \sum_j q_j (g_{j1}, \ldots, g_{jm})$ on the entire monad of z.

Corollary. If the f_j are finite on the monad, the q_j also may be chosen so as to be finite on the monad. (December 1, 1972)

K. D. Stroyan

A nonstandard characterization of mixed topologies

In this paper we generalize the characterization of the strict and Mackey topologies (on H^∞ and L^∞ respectively) to the general setting of mixed topologies. (December 3, 1973)

K. D. Stroyan

Infinitesimal relations on the space of bounded holomorphic functions

Let U denote the unit disk. On $*(\hat{H}^\infty(U))$ there are several interesting notions of infinitesimals, for example, pointwise, compact or uniform convergence can be characterized by appropriate infinitesimal relations.

Two other interesting infinitesimal relations which correspond in standard terms to mixed topologies are as follows: h and k in $*(H^\infty(U))$ are within a bounded infinitesimal provided $h(z) \sim k(z)$ for all $z \in *U$ and $h(z) \approx k(z)$ for near standard z; h and k are within a measure infinitesimal provided $h(z) \sim k(z)$ for all z and $h(z) \approx k(z)$ (as radial limits) for nearly all $|z| = 1$ (that is, except on a set of infinitesimal $d\theta$-measure). These correspond to bounded pointwise convergence and bounded L^1-convergence.

James K. Thurber and Jose Katz

Applications of fractional powers of delta functions

In calculations in electromagnetic theory, quantum scattering theory and perturbation expansions in quantum field theory, it is often desirable for such expressions as

$$\Delta\, \delta(x)\delta(x) \text{ and}$$

$$\int_{-\infty}^{\infty} \delta(x)\delta(x)\theta(x)\theta(-x)dx$$

to have a meaning, where $\delta(x)$ is the Dirac delta function and

$$\theta(x) = \begin{cases} 0 & \text{for } x < 0 \\ 1 & \text{for } x > 0 \end{cases}$$

Also sometimes it is important to distinguish between different delta functions such as

$$\lim_{\varepsilon \to +\infty} \frac{\varepsilon}{\pi(x^2+\varepsilon^2)} \quad \text{and}$$

$$\lim_{n \to +\infty} \left[\frac{n}{\pi}\right]^{1/2} e^{-nx^2}$$

And as we show in this paper it is useful to have fractional powers of the Dirac function, such as

$$\left[\delta(x)\right]^{1/2} \quad \text{and} \quad \left[\delta(x)\right]^{5/6} .$$

If the delta function is conceived solely in terms of the L. Schwarz theory of distributions, all of the above notions lead to serious difficulties and in some instances to unresolvable inconsistencies.

However A. Robinson's theory of Nonstandard Analysis can be used to make perfectly rigorous and useful sense of the above notions. We apply here nonstandard versions of delta functions to some basic calculations in classical electro-magnetic theory, quantum scattering theory, and the star-finite infinitely many body Schrodinger Equation.

In classical electromagnetic theory we are able to give a model of a point electron with a finite self-energy which is consistent with causality requirements. This leads to an interaction between charged particles of infinitesimal strength. In the infinite-star finite body problem this leads however to a finite effective interaction.

The fractional powers of delta functions as used here can be interpreted as a form of renormalization. (December 10, 1972)

Suppose X is a σ-compact, locally compact space and consider the space $F_C(X)$ of continuous functions $X \to R$ with compact support. Two of the most interesting topologies on this space are the direct limit (or inductive) topology and the δ-topology. These two distinct topologies have remarkably similar properties. This paper investigates these two topologies using the techniques of Nonstandard Analysis. The main result of this paper is that if D is an ultrafilter on ω which is a P-point and $^*F_C(X)$ is the ultrapower of $F_C(X)$ with respect to D then the direct limit and δ-topologies have the same monads. Thus, from the point of view of Nonstandard Analysis it is clear why the two topologies share so many properties. (January 25, 1973)

In this paper, we generalize the concept of a monomorphism of one superstructure into another, as previously introduced by A. Robinson and E. Zakon ("A set theoretical characterization of enlargements, [8]"). As a novelty, we consider "non-strict" monomorphisms under which an internal set may possess external elements. This generalization of the ordinary "strict" monomorphisms (including Robinson's "enlargements" as a special case), admits a particularly simple definition. One of our aims is to demonstrate that "strictness" is not needed for the validity of many important theorems. We give a self-contained presentation of all basic notions of Nonstandard Analysis as they evolve from the newly defined "monomorphisms". (February 10, 1973)

AN APPLICATION OF ELEMENTARY MODEL THEORY TO TOPOLOGICAL BOOLEAN ALGEBRAS

Andrew Adler

University of British Columbia

1. Introduction: A topological boolean algebra (t.b.a.) is a boolean algebra equipped with a unary closure operator C satisfying the usual Kuratowski axioms $C(a \cup b) = Ca \cup Cb$, $a \leq Ca$, $CCa = Ca$, and $C0 = 0$.

Topological boolean algebras provide a useful first-order approximation to the (second-order) notion of topological space. For let T be a topological space, and let B be the algebra of all subsets of T. If $a \in B$, let Ca be the closure of the set a in T. Then under the operation C, B is a t.b.a.. Of course there are t.b.a's that are very far from being topological spaces. However, the main result of Section 2 shows that every t.b.a. is a topological subalgebra of a fairly well-behaved topological space.

By suitably strengthening the Kuratowski axioms, one can get a notion that captures many more of the elementary properties of topological spaces. The appropriate axiom scheme to add says essentially that unions of definable collections exist (more or less the usual separation axioms of set theory). In this strengthened version of the Kuratowski axioms, we have been able to define many of the concepts of elementary topology (connectedness, component, inductive dimension, many others) and show that the usual elementary results hold.

In section 2, we prove a representation theorem for t.b.a.'s. In section 3, we examine in more detail those atomic t.b.a.'s that are T_3 dense-in-themselves (and in particular T_3 dense-in-themselves topological spaces).

2. Representation of Topological Boolean Algebras: Let B_Q be the t.b.a. whose underlying algebra is the algebra of all subsets of Q (the rationals) and whose

closure operator is the one induced by the usual topology on Q. In [1] McKinsey and Tarski have shown that every finite t.b.a. is isomorphic to a topological subalgebra of B_Q. We use ultraproducts to extend this representation theorem to arbitrary t.b.a.'s.

Let I be an index set, and let D be an ultrafilter on I. Let $B = B_Q{}^I/_D$. We list without proof some properties of B. These properties all follow in a straight forward way from the definition of ultrapower.

(i) B is an atomic boolean algebra. The atoms of B are the objects $^f/_D$ where f(i) is an atom (one point subset of Q) for all i. So the collection of atoms may be identified with $Q^I/_D$.

(ii) Two objects of B are equal iff they have the same atoms.

(iii) The ordinary order on Q induces a (total dense) order on the atoms of B. For any atoms p, q of B, the interval (p,q) is open in B(that is, the complement of (p,q) is a fixed point of C).

(iv) For any open object x of B, and any atom p of x, there is an open interval containing p and contained in x.

A subset X of $Q^I/_D$ is a <u>B-set</u> if there is an $x \in B$ such that X is the set of atoms of x. By (ii) there is a natural isomorphism between B (considered as a boolean algebra) and the algebra of B-sets of $Q^I/_D$. Call X open in this last algebra if the corresponding object x is open in B. Now look at the algebra of all subsets of $Q^I/_D$. If $Y \subseteq Q^I/_D$, declare Y to be open if Y is a union of open sets of B. We have defined a topology on $Q^I/_D$. From (iii) and (iv) it follows that this topology is the order topology on $Q^I/_D$. Moreover, the t.b.a. B is (isomorphic to) a topological subalgebra of the topological space $Q^I/_D$. We are now ready to prove the representation theorem.

THEOREM: Let κ be an infinite cardinal. Then there is a set I_κ (of cardinality κ) and an ultrafilter D_κ on I_κ such that any t.b.a. of cardinality κ is isomorphic to a topological subalgebra of the ordered space $Q^{I_\kappa}/_{D_\kappa}$.

Proof: Let I_κ be the set of finite subsets of κ. If $s \in I_\kappa$, let $E_s = \{t \in I_\kappa : s \subseteq t\}$. Extend the collection of sets E_s to an ultrafilter D_κ on I_κ. We show that for this choice for I_κ, D_κ, Q^{I_κ}/D_κ does the job.

For let A be a t.b.a. of cardinality κ. Then I_κ can be identified with the collection of finite subsets of A. For any $s \in I_\kappa$, define the t.b.a. A_s in the following way :

(a) As a boolean algebra, A_s is the boolean subalgebra of A generated by the elements of s.

(b) An element x of A_s is closed in A_s iff it is closed in A.

A_s is certainly finite. It is a subalgebra, but not necessarily a topological subalgebra, of A. Now by the representation theorem of McKinsey-Tarski, there is an isomorphism $\phi_s : A_s \to B_Q$. Let $\phi : A \to B_Q{}^I/D$ be defined by $\phi(x) = {}^f/_D$, where $f(s) = \phi_s(x)$ if $x \in A_s$ and (say) $f(s) = 1$ otherwise. The map ϕ is an isomorphism. We show for instance that always $\phi(Cx) = C\phi(x)$. By our choice for D_κ, for any x, $\{s : x \in s \text{ and } Cx \in s\} \in D_\kappa$. So almost everywhere $\phi_s(Cx) = C\phi_s(x)$, and the result follows. Hence A is isomorphic to a topological subalgebra of $B_Q{}^{I_\kappa}/D_\kappa$. But $B_Q{}^{I_\kappa}/D_\kappa$ is isomorphic to a topological subalgebra of the ordered space Q^{I_κ}/D_κ. This completes the proof.

So for instance every countable t.b.a. is isomorphic to a topological subalgebra of the ordered space on an η_1-set. In the next section we prove a stronger result for a much narrower class of t.b.a.'s.

3. Elementary properties of well-behaved t.b.a.'s: Let L be a first-order language for elementary topological boolean algebras. It is easy to write down definitions in L for 'atomic', for 'T_3' and for 'dense-in-itself'. Let B be an atomic T_3 dense-in-itself countable t.b.a.. B is isomorphic to a topological set algebra over the atoms of B. Just as in Section 2, by taking arbitrary unions of open objects of B, we obtain in a canonical way from B a topological space E. B is then isomorphic to a large topological subalgebra of E. Since the open

objects of B are a base for the topology of E, one can call B a <u>generating</u> topological subalgebra of the space E.

The space E is also T_3 and dense-in-itself. We show for instance that if $p \varepsilon E$ and K is a closed subset of E, where $p \notin K$, p and K may be separated in E by open sets. Since K is closed, it is an intersection of closed objects of B. Since $p \notin K$, there is a closed object Y of B such that $K \leq Y$ and $p \notin Y$. But since B is T_3, p and Y may be separated in B (and hence in E) by open sets.

Moreover, since B is a countable algebra, and so has an at most countable number of open sets, the topology of E has a countable base. Hence E is metrizable.

But E is then a countable dense in itself metrizable space. So by a result of Sierpinski ([2] p. 142) E is (homeomorphic to) the space Q. So we have proved:

THEOREM: Every countable T_3 dense-in-itself t.b.a. is isomorphic to an atomic generating topological subalgebra of the space Q.

Conversely, any atomic generating topological subalgebra of B_Q is T_3 dense-in-itself.

By the Lowenheim-Skolem theorem, every atomic T_3 dense-in-itself t.b.a. has a countable elementary (and hence atomic T_3 dense-in-itself) substructure. So we have:

THEOREM: Every atomic T_3 dense-in-itself t.b.a. is L-elementarily equivalent to a generating topological subalgebra of the space Q.

REFERENCES

1. McKinsey, J.C.C. and Tarski, A. : The algebra of topology, Annals of Math. 45 (1944) pp. 141-191.

2. Sierpinski, W. : General Topology, University of Toronto Press, Toronto 1952.

ANALYTIC SETS IN $\mathfrak{M}(D)$

Michael Behrens

Laguna Beach, California

1. Introduction. We would like to consider an application of non-standard analysis to the theory of bounded analytic functions. The classical theory of boundary values of bounded analytic functions on a planar domain D can be studied in a modern context by considering the maximal ideal space $\mathfrak{M}(D)$ of the Banach algebra $H^{\infty}(D)$ of all bounded analytic functions defined on D, endowed with the supremum norm. The space D is considered to be a subset of $\mathfrak{M}(D)$ by identifying $z \in D$ with "evaluation at z". Each function $f \in H^{\infty}(D)$ has an extension to $\mathfrak{M}(D)$ defined by $\hat{f}(m) = m(f)$, $m \in \mathfrak{M}(D)$, and the space $\mathfrak{M}(D)$ is a compact Hausdorff space when given the weakest topology making these functions continuous. Whether $\mathfrak{M}(D)$ is in fact a compactification of D is known as the Corona conjecture, and has been shown to be true for the open unit disk Δ and a few classes of infinitely connected domains [1,2]. Any theorem describing $\mathfrak{M}(D)$ is, of course, a theorem about the boundary behaviour of bounded analytic functions defined in D. Since each function $f \in H^{\infty}(D)$ is analytic on D it seems natural to ask whether there might be subsets of $\mathfrak{M}(D) - D$ on which each $\hat{f}$ is in some sense analytic. To be more precise, an analytic disk in $\mathfrak{M}(D)$ is defined to be the image of the open unit disk Δ under an injective mapping $T : \Delta \to \mathfrak{M}(D)$ such that $\hat{f} \circ T \in H^{\infty}(\Delta)$ for each $f \in H^{\infty}(D)$. An analytic set in $\mathfrak{M}(D)$ is a connected subset of $\mathfrak{M}(D)$ which is a union of analytic disks. The problem, then, is to describe the analytic subsets of $\mathfrak{M}(D) - D$, and to determine how their structure arises from the structure of D.

Central to this problem is the notion of a Gleason part. A metric is defined on $\mathfrak{M}(D)$ by $\rho(m,n) = \sup\{|m(f)| : f \in H^{\infty}(D), ||f|| \leq 1, n(f) = 0\}$. The relation $\rho(m,n) < 1$, is an equivalence relation (this is non-trivial) and the equivalence classes are known as the Gleason parts of $H^{\infty}(D)$. The Gleason part containing m will be denoted by $\mathfrak{G}(m)$. It follows immediately from Schwarz's lemma that any analytic set containing m is contained in $\mathfrak{G}(m)$. Gleason's original question, in this context, asks whether each non-trivial Gleason part of $\mathfrak{M}(D)$ is an analytic set.

In general very little is known about either the analytic sets or the Gleason parts of $\mathfrak{M}(D)$. For the open unit disk, the problem was completely solved by Hoffman [4], who showed that there are 2^{C} non-trivial Gleason parts in $\mathfrak{M}(D)$, that each of these is an analytic disk, and that the existence and placement of these disks is intimately connected with the structure of "interpolating" sequences in D . The problem for finitely connected domains is easily seen to reduce to the problem in the disk. For a few classes of infinitely connected domains, a reasonably complete description of the analytic sets is known, but even in these restricted cases the analytic structure is exceedingly complicated [1,2].

What can nonstandard analysis contribute to these problems? The immediate difficulty in studying $\mathfrak{M}(D)$, and even $\mathfrak{M}(\Delta)$, is that, even though the analytic structure of D may be quite simple, we have no "intuitive description" of $\mathfrak{M}(D)$. How do we "draw a picture" of $\mathfrak{M}(D)$ in such a way that it helps us discover and prove theorems.

In a Non-Standard model of analysis each object defined on the complex numbers C will have a nonstandard extension, which is usually denoted by the addition of an asterick e.g. C^* . A function $f : D \to C$ becomes a function $f^* : D^* \to C^*$ which agrees with f on D . The point is that as long as we use only words and concepts that are within the traditional mathematical background, then D and D^* , f and f^* , etc. cannot be distinguished [5]. Any intuitive understanding of D is equally valid for D^* . It is only when nonstandard concepts are introduced that a difference appears. The problem is to develop a fruitful interplay between new nonstandard concepts, a transferred intuitive understanding of D^* , and standard theorems on $\mathfrak{M}(D)$ which are to be proved or discovered.

2. Generalities. Let $f \in H^\infty(D)$. Then $||f||$ is a standard complex number and $|f(z)| \le ||f||$ for all $z \in D$. Since this involves only standard concepts it is also true that $|f^*(z)| \le ||f||$ for all $z \in D^*$, i.e. $||f^*||^* = ||f||$. Thus for each $z \in D^*$ and $f \in H^\infty(D)$, $f^*(z)$ is a finite element of C^* and has a standard part $°(f^*(z))$. It is easy to see that for each $z \in D^*$, and $f \in H^\infty(D)$, the mapping $f \to °(f^*(z))$ is a non-trivial homomorphism of $H^\infty(D)$ into C, i.e., defines an element of $\mathfrak{M}(D)$. We can thus define a mapping $F : D^* \to \mathfrak{M}(D)$ by $F(z)(f) = °(f^*(z))$. From now on we will follow the usual convention and write $f(z)$ for $f^*(z)$ even though $z \in D^*$.

Let S be a subset of D and let $\bar{S}$ be the closure of S in $\mathfrak{M}(D)$. Then $m \in \bar{S}$ iff for each $f_1, \ldots f_n \in H^\infty(D)$ and $\epsilon > 0$, there is a $z \in S$ such that $|f_i(z) - m(f_i)| < \epsilon$ for $i = 1, \ldots, n$. Define a relation R_m by $R_m((f,\epsilon),z)$ holds iff $z \in S$, f $H^\infty(D)$, $\epsilon > 0$ and $|f(z) - m(f)| < \epsilon$. Then $m \in \bar{S}$ iff R_m is concurrent. But this holds in the nonstandard model iff there is a $z \in S^*$ such that $R_m((f,t),z)$ holds for all standard f,ϵ, i.e., iff there is a $z \in S^*$ such that $f(z)$ is infinitesimally close to $m(f)$ for all $f \in H^\infty(D)$.

We thus have

1. For each subset $S \subset D$, $\bar{S} = F(s)$. Each sequence $\{\alpha_n : n \in N\} \subset D$ has an extension to a sequence $\{\alpha_n : n \in N^*\}$ in D^*. In this case 1. becomes

2. For each sequence $\{\alpha_n : n \in N\} \subset D$; $\overline{\{\alpha_n : n \in N\}} = \{F(\alpha_n) : n \in N^*\}$.

3. F is surjective iff the corona theorem holds for D. In particular $\mathfrak{M}(\Delta) = F(\Delta)$.

3. The case $D = \Delta$. We will now consider the open unit disk. In classical analysis two metrics, the pseudo-hyperbolic and the hyperbolic, are defined on Δ by $\rho(z,w) = \left|\frac{z - w}{1 - \bar{z}w}\right|$ and $\psi = \log\left(\frac{1 + p}{1 - p}\right)$. The connection between these metrics and the theory of bounded analytic functions is contained in a theorem of Pick (the invariant form of Schwarz's lemma) which says that an analytic mapping of the open disk into itself decreases distances in these metrics, or is a Mobius transformation, in which case it preserves distances in these metrics.

For any $z, w \in \Delta^*$, $\psi^*(z,w)$ is a nonstandard real number and hence may be infinitesmal, finite or infinite. The relations $\psi^*(z,w)$ infinitesimal and $\psi^*(z,w)$ finite are equivalence relations and the equivalence classes containing z are known respectively as the (hyperbolic) monad of z, denoted by $M(z)$ and the (hyperbolic) galaxy of z, denoted by $G(z)$.

Let e be the euclidean metric. The relation $e^*(z,w)$ infinitesmal is an equivalence relation in Δ^*. The equivalence class containing z is known as the euclidean monad of z and is denoted by $M_e(z)$. To simplify notation, if $\lambda \in \partial\Delta$, we let $M_e(\lambda) = \{z \in \Delta^* : z \approx \lambda\}$ and say that $M_e(\lambda)$ is an "exterior" euclidean monad of Δ^*. If $\lambda \in \Delta$, we say that $M_e(\lambda)$ is an "interior" euclidean monad of Δ^*. Every euclidean monad is interior or exterior.

The formula for the hyperbolic metric shows that the hyperbolic and euclidean monads and galaxies are related by: $M(z) = M_e(z)$ if $°|z| < 1$, i.e. each interior euclidean monad is also a hyperbolic monad. Each exterior euclidean monad contains a large number of hyperbolic galaxies. This follows from the existence of sequences $\{\alpha_n : n \in N\}$ such that $\lim \alpha_n = \lambda \in \partial\Delta$ but $\lim\limits_{\substack{n \neq m \\ n,n \to \infty}} \psi(\alpha_n, \alpha_m) = \infty$. For such a sequence and n,m infinite $\alpha_n, \alpha_m \in M_e(\lambda)$ but $G(\alpha_n) \neq G(\alpha_m)$. The metric space (Δ,ψ) has a transitive group of isometries, the Mobius transformations that leave Δ invariant. It follows that the hyperbolic galaxies of Δ^* all have similar structure. In fact, if $L_z(w) = \frac{z - w}{1 - \bar{z}w}$, then $L_z(M(0)) = M(z)$ and $L_z(G(0)) = G(z)$.

Assuming that one can "visualize" the principal galaxy $G(0)$ to ones own satisfaction, then it is easy to visualize the remaining hyperbolic galaxies. The trick is to visualize oneself walking towards $\partial\Delta$ along a radius in such a way that one shrinks to half ones height each time half the distance to $\partial\Delta$ is covered. Of course, the edge of Δ looks more and more like a straight line, and if vision is in some way limited one would be, for practical purposes, in a half plane. (This will be made more precise when we define the "type" of a galaxy in an infinitely connected domain, the hyperbolic galaxies in this case being of "type" a half plane.) In mathematical terms, if $z,w \in \Delta^*$ with $°|z| = 1$, then $w \in M(z)$ iff

$\dfrac{|z - w|}{1 - |z|}$ and $\dfrac{1 - |z|}{1 - |w|}$ are finite, i.e. w is not "too far away" or "too close" to $\partial\Delta^*$ using $1 - |z|$ as the unit of measure.)

Robinson [5] has shown how a number of classical problems can be considered by defining the "standard part" of a finite-valued nonstandard analytic function. Let f be a nonstandard bounded analytic function defined on $^*\Delta(f\epsilon(H^{\infty}(\Delta))^*)$ with $||f||^* \leq 1$. By Pick's Theorem, $f(M(z)) \subset M(f(z))$ and $f(G(z)) \subset G(f(z))$ for all $z\epsilon\Delta^*$. If $^\circ|f(z)| = 1$ then $f(G(0)) \subset G(f(0)) \subset M_e(^\circ(f(0)))$ so that $^\circ f$ is the constant function on Δ equal to $^\circ f(0)$. If $^\circ|f(0)| < 1$ then $f(G(0)) \subset G(f(0)) = G(0)$. Since $M_e(z) = M(z)$ for $z\epsilon G(0)$, using the triangle inequality and Pick's Theorem we have

$$\rho(^\circ f(z), {}^\circ f(w)) = {}^\circ\rho^*(f(z), f(w)) \leq \rho(z,w)$$

for all $z,w \in \Delta$. In particular $^\circ f$ is continuous.

Now $^\circ f$ is a standard function and so has a nonstandard extension to a function $(^\circ f)^*\colon \Delta^* \to C^*$. It might not be true that $(^\circ f)^* = f$ even for $z\epsilon G(0)$. For example, if $f(z) = \dfrac{a - z}{1 - \bar{a}z}$ with $a \in M_e(1)$ then $^\circ f \equiv 1$ so that $(^\circ f)^* \equiv 1$. It is true that $(^\circ f)^*(z) \simeq f(z)$ for all $z \in G(0)$ (but again not necessarily for $z \notin G(0)$ as the above example shows). Since $^\circ f$ is continuous on Δ, $(^\circ f)^*(z) \simeq (^\circ f)(^\circ z)$ for all $z \in G(0)$. From Pick's Theorem, $f(z) \simeq f(^\circ z)$ for $z \in G(0)$. Thus

$$(^\circ f)^*(z) \simeq (^\circ f)(^\circ z) = {}^\circ(f(^\circ z)) \simeq f(^\circ z) \simeq f(z)$$

for all $z \in G(0)$.

It is easy to see that $^\circ f$ is actually analytic. Let C_r be the standard circle of radius $r < 1$ so that C_r^* is the nonstandard circle of radius r. Since f is analytic in Δ^* we have that

$$f(z) = \frac{1}{2\pi i} {\int_{C_r^*}}^{*} \frac{f(\zeta)}{\zeta - z}\, d\zeta$$

for $|z|^* < r$. If $|^\circ z| < r$ then $\dfrac{1}{\zeta - z}$ is finite on C_r^* and so

$\frac{f(\zeta)}{\zeta - z} \simeq \frac{({}^\circ f)^*(\zeta)}{\zeta - z}$ there. For $z \in \Delta$, $|z| < r$, we thus have

$$({}^\circ f)(z) = {}^\circ f(z) = {}^\circ\left(\frac{1}{2\pi i}\int^*_{C_r^*} \frac{(f)(\zeta)}{\zeta - z}\,d\zeta\right)$$

$$= {}^\circ\left(\frac{1}{2\pi i}\int^*_{C_r^*} \frac{({}^\circ f)(\zeta)(\zeta)}{\zeta - z}\,d\zeta\right)$$

$$= \frac{1}{2\pi i}\int_{C_r} \frac{({}^\circ f)(\zeta)}{\zeta - z}\,d\zeta$$

and so f is analytic in C_r and hence in Δ .

An interesting and useful observation due to Robinson [5] is that, if ${}^\circ f$ is nonconstant then the value ${}^\circ f(z)$, $z \in G(0)$, is actually attained by $^*f$ in the monad $M(z)$. This follows immediately from the Argument Principle. For if ${}^\circ f$ is nonconstant and $z \in \Delta$ then, for a small standard circle $S \subset \Delta$ with center z , the value ${}^\circ f(z)$ is attained in the interior of S and at z , and the winding number of the image of S under ${}^\circ f$ is positive. But this is also the winding number of the image of S^* under $({}^\circ f)^*$ about ${}^\circ f(z)$ and hence also the winding number of the image of S^* under f about ${}^\circ f(z)$. Thus f attains the value ${}^\circ f(z)$ in the interior of S^* for all small standard circles S with center z . Since $M(z)$ is not internal, this value is also attained in $M(z)$.

Of course the standard part of any function $f \in (H^\infty(D))^*$ with $||f||^*$ finite can be defined by considering $f/||f||^*$.

By **defining** the standard part of an analytic function we have said something about its behavior on the principal galaxy (actually we have shown that bounded families of analytic functions are normal). We could just as well have defined the standard part of such a function in any hyperbolic galaxy of

Δ^*. Let $f \in H^\infty(\Delta)^*$ with $||f||^*$ finite and let $z \in \Delta^*$. Let $L_z(w) = \frac{z-w}{1-\bar{z}w}$, so that $L_z(G(0)) = G(z)$. Then $f \circ L_z \in H^\infty(\Delta)^*$ and $||f \circ L_z|| = ||f||$ so that we can define the standard function $°(f \circ L_z)$ which is an analytic function on Δ. The function $°(f \circ L_z)$ (and the function $°(f \circ L_z) \circ L_z^{-1}$) is a kind of standard part for f on $G(z)$. Now the nonstandard disk Δ^* is the union of a large number of hyperbolic galaxies and each standard $f \in H^\infty(\Delta)$ has a "standard part" (or rather f^* does) on each of these galaxies, which is an analytic function on Δ. What does this have to do with analytic sets in $\mathfrak{M}(\Delta)$? Remember that a surjective mapping $F: \Delta^* \to \mathfrak{M}(\Delta)$ was defined by $F(z)(f) = °f(z)$ and that the Gleason part in $\mathfrak{M}(\Delta)$ containing $m \in \mathfrak{M}(\Delta)$ was denoted by $\mathfrak{G}(m)$. For $z \in \Delta^*$ define a mapping $T_{F(z)}: \Delta \to \mathfrak{M}(\Delta)$ by $T_{F(z)} = F \circ L_z|\Delta$, the restriction of $F \circ L_z$ to Δ. For each $f \in H^\infty(\Delta)$, $\hat{f} \circ T_{F(z)} = °(f^* \circ L_z)$ so that each $\hat{f} \circ T_{F(z)}$ is analytic. Hoffman's result [4] shows that $T_{F(z)}$ is either constant or injective so that, in the latter case, $T_{F(z)}(\Delta) = F(G(z))$ is an analytic disk in $\mathfrak{M}(\Delta)$. Hoffman [4] also shows that for all $z \in \Delta^*$, $\mathfrak{G}(F(z)) = F(G(z))$. The analytic disks which occur in $\mathfrak{M}(\Delta)$ thus arise (under the mapping F) from the hyperbolic galaxies of Δ^*.

That each mapping $T_{F(z)}$ is either constant or injective means that there are two types of hyperbolic galaxies in Δ^*: those on which each standard analytic function has infinitesimal variation ($w \in G(z)$ implies $f^*(w) \simeq f^*(z)$ for all $f \in H^\infty(\Delta)$), which we will call flat galaxies, and those on which points in different hyperbolic monads can be separated a non-infinitesimal distance by a standard analytic function ($u,v \in G(z), \psi(u,v) \not\simeq 0$ implies that there exists an $f \in H^\infty(\Delta)$ such that $f^*(u) \not\simeq f^*(v)$) which we will call full galaxies.

To see that there are flat galaxies consider the relation defined by $\mathfrak{R}((f,\epsilon),\ell)$ holds iff $f \in H^\infty(\Delta)$, $\epsilon > 0$, ℓ is an end segment of a radius $\ell = \{se^{i\theta}, r \le s < 1\}$ and $|f(z) - f(w)| < \epsilon$ for $z,w \in \ell$. Since boundary values exist a.e on $\partial\Delta$, it follows that $\mathfrak{R}$ is concurrent and thus that there exists a nonstandard segment ℓ with $f(z) \simeq f(w)$ for $z,w \in \ell$. Such a nonstandard segment is said to be a "sticker" and a galaxy which intersects a sticker is said to be a "far out" galaxy. It is evident that the standard part of each $f \in H^{\infty}(\Delta)$ on a galaxy that intersects a sticker must be a constant, i.e., that each far out galaxy is flat.

Hoffman's work [4] also shows which galaxies are full. A Blaschke sequence in Δ is a sequence $\{\alpha_n\} \subset \Delta$ such that $\sum_n 1 - |\alpha_n| < \infty$. An interpolating sequence is a Blaschke sequence for which $\inf_n \prod_{m \neq n} \rho(\alpha_n,\alpha_m) > 0$. Each standard sequence $\{\alpha_n\} \subset \Delta$ has an extension $\{\alpha_n\}^* \subset \Delta^*$ and $\{\alpha_n\}^* = \{\alpha_n : n \in N^*\}$. The result is that $G(z)$ is full iff $G(z)$ contains a point of a standard interpolating sequence iff $M(z)$ contains a point of a standard interpolating sequence.

The mapping $F: \Delta^* \to \mathfrak{M}(\Delta)$ is surjective but it is far from being injective (e.g. $M(z) \subset F^{-1}(F(z))$) . In fact, Δ^* can have arbitrarily high cardinality. Another remark, which is a little confusing, is that if $m \in \mathfrak{M}(\Delta)$ and $r \approx 1$ with $r < 1$, then there are $z,w \in \Delta^*$ with $|z|<r<|w|$ and $F(z) = F(w) = m$ i.e. $F(\{z \in \Delta^* : |z|<r\}) = \mathfrak{M}(\Delta)$ and $F(\{z \in \Delta^* : |z|>r\}) = \mathfrak{M}(\Delta) - \Delta$. If $G(z)$ is full then there is a standard interpolating sequence $\{\alpha_n\} \subset \Delta$ and an $\omega \in N^*$ such that $\alpha_\omega \in M(z)$.

Lemmas in Hoffman can be used to show that $F^{-1}(F(z)) = \cup M(\alpha_n)$ where the union is taken over those n such that n and ω are contained in the same standard subsets of N^* , i.e., $S \subset N$ implies $n \in S^*$ iff $\omega \in S^*$. This is a relatively straightforward characterization of $F^{-1}(F(z))$ and is an important nonstandard tool.

As an example of how this might be used we will consider two theorems on fixed points in $\mathfrak{M}(\Delta)$. Let ψ be an analytic mapping of Δ into Δ , i.e., $\psi \in H^{\infty}(\Delta)$ with $||\psi|| \leq 1$. The mapping ψ has a unique continuous extension to a mapping $\tilde{\psi}: \mathfrak{M}(\Delta) \to \mathfrak{M}(\Delta)$ which is defined by $\tilde{\psi}(m)(f) = m(f \circ \psi)$. Pick's theorem (Schwarz's lemma) says that if ψ fixes two points of Δ , then ψ fixes all points of Δ . What if $\tilde{\psi}$ fixes points of $\mathfrak{M}(\Delta)$? This is not a simple question. For example, it turns out that $\widetilde{(z^2)}$ fixes only the point 0 in

$\mathfrak{M}(\Delta)$, but $\widetilde{(\frac{2z-1}{2-z})}$ fixes many, but not all of the points in the fiber at 1 . (The fiber at $\lambda \in \partial\Delta$ is the set $\mathfrak{M}_\lambda(\Delta) = \{m \in (\Delta): m(z) = \lambda\}$). In particular, even though $\widetilde{z^2}$ takes $\mathfrak{M}_1(\Delta)$ into itself, it fixes no point of this fiber.

Let G be the subset of $\mathfrak{M}(\Delta)$ consisting of those points which lie in non-trivial Gleason parts, i.e., the union of the analytic sets in $\mathfrak{M}(\Delta)$. A Stolz angle at 1 is the subset of Δ bounded by two chords at 1 . The set $\cup\{\mathfrak{M}_1(\Delta) \cap \overline{S}: S$ is a Stolz angle at $1\}$ is known as the set of non-tangential homomorphisms at 1 . It is easy to see that the set of non-tangential homomorphisms at 1 is the union of the set of Gleason parts in $\mathfrak{M}_1(\Delta)$ which intersect the closure of $(0,1)$ in $\mathfrak{M}(\Delta)$, and that each of these Gleason parts is an analytic disk. (They intersect the closure of $\{1 - \frac{1}{2^n}\}$ in $\mathfrak{M}(\Delta)$). The nonstandard characterization is given by

$$\{F(z): z \in \Delta^*, z \approx 1, \text{ and } -\frac{\pi}{2} < {}^\circ\arg(1-z) < \frac{\pi}{2}\} .$$

The theorems which we want to discuss are as follows:

Theorem 1: If $\tilde{\psi}$ *fixes two points of* G *which lie in different fibers, then* $\tilde{\psi}$ *fixes all points of* $\mathfrak{M}(\Delta)$.

Theorem 2: If $\tilde{\psi}$ *fixes a point of* $G \cap \mathfrak{M}_1(\Delta)$, *then* $\tilde{\psi}$ *fixes every non-tangential homomorphism at* 1 .

Suppose that $\tilde{\psi}$ fixes $m \in G$ and $\{\alpha_n\}$ is interpolating with $F(\alpha_\omega) = m$. Then $\psi(\alpha_\omega) \in F^{-1}(m) = \cup\{M(\alpha_n): n,\omega$ are contained in the same standard subsets of $N^*\}$. It turns out that $\psi(\alpha_\omega) \in M(\alpha_\omega)$. This is not obvious, but the proof is relatively straightforward. The other ingredient needed here is an elementary result on hyperbolic geometry, which states that, if $\psi(z) \in M(z)$ and $\psi(w) \in M(w)$, then $\psi(u) \in M(u)$ for all u on the hyperbolic line joining z and w . (This result is not true for distance decreasing mapping in euclidean geometry).

Now, if $\tilde{\psi}$ fixes points of G in different fibers of $\mathfrak{M}(\Delta)$, then there are points $z,w \in \Delta^*$ with $G(z), G(w)$ full galaxies, $\psi(z) \in M(z)$, $\psi(w) \in M(w)$ and ${}^\circ z \neq {}^\circ w$. Thus $\psi(u) \in M(w)$ for all u on the hyperbolic line joining z and w . But the hyperbolic line joining z and w intersects the principal galaxy of Δ^* and in fact, $\{{}^\circ u | u$ is on the hyperbolic line joining $z,w\} \cap \Delta$ is a standard hyperbolic line in Δ. Since ψ fixes each point on this line, ψ fixes every point in Δ and hence $\tilde{\psi}$ fixes every point in $\mathfrak{M}(\Delta)$.

Suppose $\tilde{\psi}$ fixes the point $m \in G \cap \mathfrak{M}_1(\Delta)$ and $z \in \Delta^*$ with $\psi(z) = m$. Let m_0 be a non-tangential homomorphism at 1. Then there is a $w \in \Delta^*$ with $w \approx 1$, $-\frac{\pi}{2} < {}^\circ\arg(1-w) < \frac{\pi}{2}$, and $F(w) = m_0$. We may further assume that $\frac{|1-w|}{|1-z|} \approx 0$ (it follows from elementary properties of the nonstandard model that $\mathfrak{M}_1(\Delta) = F(\{u \in \Delta^* : |1-u| < \epsilon\, |1-z|\})$, ϵ a fixed infinitesimal). Similarly there is a third point $u \in \Delta^*$ such that $F(u) = m$ and $\frac{|1-u|}{|1-w|} \approx 0$. Now $\psi(z) \in M(z)$ and $\psi(u) \in M(u)$ so $\psi(v) \in M(v)$ for all v on the hyperbolic line segment joining z and u. It is easy to see that on $G(w)$ the line segment lies infinitesimally close to the real ray $(0,1)^*$, so that $\psi(v) \in M(v)$ for all $v \in (0,1)^* \cap G(w)$. Schwarz's lemma then gives (considering $L^{-1} \circ \psi \circ L$ with L a Möbius transformation taking 0 to a point of $(0,1)^* \cap G(w)$) that $\psi(v) \in M(v)$ for all $v \in G(w)$ so that, in particular, $\psi(w) \in M(w)$ and $m_0 = F(w)$ is fixed by $\tilde{\psi}$.

The problem of giving a useful description of $F^{-1}(f(z))$ for $G(z)$ flat seems to be very difficult. Among the flat galaxies, certain ones, the "far out" galaxies, were distinguished because their existence was easy to establish. It turns out that these galaxies correspond to the Shilov boundary of H^∞; that is, the Shilov boundary $\mathcal{S}$ is equal to $\{F(z) : G(z)$ is far out$\}$.

Let E be the set of all $\lambda \in \partial\Delta^*$ such that λ is contained in no standard set of measure zero. If f and g are standard functions in L^∞ that agree almost everywhere, then $f(\lambda) = g(\lambda)$ for $\lambda \in E$. For each $\lambda \in E$, define a homomorphism $[\lambda] : L^\infty \to C$ by $[\lambda](f) = {}^\circ f(\lambda)$. Then $[\lambda]$ is well defined and $\{[\lambda] : \lambda \in E\}$ is a subset of the maximal ideal space $\mathfrak{M}_{L^\infty}$ of L^∞.

If $m \in \mathfrak{M}_{L^\infty}$ is contained in the closure of $\{[\lambda] : \lambda \in E\}$, then, for each standard $f_1, \ldots f_n \in L^\infty$, $\epsilon > 0$, and $A_1, \ldots, A_m$ sets of measure zero, there is a λ with

$$|f_i(\lambda) - m(f_i)| < \epsilon \text{ and } \lambda \notin A_j \qquad i = 1,\ldots,n$$
$$j = 1,\ldots,m .$$

Translation gives the existence of a $\lambda \in E$ with $°f(\lambda) = m(f)$ for all standard $f \in L^\infty$, i.e., $m = [\lambda]$. The set $\{[\lambda]:\lambda\in E\}$ is thus a closed subset of $\mathfrak{M}_{L^\infty}$. But if $f \in L^\infty$, then $\mu(\{z:|f(z)| > ||f||_{\text{ess sup}} - \epsilon\}) \neq 0$ for all $\epsilon > 0$. Choosing ϵ infinitesimal, it follows that every function $\hat{f} \in \hat{L}^\infty$ assumes its maximum on $\{[\lambda]:\lambda\in E\}$. Since $\hat{L}^\infty = C(\mathfrak{M}_{L^\infty})$, it must be that $\mathfrak{M}_{L^\infty} = \{[\lambda]:\lambda\in E\}$.

It is also true that $\lambda \in E$ iff there is a "sticker" in Δ^* that ends at λ. If $\lambda \notin E$, then λ is contained in a standard set of measure zero. There exist bounded analytic functions which have cluster values of modulus 0 and 1 at all points of such a set, so that no sticker can end off E. Conversely, let $\lambda \in E$. If A be a *-finite subset of $H^\infty(\Delta)^*$ that contains all of the standard functions, let $\epsilon > 0$ be infinitesimal and let B be the set of all functions in $H^\infty(\Delta)^*$ which have a radial cluster set at λ of radius less than ϵ. Clearly we can find a ray ending at λ on which each function in $A \cap B$ has variation less than 2ϵ. This ray is a sticker.

If $r_\lambda \lambda$ $(0<r_\lambda<1)$ is such that the ray from $r_\lambda \lambda$ to λ is a sticker for each $\lambda \in E$, then the mapping $[\lambda] \to F(r_\lambda \lambda)$ provides the well known mapping of $\mathfrak{M}_{L^\infty}$ onto the Shilov boundary of $H^\infty(\Delta)$.

We have seen that $m \in \mathcal{S}$ iff $F^{-1}(m)$ contains a far out galaxy. This suggests the question as to whether $F(z) \in \mathcal{S}$ iff $G(z)$ is far out. This question is particularly fascinating in that an equivalent standard formulation would be virtually unintelligable. I don't know whether the answer to this question has any interesting consequence in $\mathfrak{M}(\Delta)$.

As an example of how a simple geometric understanding of Δ^* can be useful in proving a theorem about $\mathfrak{M}(\Delta)$, consider a theorem of Max Weiss which states that the closure in $\mathfrak{M}(\Delta)$ if a convex curve in Δ which approaches 1 tangentially from above intersects each non-trivial Gleason part of $\mathfrak{M}(\Delta)$ in an oricycle. A convex curve which approaches 1 from above is a curve α such that

$\lim \alpha = 1$, $\lim \arg(\alpha-1) = \frac{\pi}{2}$ and such that any line segment joining points of α lies entirely below α. The theorem states that for any such α and any map $T_{F(z)}$ (Wermer map) $T^{-1}_{F(z)}(\bar{\alpha})$ is either empty or an oricycle, i.e. a circle in Δ tangent to $\partial\Delta$ at a single point.

Let $z \in \alpha^*$ with $z \approx 1$. It is easy to see that $T^{-1}_{F(z)}(\bar{\alpha})$ being an oricycle is equivalent to $\frac{1-|z|}{1-|w|} \approx 1$ for all $w \in \alpha^* \cap G(z)$ (in the upper half-plane an oricycle is a line $\operatorname{Im} w = \text{constant}$). Since z is non-tangential, we have that $\frac{1-|z|}{|1-z|} \approx 0$. If $\circ(\frac{1-|z|}{1-|w|}) \neq 1$, then it follows immediately from the convexity of α applied to the line through z and w that either $\lim \alpha^* \neq 1$ or α^* does not intersect $G(0)$, both impossibilities.

4. General D. The problems we have been considering become much more difficult when the domain D is infinitely connected. The machinery of Blaschke products completely breaks down and practically nothing is known of the interpolating sequences in such domains. Further, there are now several possibilities for the hyperbolic metric. These are the Poincare metric ψ, which is obtained by locally lifting the hyperbolic metric in the disk to D via the universal covering map of D, the Caratheodory metric Π, which is given by

$$\Pi(z,w) = \sup\{\psi_0(f(z),f(w)) : f \in H^\infty(D), \|f\| \le 1\}$$

(ψ_0 the hyperbolic metric in the disk), and the metric σ, which is given locally by $\frac{|dz|}{e(z,\partial D)}$ ($e(z,\partial D)$ the euclidean distance from z to ∂D). These metrics are related by $\Pi \le \psi \le \sigma$.

It turns out that all of these metrics are useful for studying analytic sets in $\mathfrak{M}(D)$. The metric σ is easily described in terms of the geometry of D and gives rise in a natural way to analytic sets in $\mathfrak{M}(D)$. The metric ψ, although in some ways more intrinsic to D than is σ, is never computable, since the universal covering map is never explicitly given. However, it turns out that, at least locally, ρ is closely related to σ, and gives rise to a phenomena which can be described as a "tacking together" of the analytic sets obtained using σ. The metric Π gives rise to more complicated phenomena.

If ρ is one of the metrics we are considering, let $M_\rho(z)$ and $G_\rho(z)$ be the monad and galaxy of D^* that contains $z \in D^*$. From $\Pi \le \psi \le \sigma$ it follows that $M_\sigma(z) \subset M_\psi(z) \subset M_\Pi(z)$ and $G_\sigma(z) \subset G_\psi(z) \subset G_\Pi(z)$ for all $z \in D^*$. Also $M_\sigma(z) = M_\psi(z) = M_\Pi(z) = \{w \in D^* : |z-w| \approx 0\}$ for $z \in D$ (we assume $H^\infty(D)$ contains non-constant functions), and $G_\sigma(z) = G_\psi(z)$ for $z \in D$. It need not be the case that $G_\sigma(z) = G_\Pi(z)$ for $z \in D$. Consider the example $D = \Delta - \{0\}$. Here $G_\Pi(\frac{1}{2}) = \{z \in \Delta^* : 0 < {}^\circ|z| < 1\}$. If $0 \ne z \approx 0$, then $M_\Pi(z) = \{w \in \Delta^* : 0 \ne w \approx 1\}$, $M_\sigma(z) = \{w \in D^* | \frac{|z-w|}{|z|} \approx 0\}$, and $G_\sigma(z) = \{w \in D^* | \frac{|w|}{|z|}$ is finite and not infinitesimal$\}$. The sets $M_\psi(z)$ and $G_\psi(z)$ are each the union of a large number of σ-galaxies. This follows from the fact that if $\epsilon > 0$ is fixed then $\lim_{z \to 0} \psi(z, \epsilon z) = 0$. This phenomena relating the σ and ψ monads and galaxies is actually quite general.

We define the type of a σ-galaxy $G_\sigma(z)$ to be a planar domain $\mathcal{E} \subset C$ if there is a nonstandard linear mapping $L(w) = aw + b$ such that $L(G^p_\sigma(\mathcal{E})) = G_\sigma(z)$, where $G^p_\sigma(\mathcal{E})$ is the principal σ-galaxy of $\mathcal{E}$ i.e. the σ-galaxy containing $\mathcal{E}$. Let $z \in D^*$ and let $L_1(w) = \frac{w-z}{e(z, \partial D)}$. Then $L_1(D^*)$ is a nonstandard domain in C^*. Define a domain $\mathcal{E}_1$ by: $w \in \mathcal{E}_1$ iff for some standard $\epsilon > 0$ $\{u \in C^* : |u-w| < \epsilon\} \subset L_1(D^*)$ and let $\mathcal{E}$ be the component of $\mathcal{E}_1$ containing 0. Then $\mathcal{E}$ is a standard domain and $G_\sigma(z)$ is of type $\mathcal{E}$. In the above example each $G_\sigma(z)$ is of type $C - \{0\}$ (if $z \approx 0$) $\Delta - \{0\}$ (the principal σ-galaxy) or a half-plane (if $|z| \approx 1$).

The following result, though surprising, has an elementary proof.

Theorem 3: If $G_\sigma(z)$ is of type $C - \{0\}$, then $M_\psi(z) \supset G_\sigma(z)$ and $M_\Pi(z) \supset G_\psi(z)$.

If $G_\sigma(z)$ is not of type $C - \{0\}$, then $M_\psi(z) = M_\sigma(z)$ and $G_\psi(z) = G_\sigma(z)$. Further, in this case, there is a standard constant κ such that $\kappa\sigma \le \psi$ in $M_\sigma(z)$.

We will consider two examples which illustrate this result.

Let c_n be a real sequence that converges to 0 and let r_n be such that $\Sigma \frac{r_n}{c_n} < \infty$. Further assume that $0<a<\frac{c_{n+1}}{c_n}<b<1$ for all n and some constants a,b. Let Δ_n be the closed disk with center c_n and radius r_n and let $D_1 = (\Delta-\{0\}) - \cup\Delta n$. The interesting galaxies occur when $z \approx 0$. There are several possibilities. If $z \approx 0$ and $°(\arg z) \neq 0$, then $G_\sigma(z)$ is of type $C - \{\alpha_n : n=0, \pm 1, \pm 2, \ldots\}$ where α_n is real and $a < \frac{\alpha_{n+1}}{\alpha_n} < b$. It follows from the theorem that $G_\psi(z) = G_\sigma(z)$. Also $M_\psi(z) = M_\sigma(z) = \{w \in \Delta^* | \frac{|z-w|}{|z|} \approx 0\}$. If $\frac{|z-c_n|}{|z|} \approx 0$ for some n but $\frac{|z-c_n|}{r_n}$ is infinite, then $G_\sigma(z)$ is of type $C - \{0\}$, so that $G_\sigma(z) \subset M_\psi(z)$. Also $G_\sigma(z)$ is the union of a number of such σ-galaxies. If $\frac{|z-c_n|}{r_n}$ is finite but not infinitesimally close to 1, then $G_\sigma(z)$ is of type $\{z \in C | \, |z| > 1\}$. Thus $G_\psi(z) = G_\sigma(z)$ and $M_\psi(z) = G_\sigma(z) = \{w \in \Delta^* | \frac{|w-z|}{r_n} \approx 0\}$. If $\frac{|z-c_n|}{r_n} \approx 1$, then $G_\sigma(z)$ is of type a half-plane, $G_\psi(z) = G_\sigma(z)$ and $M_\psi(z) = M_\sigma(z) = \{w \in D^* | \frac{|w-z|}{|z-c_n|-r_n} \approx 0\}$.

If $f \in H^\infty(D_1)$ and n is infinite, then

$$f = \frac{1}{2\pi i}\int_{\partial\Gamma_n} \frac{f(\xi)d\xi}{\xi-z} + \left(f - \frac{1}{2\pi i}\int_{\partial\Gamma_n} \frac{f(\xi)d}{\xi -z}\right) = f_1 + f_2$$

where f_1 is analytic on $C^* - \Delta_n$ and f_2 is analytic on $D_1^* \cup \Delta_n$. That f_1 is infinitesimal for $\frac{|z-c_n|}{r_n}$ infinite is immediate. That f_2 has infinitesimal variation on $\{z \in \Delta^* | \frac{|z-c_n|}{|z|} \approx 0\}$ follows from the fact that each finite valued analytic function in Δ^* has a "standard part". We can conclude that the set $\{z \in D^* | \frac{|z-c_n|}{r_n}$ is infinite and $\frac{|z-c_n|}{|z|}$ is infinitesimal$\}$ is contained in a single Π-monad. More is true. For if $G_\sigma(-c_n)$ is of type $\mathcal{E} = C - \{\alpha_n | n=0,\pm 1,..\}$ and $L: G_\sigma^p(\mathcal{E}) \to G_\sigma(-c_n)$, then $f \circ L$ is a finite valued analytic function so has a standard part $°(f \circ L)$ which is an element of $H^\infty(\mathcal{E})$ and is hence constant.

A result of Zalcman [6] states, for domains similar to D_1, that $\lim_{\substack{r\to 0\\ r<0}} f(z)$ exists for all $f \in H^\infty(D_1)$. If we denote this limit by $m_0(f)$, then, in particular, the standard part of $f \circ L$ on $\mathcal{E}$ satisfies $°(f°L) \equiv m_0(f)$. We conclude that there is a single Π-monad $M_\pi(z)$ such that

$$M_\pi(z) = \{z\in D_1^\infty | z \approx 0 \text{ and } \frac{|z-c_n|}{r_n} \text{ is infinite for all infinite } n\}.$$

The remaining σ-monads are Π-monads. Also, the principal Π-galaxy is equal to

$$G_\sigma^p(D_1) \cup \{z\in D_1^* | z \approx 0 \text{ and } \frac{|z-c_n|}{r_n} \not\approx 1 \text{ for all } n\}.$$

The second example that we will consider is given by $D_2 = (C-\{0\}) - \cup\{\Delta_n : n=0,\pm 1,\pm 2,\dots\}$, where Δ_n is the closed disk with radius 2^{n-3} and center 2^n. It is easy to see that each σ-galaxy of D_2^* is either of type D_2 or of type a half-plane and that $M_\sigma(z) = M_\psi(z) = M_\pi(z)$ and $G_\sigma(z) = G_\psi(z) = G_\pi(z)$ for all $z \in D_2^*$.

If the mapping $F: D^* \to \mathfrak{M}(D)$ is constant on $G_\sigma(z)$, we will say that $G_\sigma(z)$ is a flat σ-galaxy. If $G_\sigma(z)$ is not flat, then we will say that $F(G_\sigma(z))$ is a natural analytic set in $\mathfrak{M}(D)$. This is justified by the fact that if $G_\sigma(z)$ is of type $\mathcal{E}$ and $L(G_\sigma^p(\mathcal{E})) = G_\sigma(z)$, then $F \circ (L|\mathcal{E})$ is an analytic map of $\mathcal{E}$ into $\mathfrak{M}(D)$. For $f \in H^\infty(D)$, $\hat{f} \circ (F\circ(L|\mathcal{E})) = °(f\circ L) \in H^\infty(\mathcal{E})$. If $F(u) \neq F(v)$ for $u,v \in G_\sigma(z)$ with $\sigma(u,v) \not\approx 0$, then $G_\sigma(z)$ is said to be full. We conjecture that every σ-galaxy is either flat or full.

In the example D_1, the galaxies of type $\{z\in C | \, |z| > 1\}$ are all full. (Consider $\sum \frac{z-c_n}{r_n}$). There are thus a large number of disks in $\mathfrak{M}(D_2)$ which have their centers identified to the homomorphism m_0 ($m_0(f) = \lim_{\substack{r\to 0\\ r<0}} f(r)$). Enough has been said that this should be clear. It turns out [1] that a set $(\beta N-N) \times \Delta$, with $(\beta N-N) \times \{0\}$ identified to m_0, exists in $\mathfrak{M}(D_1)$. Here βN is the Stone-Cěch compactification of the integers.

In the example D_2, every σ-galaxy of type D_2 is full and it turns out that the resulting analytic copies of D_2 in $\mathfrak{M}(D_2)$ are Gleason parts. The remaining Gleason parts are all point parts or analytic disks.

There are domains D for which $\mathfrak{M}(D)$ contains analytic sets which are not natural analytic sets. Let α_n be a sequence in Δ which accumulates only on $\partial\Delta$ for which $\lim\limits_{|z|\to 1} \psi(z,\{\alpha_n\}) = 0$. Let r_n be such that $r_n \to 0$ very fast (such that $\sum \frac{r_n}{e(\alpha_n, \bigcup_{n\neq m} \{\alpha_m\}} < \infty$). Let $\{\Delta_n\}$ be disjoint closed disks with centers α_n and radii r_n and let $D_3 = \Delta - \cup\Delta_n$. The σ-galaxies $G_\sigma(z)$ with $|z-\alpha_n| \approx r_n$ for some infinite n are of type a half-plane. The remaining $G_\sigma(z)$ witn $\frac{|z-\alpha_n|}{r_n}$ finite for some infinite n are of type $\{z\in C|\ |z| > 1\}$ and are full σ-galaxies. The $G_\sigma(z)$ with $\frac{|z-\alpha_n|}{r_n}$ infinite but $\frac{|z-\alpha_n|}{|z-\alpha_m|} \approx 0$ for all $m \neq n$, are of type $C - \{0\}$. The remaining σ-galaxies may be of any type whatever (except C) and are all flat. The domain D_3 may actually be chosen so that σ-galaxies of all planar type occur in D_3^*. Let $G(z,\Delta)$ be a hyperbolic galaxy in Δ^*. It follows from the fact that each $f \in H^\infty(D_3)$ has a representation $f = f_1 + (\sum_n \frac{1}{2\pi i} \int_{\partial\Delta_n} \frac{f(\xi)d\xi}{\xi - z})$ where $f_1 \in H^\infty(\Delta)$ and that $H^\infty(\Delta) \subset H^\infty(D_3)$, that $\{w\in G(z,\Delta)|\ \frac{|w-\alpha_n|}{r_n}$ is infinite for all $n\}$ is a Π-galaxy in D_3^*. The Π-monads contained in this Π-galaxy are equal to the intersection of the hyperbolic monads of $G(z,\Delta)$ intersected with the set $\{w\in\Delta^*|\ \frac{|w-\alpha_n|}{r_n}$ is infinite for all $n\}$. This Π-galaxy is full in the sense that the mapping F takes points of the different Π-monads of this Π-galaxy into different points of $\mathfrak{M}(D_3)$ (since $H^\infty(\Delta)\subset H^\infty(D_3)$). The resulting analytic disk in $\mathfrak{M}(D_3)$ is not a natural analytic set. It turns out that the resulting Gleason part in $\mathfrak{M}(D_3)$ is equal to an analytic disk with the origins of 2^C different analytic disks identified to each point of this disk (different disks for different points) (see [2], where this Gleason part is said to be a "congested garden").

It is possible to develop a general theory, using analytic capacity, to describe when Π-galaxies such as occur in D_3^* arise. The pertinent fact in the domain D_3^* is that $\frac{\gamma(\Delta(w,1-|w|)-D_3^*)}{(1-|z|)^2} \approx 0$ for $w \in G(z,\Delta))$, where γ is the analytic capacity and $\Delta(w,1-|w|)$ is the disk with center w and radius $1-|w|$.

As an example of the kind of standard theorems obtained using these methods we state the following.

Theorem 4: Let $D \subset \Delta$ *be a domain and let* $\Phi: \mathfrak{M}(D) \to \mathfrak{M}(\Delta)$ *be the natural restriction mapping. Then* Φ *has an analytic cross-section over every non-trivial Gleason part in* $\mathfrak{M}(\Delta)$ *(except* Δ *) iff*

$$\lim_{|z|\to 1} \frac{\gamma(\Delta(z,r(1-|z|))-D)}{(1-|z|)^2} = 0$$

$(0<r<1$ *a fixed real number).*

An analytic cross-section over the Gleason part $\mathcal{G}(m)$ is a map $\Psi: \mathcal{G}(m) \to \mathfrak{M}(D)$ such that $\Phi \circ \Psi$ is the identity map on $\mathcal{G}(m)$ and $\hat{f} \circ \Psi \circ L \in H^\infty(\Delta)$ for every $f \in H^\infty(D)$, where L is a Wermer map for $\mathcal{G}(m)$ ($T_{F(z)}$ where $F(z) = m$).

In the examples that we have been considering it turns out that $\mathcal{G}(F(z)) = F(G_\Pi(z))$. Since this is so fundamental, we state it as a conjecture:

<u>Conjecture</u>: It is always true that $\mathcal{G}(F(z)) = F(G_\Pi(z))$.

Bibliography

1. M. Behrens, The corona conjecture for a class of infinitely connected domains, Bull. Amer. Math. Soc. 76 (1970), 387-391.

2. ——————, The maximal ideal space of algebras of bounded analytic functions on infinitely connected domains, Trans. Amer. Math. Soc. 161 (1971), 359-379.

3. K. Hoffman, Banach spaces of analytic functions, Prentice-Hall Series in Modern Analysis, Prentice-Hall, Englewood Cliffs, N.J. 1962 MR 24 #A 2844.

4. ——————, Bounded analytic functions and Gleason parts, Ann. of Math. (2) 86 (1967), 74-111. MR 35 #5945.

5. A. Robinson, Non-standard analysis, Studies in Logic, North-Holland, Amsterdam, 1966.

6. L. Zalcman, Bounded analytic functions on domains of infinite connectivity, Trans. Amer. Math. Soc. 144 (1969), 241-270.

BOUNDARY VALUE THEOREMS FOR MEROMORPHIC FUNCTIONS DEFINED IN THE OPEN UNIT DISK

Michael Behrens

Laguna Beach, California

1. Introduction

In [7], [8] and [9], Abraham Robinson has studied the classical theory of a single complex variable using the methods of nonstandard analysis. In the spirit of those papers, we would like to show that several theorems from the classical theory of cluster sets are easily accessible to the methods of nonstandard analysis. In particular, we will consider a rather deep theorem due to Gross [5]. This theorem was discovered by Gross in 1916, although the proof that he gave is obscure. A more acceptable proof was given by Doob [4] although [4] is still somewhat difficult.

To state Gross's theorem we need several definitions. Let f be a meromorphic function defined in the open unit disk Δ . The range of values of f at λ , $R(f,\lambda)$, is the set of values that are assumed by f infinitely often in every neighborhood of λ , i.e.,

$$R(f,\lambda) = \{b: \exists \{\alpha_n\} \subset \Delta, \lim \alpha_n = 1, \text{ and } f(\alpha_n) = b \text{ for every } n\}$$

The nonstandard characterization of this set is given by

$$R(f,\lambda) = \{f(z): z \in M_e(\lambda) \cap \Delta^* \text{ and } f(z) \text{ is standard}\}$$

where $M_e(1)$ is the exterior euclidean monad of 1 defined in the previous paper. We will denote $R(f,1)$ by $R(f)$. A set similar to $R(f,\lambda)$ but usually easier to analyse is the cluster set of f at λ through $E \subset \Delta$, defined by

$$Cl(f,E,\lambda) = \{b: \exists\{\alpha_n\} \subset E\ ,\ \lim \alpha_n = \lambda \text{ and } \lim f(\alpha_n) = b\}$$

whose nonstandard characterization is given by

$$Cl(f,E,\lambda) = \{°f(z): z \in E^* \cap M_e(\lambda)\}\ .$$

The cluster set of f through E is defined by

$$Cl(f,E) = \bigcup_{|\lambda|=1} Cl(f,E,\lambda)\ .$$

A principal value of f at 1 is a complex number which is in the cluster set of f at 1 through every curve in Δ which approaches 1. Finally, a value b is a nontangential cluster value of f at 1 if there is a sequence $\{\alpha_n\}$ in Δ such that $\lim \alpha_n = 1$, $\{\frac{1-\alpha_n}{1-|\alpha_n|}\}$ is a bounded sequence, and $\lim f(\alpha_n) = b$.

Gross's Theorem: Let f be meromorphic in Δ. Then every nontangential cluster value of f at 1 which is a limit point of the complement of $R(f)$ is a principal value of f at 1.

The advantage of the nonstandard viewpoint is that we can consider that portion of a standard curve approaching 1 in Δ which lies in the monad of 1 in Δ^*. By mapping Δ^* conformally onto a domain bounded by a segment of this curve and a semi-circle (which also lies in the monad of Δ), and composing with the nonstandard extension of f, the theorem is reduced to a tractable result on nonstandard meromorphic functions defined in Δ^*.

2. Nonstandard Theorems

The starting point in Robinson's theory of complex variables is the fact that each nonstandard function f which is defined and analytic in the nonstandard open disk Δ^*, and which is bounded by a (standard) finite number there, has a "standard part" $°f$ which is analytic in Δ (9, p. 155). By the nonstandard version of Fatou's theorem, the function f has finite boundary values almost everywhere on $\partial\Delta^*$. We would like to consider some theorems which relate the behavior of these boundary values in $\partial\Delta^*$ to the question of when $^\circ f$ is constant and the question of what standard values are assumed by f in Δ^*. A relevant theorem of Robinson (9, page 158) states that if $^\circ f$ is non-constant, then the value $(^\circ f)(\lambda)$ is assumed by f in the euclidean monad of $\lambda \in \Delta$.

The following theorem is similar to the Riesz uniqueness theorem [6].

Theorem 1: Let f be a holomorphic function defined in Δ^* for which $\|f\|_\infty$ is finite. Suppose that the boundary values of f on $\partial\Delta^*$ satisfy $^\circ(f(\lambda)) = \alpha$ for $\lambda \in E \subset \partial\Delta^*$, where $^\circ(\mu E) > 0$. Then, $^\circ f \equiv \alpha$. (Here μ is normalized Lebesgue measure on $\partial\Delta$.)

Proof. It suffices to show that $^\circ f(0) = \alpha$ since the function $f \circ L$ -- L being a standard Mobius transformation -- also satisfies the hypothesis of the theorem. Since $f(\lambda) - \alpha$ is infinitesimal on E, $\varepsilon = \sup\{|f(\lambda) - \alpha| : \lambda \in E\}$ must also be infinitesimal. Choose A infinite such that $\varepsilon e^{\frac{A}{\mu(E)}}$ is infinitesimal. Let h be the harmonic function in Δ^* with boundary values equal to $\frac{A}{\mu(E)}$ on E and equal to $\frac{A}{\mu(E)-1}$ on $\partial\Delta^*\setminus E$. Let g be a harmonic conjugate for h and let $\phi = e^{h+ig}$. Then, $|\phi(0)| = e^{h(0)} = e^0 = 1$. Hence,

$$^\circ|f(0) - \alpha| = {}^\circ|(f(0)-\alpha)\phi(0)| = {}^\circ\left|\int_{\partial\Delta^*} \frac{(f-\alpha)\phi}{z}\,d\mu\right|$$

$$\leq {}^\circ\left|\int_E \frac{(f-\alpha)\phi}{z}\,d\mu\right| + {}^\circ\left|\int_{\partial\Delta^*\setminus E} \frac{(f-\alpha)\phi}{z}\,d\mu\right|$$

$$\leq {}^\circ\left(e^{\frac{A}{\mu(E)}}\mu(E)\sup\{|f(\lambda)-\alpha| : \lambda\in E\}\right) + {}^\circ\left(e^{\frac{A}{\mu E-1}}(1-\mu(E))(\|f\|_\infty + |\alpha|)\right) = 0 .$$

Of course, it may be that $^\circ f \equiv \alpha$, but that f is not infinitesimally close to α on any set of $\partial\Delta^*$ whose measure is not infinitesimal. An example of such a function is

$f(z) = e^{m\left(\frac{1+z}{1-z}\right)}$, where m is infinite. In this case, $^\circ f \equiv 0$, but $|f|$

is equal to 1 a.e. on $\partial\Delta^*$. This occurs because f winds around 0 infinitely often on $\partial\Delta^*$. □

<u>*Theorem 2:*</u> *Let f be an n.s. finite-valued function defined and continuous on the closed n.s. disk, and analytic on the open n.s. disk. Suppose further that f does not assume the value 0 in the closed n.s. disk. Then,* $^\circ f \equiv 0$

iff log f is infinite on a set E satisfying $^\circ\mu E > 0$ *for every determination of the logarithm;*

iff log f is infinite on a set E satisfying $^\circ\mu E = 1$ *for every determination of the logarithm.*

<u>Proof</u>. We will assume that $|f| \le 1$. Suppose that $^\circ f \equiv 0$. Since $|\log f| \ge |\log|f||$, log f is infinite for all $z \in \Delta^*$ with $^\circ|z| < 1$ for each determination of the logarithm. The function $\phi = \frac{\log f + 1}{\log f - 1}$ thus satisfies $^\circ\phi(0) = 1$. But, $|\phi| \le 1$ and $\phi(0) = \int_{\partial\Delta} \phi d\mu$. If $|\phi| < r < 1$ (r standard) on a set $E \subset \partial\Delta^*$ with $^\circ(\mu E) > 0$, then $^\circ|\phi(0)| = {}^\circ|\int_{\partial\Delta} \phi d\mu|$

$= {}^\circ|\int_E \phi d\mu + \int_{\partial\Delta\setminus E} \phi d\mu| \le r(\mu E) + {}^\circ\mu(\partial\Delta^*\setminus E) < 1$. It follows that $^\circ|\phi| = 1$ on a set E such that $^\circ(\mu E) = 1$. Conversely, if log f is infinite on a set E such that $^\circ\mu E > 0$, then $\phi = \frac{\log f + 1}{\log f - 1}$ will be infinitesimally close to 1 on E and so, by Theorem 1, $^\circ\phi \equiv 1$. This means that log f is infinite for each $z \in \Delta^*$ with $^\circ|z| < 1$. Now f is finite valued so that if there is some point $z_0 \in \Delta^*$, with $^\circ|z| < 1$ for which $^\circ f(z_0) \ne 0$, then we may certainly choose a branch of the logarithm for which log $f(z_0)$ is finite. Thus, $^\circ f \equiv 0$. □

The next theorem identifies the location of some standard values that are assumed by f in Δ^*.

Theorem 3: *Let* f *be an n.s. finite-valued function defined and continuous on the closed n.s. disk and analytic on the n.s. open disk. Let* $J \subset \partial\Delta^*$ *with* $°(\mu J) > 0$. *If* $\alpha \in (°f)(\Delta) \setminus °(f(J))$ (α *standard) is not assumed by* f *in* Δ^* *, then every (standard) value in the component of the complement of* $°(f(J))$ *containing* α *is assumed by* f *in* Δ^* .

Proof. Suppose β is not assumed by f in Δ^* and is in the same component of the complement of $°(f(J))$ as is α . Let A be a (standard) arc joining α and β contained in this component. Let $\log \frac{w-\alpha}{w-\beta}$ be a branch of the logarithm defined on the complement of A in the Riemann sphere. The (standard) function $\log \frac{w-\alpha}{w-\beta}$ is bounded on $°(f(J))$, say, by M . Then, if $\log(f(z)-\beta)$ is any branch of the log, there is a branch of $\log(f(z)-\alpha)$ such that $|\log(f(z)-\alpha) - \log(f(z)-\beta)| < M$ for $z \in J$. Now, $°f \equiv \alpha$ so that $\log(f(z)-\alpha)$ is infinite on a set E with $°\mu E = 2\pi$. But then, $\log(f(z)-\beta)$ is also infinite on $E \cap J$ and $°(\mu(E\cap J)) = °\mu(J) > 0$ so that $°f \equiv \beta$. Since $\alpha \neq \beta$, this is a contradiction. □

The next theorem extends Theorem 3 to nonfinite functions.

Theorem 4: *Let* f *be a continuous mapping of the closed nonstandard disk into the nonstandard Riemann sphere which is meromorphic on* Δ^* . *Let* $J \subset \partial\Delta^*$ *satisfy* $°(\mu J) > 0$. *If* $\alpha \in °f(\Delta)\setminus°(f(J))$ *is not assumed by* f *in* Δ^* *, then there is at most one other standard point in the component of the complement of* $°(f(J))$ *containing* α *which is not assumed by* f *in* Δ^* . *If two points of this component are omitted by* f *in* Δ^* *, then every other (standard) point of the Riemann sphere is assumed by* f *on* Δ^* .

Proof.

(a) Suppose that besides α , the values a, b, and c are also omitted by f in Δ^* . Let Ψ be the (standard) modular function taking Δ onto $C \cup \{\infty\}\setminus\{a,b,c\}$. We may choose a branch ϕ of $\Psi^{-1} \circ f$ for which $\phi(\alpha) = \alpha'$ is a standard number in Δ . Suppose β is in the same component of the complement of $°f(J)$ as α and is not assumed by f in Δ^* . Let Γ

be a standard curve joining α and β which lies entirely in this component. Then, we may find a standard point $\beta' \in \Delta$ and a standard curve Γ' in Δ joining α' and β' such that $\Psi \circ \Gamma' = \Gamma$. Now, Γ' lies entirely in the complement of $°(\phi(J))$ since if $z \in \Gamma'$ and $z \simeq w$ with $w \in \phi(J)$, then $\Psi(z) \in \Gamma$ and $\Psi(z) \simeq \Psi(w) \subset f(J)$ (by the continuity of Ψ). By Theorem 3 the value β' is assumed by ϕ and so the value $\beta = \Psi(\beta')$ is assumed by f.

(b) Suppose now that f does not assume the values α, β, and γ, where α and β are in the same component of the complement of $°(f(J))$. We will obtain a contradiction. Assume that $\alpha = 0$, $\beta = \infty$, and $\gamma = 1$. Let ψ_0 be a meromorphic function such that $\psi_0^3 = f$, i.e., a branch of $f^{1/3}$. Now, $°(\psi_0(J))$ is a compact subset of C not containing 0. Let z_0, z_1 be such that $|z_0| < \inf|\psi_0(J)|$ and $|z_1| > \sup|\psi_0(J)|$. Let Γ be a standard arc joining z_0^3 and z_1^3 which lies entirely in the complement of $°(\psi_0(J))$. Then, there is a standard arc Γ' joining z_0 and some determination of $(z_1^3)^{1/3}$ such that $(\Gamma')^3 = \Gamma$. Then, Γ' is contained in the complement of $°(\psi_0(J))$. This shows that 0 and ∞ lie in the same component of the complement of $°(\psi_0(J))$. But this contradicts part (a) since ψ_0 omits the values $e^{\frac{2\pi i}{3}}$, $e^{\frac{4\pi i}{3}}$, and 1. □

3. <u>Boundary Value Theorems</u>

We will now prove several well-known boundary value theorems. We begin with a theorem of Doob [2]. Let E be a subset of Δ. We say that a set $M \subset \partial\Delta$ has $\{{upper \atop lower}\}$ E-density δ if $\lim\limits_{\mu J \to 0} \{{\sup \atop \inf}\} \frac{\mu(M \cap J)}{\mu(J)} = \delta$, where μ is normalized Lebesque measure and J ranges over all intervals of the form $\{e^{i\theta} | \theta_0 - r < \theta < \theta_0 + r\}$, where $(1-r)e^{i\theta_0} \in E$. For $E \subset \Delta$, we define a "stolz angle" about E to be a set $\Lambda_p(E) = \{z \in \Delta | \psi(z,E) < p\}$, where ψ is the hyperbolic metric and p is a standard positive real number. We say that α is a cluster value of f through E if $\alpha \in Cl(f,E)$ and that α is the nontangential limiting value of f along E if $Cl(f,\Lambda_p(E)) = \{\alpha\}$ for every $p > 0$.

<u>*Doob's Theorem:*</u> *Let f be a standard bounded analytic function defined on Δ . If, for each standard $\varepsilon > 0$, the set $\{\lambda \in \partial\Delta \mid |f(\lambda)-\alpha|<\varepsilon\}$ has upper (lower) E-density greater than some fixed $\delta > 0$, then α is the nontangential limiting value of f along E (is a cluster of f along E). If $|\alpha| = \|f\|_\infty$, then the converse holds with $\delta = 1$.*

<u>Proof</u>. In the nonstandard language, the fact that α is the nontangential limiting value for f along E is equivalent to the condition $°(f(z)) = \alpha$ for all $z \in \Delta^*$ with $\psi(z,E^*)$ finite and $°|z| = 1$. Let $L_z(w) = \dfrac{z-w}{1-\bar{z}w}$. Since the Mobius transformations of this form $(z \in \Delta^*)$ are transitive on Δ^* and preserve hyperbolic distance, this latter condition is equivalent to $°(f \circ L_z) \equiv \alpha$ for all $z \in E^*$ with $°|z| = 1$. Suppose $A_\epsilon = \{\lambda \in \partial\Delta : |f(\lambda)-\alpha| < \epsilon\}$ has upper E-density greater than δ for every $\epsilon > 0$, and $z = (1-r)e^{i\theta_0} \in E^*$ with $r \simeq 0$. Then $\dfrac{\mu(A_\epsilon \cap J)}{\mu(J)} \geq \delta$ for all standard $\epsilon > 0$ where $J = \{e^{i\theta} : \theta_0 - r < \theta < \theta_0 + r\}$. A general property of the nonstandard model guarantees that $\dfrac{\mu(A_\epsilon \cap J)}{\mu(J)} \geq \delta$ for some infinitesimal ϵ . Noticing that, under L_z^{-1}, J goes into an arc which is infinitesimally close to a half-circle, and that the ratio of the maximum and minimum of the derivative of L_z^{-1} on J is 2 , we have that $f \circ L_z$ has boundary values infinitesimally close to α on a set of measure greater than $^{\delta\pi}/_2$. By Theorem 1, $(f \circ L_z) \equiv \alpha$.

If we replace upper E-density by lower E-density then we conclude that $°(f \circ L_z) = \alpha$ for some $z \in E^*$ with $°|z| = 1$. In particular $°(f(z)) = \alpha$ and α is in the cluster set of f along E .

If $°(f \circ L_z) \equiv \alpha$ with $|\alpha| = \|f\|_\infty$, then it follows from the relation $\alpha = °(f(0)) = °\left(\int (f \circ L_z) d_\mu\right)$ that $f \circ L_z$ has boundary values infinitesimally close to α on a set with measure infinitesimally close to 1 . Mapping the boundary by L_z^{-1} gives the converse. □

We need the following lemma on conformal mappings.

Lemma 1. Let D_1 be the domain bounded by $[-1,1]$ and $\Gamma_1 = \{e^{i\theta}: 0 \le \theta \le \pi\}$. Let σ be a simple curve joining r, ${}^\circ r = 1$, and s, ${}^\circ s = -1$, which lies infinitesimally close to Γ_1, i.e., such that ${}^\circ \operatorname{Im}\sigma(t) \ge 0$ and ${}^\circ|\sigma(t)| = 1$ for all $t \in (0,1)$. Let D_2 be the simply connected region bounded by $[r,s]$ and $\Gamma_3 = \sigma([0,1])$. Let f be a continuous mapping of $\bar{D}_1$, into $\bar{D}_2$ such that the restriction of f to D_1 is a conformal mapping of D_1 onto D_2. Suppose further that f is normalized so that $f(1) = r, f(-1) = s$ and $f(0) = 0$. Then, for all $z \in [0,1] \cup D_1$ with ${}^\circ|z| < 1$, we have ${}^\circ f(z) = {}^\circ z$.

Proof. Let D_3 be the region bounded by σ and the reflection of σ in the real axis. Since f is real on $[-1,1]$, we may extend f to a mapping of $*\Delta$ into D_3 by defining $f(\bar{z}) = \overline{f(z)}$. Now the standard part of f satisfies ${}^\circ f(0) = 0$ and $|{}^\circ f| \le 1$. Hence, by Schwarz's theorem, $|{}^\circ f(z)| \le {}^\circ|z|$ for all $z \in \Delta*$ for which ${}^\circ|z| < 1$. If $r < 1$ is standard, then the nonstandard disk $*\{z|\ |z| < r\}$ is contained in D_3. Since f is one-to- one on $*\Delta$, the inverse of f on $*\{z|\ |z| < r\}$ is defined. By applying Schwarz's lemma to the standard part of this function, we have that $|{}^\circ f^{-1}(z)| \le \frac{|{}^\circ z|}{r}$ for all z with ${}^\circ|z| < r$. Since this is true for every standard $r < 1$, we have that ${}^\circ|f^{-1}(z)| \le |{}^\circ z|$ for all $z \in *\Delta$ with ${}^\circ|z| < 1$. We thus have that $|{}^\circ f(z) = |{}^\circ z|$ for all $z \in \Delta*$ with ${}^\circ|z| < 1$ and, since ${}^\circ f$ is real on $[-1,1]$, that ${}^\circ f(z) = {}^\circ z$ for all such z. □

Lindelöf's Theorem: Let f be a bounded analytic function in Δ. If $f(z)$ approaches λ as z approaches 1 along some curve σ in Δ which approaches 1, then $f(z)$ approaches λ as z approaches 1 nontangentially.

Proof. We may assume that σ is a simple curve. Let $r \in (0,1)^*$ with ${}^\circ r = 1$. We want to show that ${}^\circ(f \circ L_r) \equiv \lambda$.

Consider $L_r^{-1} \circ \sigma*$. It is easy to see that the standard part of the image of this curve, $\{{}^\circ(L_r^{-1}(\sigma*(t))): t \in [0,1)^*\}$ is a closed connected set containing 1. Thus if $L_r^{-1}\circ\sigma^*$ passes through $\{z: {}^\circ|z| < 1\}$, then ${}^\circ(f\circ L_r)$ is equal to λ on an infinite connected set and, being analytic, is identically equal to λ.

If $L_r^{-1}\circ\sigma^*$ does not pass through $\{z: {}^\circ|z| < 1\}$, then let D be the simply connected domain bounded by the real axis and $L_r^{-1}\circ\sigma^*$. By Lemma 1 there is a conformal mapping ϕ of the domain bounded by the real axis and $\partial\Delta$ on D such that ${}^\circ\phi(z) = {}^\circ z$ if ${}^\circ|z| < 1$ and points of $\partial\Delta$ go into points of $\sigma^* \cap \partial D$. By applying Theorem 1, we conclude that ${}^\circ(f\circ L_r\circ\phi)(z) = \lambda$ for z with ${}^\circ|z| < 1$ and $\mathrm{Im} z > 0$. Thus ${}^\circ(f\circ L_r)(z) = {}^\circ(f\circ L_r)(\phi(z)) = \lambda$ for $z \in \Delta$. □

The set R(f) is, in the nonstandard model, equal to the set of standard values that are assumed by f in the euclidean monad of 1. If ${}^\circ z = 1$ and ${}^\circ(f\circ L_z)$ is nonconstant, then, by the theorem of Robinson mentioned above ([9]p.158), ${}^\circ(f\circ L_z)(\Delta)$ is contained in the interior of R(f). In particular, if a is an accumulation point of the complement of R(f) and ${}^\circ(f(z)) = a$ with ${}^\circ z = 1$, then ${}^\circ(f\circ L_z) \equiv a$.

Proof of Gross's Theorem

The value a is a nontangential cluster value of f at 1 if there is a sequence $\{z_n\}$ in Δ which approaches 1 in a Stolz angle (i.e., for which $\lim z_n = 1$ and $\{\frac{|1-z_n|}{1-|z_n|}\}$ is bounded) such that $\lim f(z_n) = a$. In the nonstandard model this is equivalent to the existence of a z_o with ${}^\circ z_o = 1$, $\frac{|1-z_o|}{1-|z_o|}$ finite, and ${}^\circ(f(z_o)) = a$. If a is also an accumulation point of the complement of R(f), then by what was said above, ${}^\circ(f\circ L_{z_o}) \equiv a$.

Let σ be a curve in Δ that approaches 1. Suppose that a is not a cluster value of f along σ. Then there is a standard $\varepsilon > 0$ such that $C(f(z, a) > \varepsilon$ for all z which lie on the segment of σ^* which lies in the euclidean monad of 1, i.e., $C(f(\sigma(t)),a) > \varepsilon$ for $t \in [0,1)$, $t \approx 1$ (C the chordal metric). In particular, it follows from the fact that $(f\circ L_{z_o}) \equiv a$, that if

$|1-\sigma(t)| < 2|1-z_o|$, then $\frac{1-|\sigma(t)|}{1-|z_o|}$ is infinitesimal (since the hyperbolic distance from σ^* to z_o must be infinite). We may assume that $\frac{|\mathrm{Im} z_o|}{1-|z_o|} > 3$ (replacing z_o by $z_o+3i(1-|z_o|)$, if necessary). Let Γ_o be the semi-circle through z_o which intersects $\partial\Delta^*$ at right angles. Let D_o be the domain bounded by Γ_o and $\partial\Delta^*$ in the euclidean monad of 1, and let $\bar{D}_o$ be the reflection of D_o in the real axis.

Since σ approaches 1, at least one of the two sets $D_o \setminus \sigma^*$, $\bar{D}_o \setminus \sigma^*$ must be disconnected. We may assume that it is $D_o \setminus \sigma^*$. A simple argument shows that there is a segment $\sigma^*([s,t])$ such that $D_o \setminus \sigma^*([s,t])$ is disconnected and $\sigma^*([s,t]) \cap \Gamma_o = \{\sigma^*(s), \sigma^*(t)\}$. Let D be the component of $D_o \setminus \sigma^*([s,t])$ that contains z_o in its boundary. If the value a is not assumed in $\Gamma_o \cup D$, then (an obvious variation on) Lemma 1 and Theorem 4 lead to a contradiction. The value a can be assumed by f only a *-finite number of times in $\Gamma_o \cup D$. If f assumes the value a in $\Gamma_o \cup D$, let z_1 be such that $f(z_1) = a$ and $1-|z_1| = \min\{1-|z| : z \in \Gamma_o \cup D \text{ and } f(z) = a\}$, and let $z_2 = z_1 + \frac{1}{2}(1-|z_1|)$. Obtaining a domain bounded by a semi-circle through z_2 and a segment of σ^*, we again arrive at a contradiction via Lemma 1 and Theorem 4. □

References

1. E. R. Collingwood and A. J. Lohwater, The Theory of Cluster Sets, (Cambridge University Press, Cambridge, 1966).

2. J. L. Doob, "The Boundary Values of Analytic Functions," Trans. Amer. Math. Soc., 34 (1932), 153-170.

3. _________, "The Boundary Values of Analytic Functions," Trans. Amer. Math. Soc., 35, (1933), 418-451.

4. _________, "One-sided Cluster Value Theorems," Proc. London Math. Soc. (3) 13 (1963), 461-470.

5. W. Gross, "Zum Verhalten der Konformen Abbildung am Rande," Math. Z. 3 (1919), 43-64.

6. F. Riesz and M. Riesz, "Über die Randwerte einer Analytischen Funktion," 4. Cong. Scand. Math. Stockholm, (1916), 27-44.

7. A. Robinson, "Complex Function Theory over Non-archimedean Fields," Technical-scientific note, No.30, U.S.A.F. contract No. 61 (052)-187, Jerusalem.

8. _________, "On the theory of normal families," Acta Philosophica Fennica, fasc. 18 (Rolf Nevanlinna anniversary volume), 159-184.

9. _________, Nonstandard Analysis. North-Holland, Amsterdam, 1966.

A LOCAL INVERSE FUNCTION THEOREM

Michael Behrens

Laguna Beach, California

We would like to give a purely local version of the inverse function theorem. The definition of uniform differentiability that we use arises naturally in a nonstandard treatment of this theorem. Although the theorem that we prove may be new, our main purpose is in suggesting a nonstandard treatment of the inverse function theorem which might be appropriate for an elementary course in Real Analysis.

Definition (standard). Let f be a mapping of a neighborhood of $x \in R_n$ into R_m. Then f is said to be uniformly differentiable at x with derivative A (A a linear mapping of R_n into R_m) if for each $\varepsilon > 0$ there is a neighborhood V_ε of x such that

$$\frac{|f(u) - f(v) - A(u-v)|}{|u-v|} < \varepsilon$$

for all $u,v \in V_\varepsilon$. (Here $|\cdot|$ denotes the usual Euclidean norm).

Local inverse function theorem (standard)

Let f map a neighborhood V of $x \in R_n$ into R_n. If f is uniformly differentiable at x with invertible derivative A, then f maps a neighborhood $V \subset U$ of x injectively onto a neighborhood of f(x) and the inverse function f^{-1} defined on f(V) is uniformly differentiable at x with derivative A^{-1}.

Before translating these statements into the nonstandard model we need some notation. Two nonstandard real numbers are said to be infinitesimally close together, $x \sim y$, if x-y is infinitesimal. Two vectors $x,y \in R^*_n$ will be said to be infinitesimally close together mod r, $x \tilde{\sim} y$ (mod r), if $r \in R^*$ and

$\frac{|x-y|}{r}$ is infinitesimal. This corresponds to the idea of using r as a unit of length. We will use obvious properties of the relation without proof. For example, it is convenient to know that if $s \in R^*$ and $\frac{r}{s}$ is finite but not infinitesimal, then $x \approx y \pmod r$ iff $x \approx y \pmod s$.

For each $x \in R_n$, let $M(x)$, be the euclidean monad of x, $M(x) = \{y \in R_n^*: x \approx y \pmod 1\}$. Let f be a nonstandard mapping of M(x) into R_m^* and let A be a (standard) linear mapping of R_n into R_m.

Definition (nonstandard).

The mapping $f:M(x) \to R_m^*$ is said to be uniformly differentiable at x with derivative A if

$$f(u) \approx f(v) + A(u-v) \pmod{|u-v|}$$

for all $u,v \in M(x)$.

Local inverse function theorem (nonstandard)

Let $f:M(x) \to R_n^*$ be an internal function which is uniformly differentiable with invertible derivative A. Then f defines an injective mapping of M(x) onto M(f(x)) and the resulting inverse mapping $f^{-1}:M(f(x)) \to M(x)$ is uniformly differentiable with derivative A^{-1}.

Proof. If A is invertible, then $\frac{|A(h)|}{|h|} \geq \|A^{-1}\|^{-1}$, so that $f(y+h)-f(y) \approx A(h) \not\approx 0$ mod $(|h|)$ for all $y \in M(x)$, $h \in M(0)$, $h \neq 0$. In particular, $f(y+h) \neq f(y)$ for all $y \in M(x)$, $h \in M(0)$, $h \neq 0$; i.e., f is one-to-one in M(x).

Suppose $u \in M(f(x)) \setminus f(M(x))$. The closed ball $S(x, 3|u-f(x)|\, \|A^{-1}\|)$ of radius $3|u-f(x)|\, \|A^{-1}\|$ is compact and f is continuous on M(x) (that is, on every internal subset of M(x)), so that the set $f(S(x, 3|u-f(x)|\, \|A^{-1}\|))$ is compact. Let $v \in f(S(x, 3|u-f(x)|\, \|A^{-1}\|))$ be such that $|u-v| = \mathrm{dist}(u, f(S(x, 3|u-f(x)|\, \|A^{-1}\|)))$. In particular, $|u-v| \leq |u-f(x)|$. If $y \in M(x)$ with $|y-x| \geq 3|u-f(x)|\, \|A^{-1}\|$, then $f(y)-f(x) \approx A(y-x)$ mod $(|y-x|)$ and $|A(y-x)| \geq \|A^{-1}\|^{-1}|y-x| \geq |3|u-f(x)||$ so that $|f(y)-f(x)| \geq 2|u-f(x)|$, and the set $S(u, |u-v|) \cap F(M(x))$ is empty. But this is certainly not the case. For, choosing $y \in M(x)$ with $f(y) = v$, and letting $h = A^{-1}(u-v)$, and $z = y+h$, we have that $f(z) \approx f(y) + A(h) = u$ (mod $(|u-v|)$).

Another proof can be given by defining a sequence in $M(x)$ inductively by $x_{n+1} = x_n + A^{-1}(u-f(x_n))$. Since $f(x_{n+1}) \approx u \bmod (|u-f(x_n)|)$ implies that $|u-f(x_{n+1})| \le \frac{1}{2}|u-f(x_n)|$, and since $|x_n - x_{n+1}| \le \| A^{-1} \| \, |u-f(x_n)|$, it must be that $\{x_n\}$ is a Cauchy sequence and $\{f(x_n)\}$ converges to u. Since $\{x_n\}$ is internal, and R_n^* is complete, $\{x_n\}$ converges to some $y \in M(x)$ so that $f(y) = u$.

To see that f^{-1} is uniformly differentiable at x, let $u, u+h \in M(f(x))$ and let $u = f(y)$ and $u+h = f(y+k)$. Then $u+h = f(y+k) \approx f(y)+A(k) = u+A(k) \bmod (|k|)$ so that $h \approx A(k) \bmod (|k|)$. This implies that $A^{-1}(h) \approx k \bmod (|h|)$, i.e. that $f^{-1}(u+h) \approx f^{-1}(u)+A^{-1}(h) \bmod (|h|)$. □

Note 1. It is an easy exercise to see that a function $f:V \to R^m$ defined on an open set $V \subset R^n$ is continuously differentiable in V iff it is uniformly differentiable at each point of V.

Note 2. It is easy to give examples of functions which are uniformly differentiable at a point but not differentiable at all points of any neighborhood of that point. For example, let $f:R \to R$ be such that f is continuous, $f(\pm \frac{1}{n}) = \frac{1}{2}n$, $n \ge 1$, and such that f is linear except at the points $0, \pm 1, \pm \frac{1}{2}, \ldots$.

Note 3. Our nonstandard definition of uniform differentiability is suitable for the study of other theorems on vector valued functions. For instance, a simple and intuitive proof of the formula for change of variable in an integral involving the Jacobian can be given by partitioning the domain into infinitesimal boxes, and a construction of integral curves for continuously differentiable vector fields can be given by fitting together infinitesimal line segments in each monad.

Bibliography

A. Robinson, Nonstandard Analysis, North-Holland (Amsterdam, 1966).

W. Rudin, Principles of Mathematical Analysis, McGraw-Hill (1964).

NONSTANDARD TOPOLOGICAL VECTOR SPACES*

Steven F. Bellenot

Claremont Graduate School

We prove a well known theorem on the uniqueness of finite dimensional Hausdorff topological vector spaces with the aid of nonstandard analysis. Standard proofs are based on duality, either directly (Robertson and Robertson [2, Prop. 11, p. 37]) or in showing the continuity of certain linear maps (Schaefer [5, Th. 3.2, p. 21] or Treves [6, Th. 9.1, p. 79]). Nonstandardly, we show that every such topology has the same monad at the origin. Although this is hardly a deep result it does illustrate the simplifying power of nonstandard analysis. We shall assume familiarity with nonstandard analysis via Robinson [3] or Robinson and Zakon [4].

1. Preliminaries

A TVS (topological vector space) is a vector space E over Φ (which will always be either the reals or the complexes) with a topology such that both vector addition: $E \times E \to E$ and the scalar multiplication: $\Phi \times E \to E$ are continuous maps. For $x \in E$ we define the monad of x to be $\mu(x) = \bigcap {}^*U$ as U runs over a neighborhood basis for x. $\mu(x)$ is independent of the choice of the neighborhood basis. For notation's sake we will refer to $\mu(0)$ (0 the origin of E) as just μ. The following statements have proofs similar to results on topological groups in Parikh [1, pp. 280-2]: let $x,y \in E$, $\lambda \in \Phi$, then:

(A) $\mu(x) + \mu(y) = \mu(x) + y = \mu(x+y) = x + y + \mu$.

(B) $\delta(\lambda)x \subset \delta(\lambda)\mu(x) = \lambda\mu(x) = \mu(\lambda x)$ (where $\delta(\lambda)$ is the monad of λ with respect to the usual topology on Φ.)

(C) The topology is Hausdorff iff $\mu \cap E = \{0\}$.

(D) If ξ, η are two TVS topologies on E with monads μ and ω respectively, then $\xi = \eta$ iff $\mu = \omega$.

---

*This work was supported in part by NSF Research Grant GP-20838.

From (D) it is clear that the topology of a TVS is completely determined by the monad of the origin (translation invariance of the topology). Let $\mathcal{U}$ be a filter base of subsets of E about the origin and further let μ be its monad (i.e., $\mu = \cap^* U \quad (U \in \mathcal{U})$).

Proposition 1. $\mathcal{U}$ is a neighborhood base of the origin for a TVS structure on E iff the following are true:

(a) $x \in E$ and λ an infinitesimal element of $^*\Phi$ imply $\lambda x \in \mu$.

(b) μ is a vector space over Φ.

(c) $\delta(0)\mu \subset \mu$.

Proof: If $\mathcal{U}$ is a neighborhood basis of the origin then clearly (A) and (B) above imply (a), (b) and (c).

Conversely, we define a topology on E by translating $\mathcal{U}$ to form a neighborhood basis, $\mathcal{U}_x$, at $x \in E$; that is $\mathcal{U}_x = x + V\colon V \in \mathcal{U}\}$ (Treves [6, pp. 23-4]). Clearly, for x an element of E, $\mu(x) = x + \mu$. To complete the proof we need to show the continuity of scalar multiplication and vector addition. We need to show that for x,y in E and λ an element of ϕ, then $\delta(\lambda)\mu(x) \subset \mu(\lambda x)$ and $\mu(x) + \mu(y) \subset \mu(x+y)$ (Robinson [3, Th. 4.2.7, p. 98]).

Now $\delta(\lambda)\mu(x) = (\lambda+\delta(0))(x+\mu) = \lambda x + x\delta(0) + \lambda\mu + \delta(0)\mu$. By (a), $x\delta(0) \subset \mu$; by (b), $\lambda\mu \subset \mu$; and by (c), $\delta(0)\mu \subset \mu$. Thus, by applying (b) again we have $\delta(\lambda)\mu(x) \subset \lambda x + \mu + \mu + \mu = \lambda x + \mu = \mu(\lambda x)$; and the continuity of scalar multiplication. The continuity of vector addition follows similarly, and we are done. (Compare Parikh [1, Th. 1, p. 280].)

2. Finite-Dimensional TVS's

Theorem 1: For each finite dimension, there is only one Hausdorff TVS topology.

Proof: Let E be a finite dimension vector space with basis $\{x_1,x_2,\ldots x_n\}$. Note that $\{x_1,x_2,\ldots x_n\}$ is also a *basis for $^*E$. Suppose μ is the monad of a Hausdorff TVS topology on E. By Proposition 1(a) $\lambda x_i \in \mu$ if λ is infinitesimal, $i = 1,2,\ldots n$. Since μ is a vector space, it contains the set

$\{\Sigma_{i=1}^{n} \lambda_i x_i :$ all λ_i infinitesimal$\}$. Actually the sets are equal. For, if not, there would exist, in μ, an element of the form $\Sigma_{i=1}^{n} \lambda_i x_i$, where not all the λ_i are infinitesimal. We can assume all the λ_i are finite, multiplying by the reciprocal of max $|\lambda_i|$ if necessary. Then, by (B), all the λ_i can be taken to be finite standard scalars, at least one non-zero. This contradicts the fact that the topology is Hausdorff via (C). Thus any two Hausdorff TVS topologies on E must have the same monad, hence, by (D), they are identical. (Compare Schaefer [5, Th. 3.2, p. 21].)

REFERENCES

1. R. Parikh, "A Nonstandard Theory of Topological Groups," Applications of Model Theory to Algebra, Analysis, and Probability, (W.A.J. Luxemburg, editor) Holt, Rinehart and Winston, New York, 1969.

2. A. P. Robertson and W. J. Robertson, Topological Vector Spaces, Cambridge Tracts in Mathematics and Mathematical Physics 53, Cambridge Univ. Press, Cambridge, 1964.

3. A. Robinson, Non-Standard Analysis, Studies in Logic and the Foundations of Mathematics, North Holland, Amsterdam, 1966.

4. A. Robinson and E. Zakon, "A Set-Theoretical Characterization of Enlargements." Applications of Model Theory to Algebra, Analysis and Probability (W.A.J. Luxemburg, editor) Holt, Rinehart and Winston, New York, 1969.

5. H. Schaefer, Topological Vector Spaces, MacMillan, New York, 1966.

6. F. Treves, Topological Vector Spaces, Distributions, and Kernels, Pure and Applied Mathematics 25, Academic Press, New York, 1967.

A NON-STANDARD INTEGRATION THEORY FOR UNBOUNDED FUNCTIONS

Allen R. Bernstein
University of Maryland

and

Peter A. Loeb[1]
Yale University and
University of Illinois, Urbana

1. INTRODUCTION

This paper contains a development of the non-standard theory of integration based on the approach taken in [8]-[10].

Here, however, instead of truncating the extension $*f$ of an unbounded integrable function f (i.e. replacing $*f$ with $-\omega \vee *f \wedge \omega$ where ω is a suitable infinite integer) as in [8] and [9], we restrict $*f$ to a fixed proper subset A of the extension of the measure space. The latter process, being independent of the choice of f, is linear. However, the set A does depend on the choice of the measure whereas the truncation process in [9] is not dependent on the measure.

The non-standard theory of measure has been examined by several authors. Robinson's book [11] contains a brief account of the extension of Lebesgue integration from the standard to the non-standard world. A different approach was taken in [6] where Lebesgue measure was defined as a counting measure with the condition among other things that the measure of a singleton

[1] This work was supported by N.S.F. Grant NSF GP 14785 and a grant from the University of Illinois Center for Advanced Study.

real number be a non-zero infinitesimal. Another approach is taken in [8]-[10] which develops general measure theory concentrated on sets instead of points. It is this latter approach that we shall extend here.

An integration theory based on the approach in [6] may be found in [3], and in Henson's work [7].

We shall continue to use in this paper the notation established by one of the authors in [10].

2. INTEGRATION

Let $(X,\mathfrak{M},\mu)$ be a measure space. We will assume that X is σ-finite with respect to μ so that we may, by restricting integration to a subset Y of $*X$, obtain the proper values for the integrals of all standard positive $\mathfrak{M}$-measurable functions. We will show, however, that it is not necessary to assume that X is σ-finite if one only wants to consider the integrable functions.

Assume that $X = \bigcup_{n=1}^{\infty} X_n$ with $\mu(X_n) < \infty$ and $X_n \subset X_{n+1}$ for $n = 1,2,\ldots$. (If $\mu(X) < \infty$, let $X_n = X$ for all $n\in N$.) Let M_{R+} be the non-negative measurable <u>real-valued</u> functions on $(X,\mathfrak{M})$, L^+ be the μ-integrable functions in M_{R+} and $M^{\infty} = M_{R+} - L^+$. Let M_F be a *finite subset of $*M_{R+}$ such that if $h\in M_{R+}$ then $*h\in M_F$. For brevity, we often write $\int h$ for $\int h d\mu$ or $\int h\, d*\mu$. Given a partition $P\in *\mathcal{P}$ with index set I, and a set $C\in *\mathfrak{M}$, we let $I(C) = \{i\in I: A_i\in P,\ A_i \subset C, \text{ and } \mu(A_i) > 0\}$; $|I(C)|$ denotes the number (in $*N$) of elements in $I(C)$.

1. THEOREM. For a given choice of $\gamma\in m^+(0)$ and $\omega_0\in *N-N$, there exist integers η and β in $*N - N$ such that

(i) if X_η is the set corresponding to η in the collection $*\{X_n\}$, then there is an internal set $Y \subset X_\eta$ with the property

$$*\mu(Y) > *\mu(X_\eta) - \gamma$$

and the following additional properties:

(ii) For each $f \in L^+$, $\int_Y *f\, d*\mu > \int_X f d\mu - \gamma$.

(iii) For each $g \in M^\infty$, $\int_Y *g\, d*\mu > \omega_0$.

(iv) For each $h \in M_F$, $\sup_Y h \leq \beta$.

Note that $X_\eta = *X$ if $\mu(X) < \infty$.

<u>Proof</u>: Let f_s be the sum of the functions in $M_F \cap *L^+$; then $f_s \in *L^+$. Choose $\eta \in *N-N$ so that $\int_{X_\eta} f_s > \int_{*X} f_s d*\mu - \frac{\gamma}{2}$ and $\int_{X_\eta} g > \omega_0$ for each $g \in M_F \cap *M^\infty$. Given $h \in M_F \cup \{f_s\}$, $X_\eta = \bigcup_{n=1}^{\infty} \{x \in X_\eta : h(x) \leq n,\ n \in *N\}$. Since $*\mu(X_\eta) \in *R$, we have

(1) $\lim_{\substack{n \to \infty \\ n \in *N}} *\mu(\{x \in X_\eta : h(x) > n\}) = 0$. We also have

(2) $\lim_{\substack{n \to \infty \\ n \in *N}} \int_{\{x \in A:\ h(x) \leq n\}} g\, d*\mu = \int_A g\, d*\mu$

for each internal set $A \subset X_\eta$ and each $g \in M_F \cup \{f_s\}$. Therefore, we may choose $Z_0 \subset X_\eta$ so that f_s is bounded on

$X_\eta - Z_0$, $*\mu(Z_0) < \frac{1}{2}\gamma$, $\int_{X_\eta - Z_0} f_s > \left(\int_{X_\eta} f_s\right) - \frac{\gamma}{4}$, and

$$\int_{X_\eta - Z_0} g > \omega_0$$

for each $g \in M_F \cap *M^\infty$. Let $\{g_1, g_2, \ldots, g_\nu\}$ be an enumeration of $M_F \cap *M^\infty$. Given Z_{i-1} for $1 \leq i \leq \nu$, we choose $Z_i \subset X_\eta$ so that g_i is bounded on $X_\eta - Z_i$, $*\mu(Z_i) < \frac{1}{2^{i+1}}\gamma$,

$$\int_{X_\eta - Z_i} f_s > \left(\int_{X_\eta} f_s\right) - \frac{1}{2^{i+2}}\gamma \text{ , and } \int_{X_\eta - \bigcup_{j=0}^{i} Z_j} g > \omega_0$$

for each $g \in M_F \cap *M^\infty$. Now set $Y = X_\eta - \bigcup_{i=0}^{\nu} Z_i$, and we are done.

Note: If we only wish to find a subset $Y \subset *X$ such that $*\mu(Y) < +\infty$, the function f_s is bounded on Y by some number $\beta \in *N$, and Condition (ii) holds for a given $\gamma \in m^+(0)$, then we do not need to assume that X is σ-finite. Simply let $A = *X$ and $h = g = f_s$ in Equation 2. The existence of Y follows.

2. THEOREM. Given γ, ω_0, η, β and Y as above, there is a partition $P_0 \in *\mathcal{P}_0$ such that Y is exactly the union of sets from P_0 and for any partition $P \geq P_0$ in $*\mathcal{P}$, we have the following properties in terms of the index set I for P and an arbitrary choice of points $x_i \in A_i$, $i \in I(Y)$:

$$\text{(i)} \quad \left| \int_B f d\mu - \sum_{i \in I\,(Y \cap *B)} *f(x_i)\, *\mu(A_i) \right| < 2\gamma$$

for each $f \in L^+$ and $B \in \mathcal{M}$, and

(ii) $\sum_{i\in I(Y)} {}^*g(x_i)\ {}^*\mu(A_i) > \omega_0 - \gamma$

for each $g\in M^\infty$.

Proof: Each $h\in M_F$ is bounded by β on Y. Therefore, we can find a partition $P_0\in{}^*\mathcal{P}_0$ so that $P_0 \geq \{Y, {}^*X-Y\}$ and for each set $C\in P_0$ with $C\subset Y$ and each $h\in M_F$ we have $\sup_C h - \inf_C h < \frac{\gamma}{{}^*\mu(Y)}$. Now take any partition $P \geq P_0$ with index set I and choose a point $x_i\in A_i$ for each $i\in I(Y)$. Given any $h\in M_F$, we have

$$\int_Y h\ d{}^*\mu = \sum_{i\in I(Y)} (h(x_i)+\delta_i){}^*\mu(A_i)$$

where $|\delta_i| < \frac{\gamma}{{}^*\mu(Y)}$ for every $i\in I(Y)$. Thus

$$\left|\int_Y h\ d{}^*\mu - \sum_{i\in I(Y)} h(x_i)\ {}^*\mu(A_i)\right| < \frac{\gamma}{{}^*\mu(Y)} \sum_{i\in I(Y)} {}^*\mu(A_i) = \gamma.$$

If $f\in L^+$, and $B\in\mathfrak{M}$, we let $h = {}^*f\cdot\chi_{{}^*B}$ where $\chi_{{}^*B}$ denotes the characteristic function of ${}^*B$. The rest is clear.

3. COROLLARY. If μ is a non-atomic measure, we may choose the partition $P \geq P_0$ with index set I so that for any choice of $x_i\in A_i$, $i\in I(Y)$, and any $f\in L^+$, $g\in M^\infty$, and $B\in\mathfrak{M}$, we have

$$\left| \int_B f d\mu - \frac{*\mu(Y)}{|I(Y)|} \cdot \sum_{i \in I(Y \cap *B)} *f(x_i) \right| < 3\gamma$$

and

$$\frac{*\mu(Y)}{|I(Y)|} \sum_{i \in I(Y)} *g(x_i) > \omega_0 - 2\gamma.$$

Proof: Recall that β is an upper bound for the functions in M_F on Y. By Theorem 5 of [10], there is a partition $P \geq P_0$ with index set I such that for each $i \in I(Y)$,

$$\left| *\mu(A_i) - \frac{*\mu(Y)}{|I(Y)|} \right| < \frac{\gamma}{|I(Y)| \cdot \beta}.$$

Therefore, given $h \in M_F$ and $B \in \mathfrak{M}$, we have

$$\left| \sum_{i \in I(Y \cap *B)} h(x_i) *\mu(A_i) - \frac{*\mu(Y)}{|I(Y)|} \cdot \sum_{i \in I(Y \cap *B)} h(x_i) \right|$$

$$\leq \sum_{i \in I(Y \cap *B)} h(x_i) \left| *\mu(A_i) - \frac{*\mu(Y)}{|I(Y)|} \right|$$

$$< \frac{\gamma}{|I(Y)| \cdot \beta} \cdot \sum_{i \in I(Y \cap *B)} h(x_i) \leq \gamma.$$

4. COROLLARY. If $\mu(X) = 1$, then for any $f \in L^+$, $g \in M^\infty$ and $B \in \mathfrak{M}$, we have

$$\int_B f d\mu \simeq \frac{1}{|I(*X)|} \sum_{i \in I(*B)} *f(x_i)$$

and

$$\frac{1}{|I(^*X)|} \sum_{i \in I(^*X)} {}^*g(x_i) > \omega_0 - 1.$$

3. PROJECTIONS IN $L_1(X,\mathfrak{M},\mu)$

We conclude this paper with an application of the integration theory developed here to operator theory on $L_1(X,\mathfrak{M},\mu)$. By $L_1(X,\mathfrak{M},\mu)$ we mean the Banach space of real-valued integrable functions on $(X,\mathfrak{M},\mu)$ with the norm $||f|| = \int_X |f|d\mu$, the complex case could be handled equally as well.

The reader may recall previous applications of non-standard analysis to the theory of linear operators on Hilbert space (cf. [1], [2], [4], [5], [11]). All of these applications proceded by first reducing a problem concerning a Hilbert space H to one concerning a suitable *finite-dimensional subspace H_ω of $^*H$ and then lifting the result back up to H. This could be done because of the existence of a linear, idempotent operator T with finite norm from $^*H$ onto H_ω such that $T^*h \sim {}^*h$ for standard h, where $T^*h \sim {}^*h$ means the norm of $T^*h - {}^*h$ is infinitesimal.

The problem of extending such a method to deal with operators on a Banach space often hinges precisely on obtaining analogous projection operators onto suitable *finite-dimensional subspaces. For example in [2], by shifting to an equivalent norm and assuming an operator had no invariant subspace it was possible to define a bounded, idempotent, semi-linear projection T which was used in proving a theorem concerning invariant subspaces. Also if one is able to establish the existence of a suitable type of basis for a separable Banach space then it is possible to define projections there having all the desired properties.

What we wish to show here is that our development of integration theory provides a very natural way to define suitable projection operators on $L_1(X,\mathfrak{M},\mu)$ for any measure space $(X,\mathfrak{M},\mu)$. (Recall the remark following Theorem 1). These projections may then be used, as in the Hilbert space case, to study questions concerning bound linear operators by using suitable finite-dimensional results. Observe that since we have concentrated measures on sets, we can obtain these projection operators with finite norms.

5. THEOREM. Let $(X,\mathfrak{M},\mu)$ be a measure space and let $\mathfrak{B} = L_1(X,\mathfrak{M},\mu)$. Then there is a *finite-dimensional subspace $\mathfrak{B}_\omega$ of $^*\mathfrak{B}$ and an operator T from $^*\mathfrak{B}$ onto $\mathfrak{B}_\omega$ such that

(i) T is linear, idempotent, and has finite norms,

(ii) $T^*f \sim {}^*f$ for any $f \in \mathfrak{B}$.

<u>Proof</u>: Given $\gamma \simeq 0$, $\gamma > 0$, let Y be the set given by Theorem 1 and the remark following Theorem 1, and let P_0 be the partition defined in the proof of Theorem 2. Assume P_0 has index set I, and let $\mathfrak{B}_\omega$ be the space of simple functions which are constant on each $A \in P_0$ and vanish on $X-Y$. Clearly, $\mathfrak{B}_\omega \subset {}^*\mathfrak{B}$. For each function $f \in {}^*\mathfrak{B}$, set $T(f) = \sum_{i \in I(Y)} a_i \chi_{A_i}$,

where for each $i \in I(Y)$ χ_{A_i} is the characteristic function of $A_i \in P_0$ and $a_i = \int_{A_i} f d^*\mu / {}^*\mu(A_i)$. Clearly, T is linear and idempotent. If $f \in {}^*\mathfrak{B}$ is nonnegative,

$||T(f)|| = \int_Y fd{*\mu} \leq \int_X fd{*\mu} = ||f||$, so for arbitrary $f \in {*\mathcal{B}}$,

$||T(f)|| \leq ||T(f^+)|| + ||T(f^-)|| \leq ||f^+|| + ||f^-|| \leq 2||f||$,
whence $||T|| \leq 2$. Finally, the fact that for any $f \in \mathcal{B}$, $T(f) \sim f$, follows from the fact that for $f \geq 0$, $f \in \mathcal{B}$, we have

$$\int_X |{*f} - T({*f})| d\mu$$

$$= \int_{X-Y} {*f}d{*\mu} + \sum_{i \in I(Y)} \int_{A_i} \left| {*f}(x) - \left(\int_{A_i} {*f}(x)d{*\mu} / {*\mu}(A_i) \right) \right| d{*\mu}$$

$$\leq \gamma + {*\mu}(Y) \cdot [\max_{i \in I(Y)} (\sup_{A_i} {*f} - \inf_{A_i} {*f})] < 2\gamma \simeq 0,$$

by the choice of the partition P_0 and Theorems 1 and 2.

REFERENCES

[1] A. BERNSTEIN, Invariant subspaces for linear operators, doctoral dissertation, U.C.L.A. 1965, unpublished.

[2] ——————, Invariant subspaces of polynomially compact operators on Banach space, Pac. J. of Math., 21(1967), pp. 445-463.

[3] ——————, A non-standard integration theory for unbounded functions, University of Maryland Technical Report, TR 71-50, 1971.

[4] ——————, Invariant subspaces for certain commuting operators on Hilbert space, Annals of Math., 95(1972), pp. 253-260.

[5] A. BERNSTEIN and A. ROBINSON, Solution of an invariant subspace problem of K. T. Smith and P. R. Halmos, Pac. J. of Math., 16(1966), pp. 421-431.

[6] A. BERNSTEIN and F. WATTENBERG, Non-Standard Measure Theory, Applications of Model Theory to Algebra, Analysis and Probability, ed. W. A. J. Luxemburg, Holt, Rinehart and Winston, 1969.

[7] C. HENSON, On the non-standard representation of measures, Trans. Amer. Math. Soc., 172, October, 1972, pp. 437-446.

[8] P. A. LOEB, A non-standard representation of measurable spaces and L_∞, Bull. Amer. Math. Soc. 77, No. 4, July 1971, pp. 540-544.

[9] ———, A non-standard representation of measurable spaces, L_∞ and L_∞^*, Contributions to Non-Standard Analysis, Ed. by W. A. J. Lexemburg and A. Robinson, North-Holland, 1972, pp. 65-80.

[10] ———, A non-standard representation of Borel measures and σ-finite measures, this volume.

[11] A. ROBINSON, Non-Standard Analysis, North Holland, Amsterdam, 1966.

CARDINALITY-DEPENDENT PROPERTIES OF TOPOLOGICAL SPACES

Allen R. Bernstein
University of Maryland
and
Frank Wattenberg
University of Massachusetts

I. INTRODUCTION

The non-standard theory of topological spaces has received considerable attention by a number of authors. One of the reasons for this is the striking way in which many of the basic notions of topology may be characterized by passage to a non-standard model. These characterizations take the form of an equivalence between a standard property and a simpler or more intuitive property in an appropriate non-standard model. In many cases the passage from the standard to the non-standard world is valid in an arbitrary non-standard model while the converse direction requires that the model be an enlargement. For example, the properties of being compact, Hausdorff, open, continuous, etc. all follow this pattern [6].

There are certain other properties, however, which have not been amenable to such treatment. In particular those properties whose definitions mention a particular infinite cardinality are in this category. A perfect example of this is the property of being a Baire space since its definition depends on the cardinal $\aleph_0$. The usual techniques of non-standard analysis have been ineffective in dealing with such properties. Arbitrary non-standard models, enlargements, or even saturated models simply appear too insensitive to this type of cardinality condition to yield successful non-standard characterizations.

The approach we take in this paper is to investigate certain cardinality-dependent properties by working with non-standard models which are weaker than those generally used. Specifically, in order to investigate a property which involves cardinality α, we work with ultrapowers with respect to α.

In section 2 we give a characterization of α-Baire spaces in terms of ultraproducts over sets of cardinality α. Thus in particular, Baire spaces can be characterized by looking at countably-indexed ultrapowers. In section 3 we examine certain equivalence conditions between topological spaces which rely on the cardinal $\aleph_0$ and explore their relationship with countably indexed ultrapowers.

The reader may wish to consult [1] and [12] both of which focus on the extension of topological spaces to countably indexed ultrapowers. However, except for the rudiments of the theory of non-standard analysis and ultrapowers this paper is self-contained.

II. α-BAIRE SPACES

Let (X,σ) denote a topological space where X is the set of points and σ is the set of open sets. We denote by C_σ the set of closed sets of (X,σ). If $A \subseteq X$ then $c\ell_\sigma(A)$ denotes the closure of A in (X,σ).

Now let $\mathcal{M}$ be the complete higher order structure of X so $\mathcal{M}$ contains all objects of finite type built up using X as the set of atoms. Let D be an ultrafilter on a set I. We denote by ${}^D\mathcal{M}$ the D-power of $\mathcal{M}$ (see [3] for a suitable framework in which to form this "higher-order" ultrapower). For any object Z in $\mathcal{M}$ we denote by ${}^DZ$ the corresponding object in ${}^D\mathcal{M}$ denoted by the same symbol of the formal language. For $x \in {}^DX$, ${}^D\mu_\sigma(x)$ denotes the σ-monad of x, ${}^D\mu_\sigma(x) = \bigcap\{{}^D\mathcal{V} \mid x \in \mathcal{V} \in \sigma\}$. We denote by $\mathcal{V}(I)$ the set of ultrafilters on I and by $Npu(I)$ the set of non-principle ultrafilters on I. If I is the set $S_\omega(J)$ of all finite subsets of a set J then there exist ultrafilters on I which contain every set of the form $\hat{j} = \{i \in I \mid j \in i\}$ where $j \in J$. An ultrafilter is <u>regular</u> if it is isomorphic to such an ultrafilter. We denote by ω the set of non-negative integers.

<u>Definition</u>. Let α be an infinite cardinal. (X,σ) is <u>α-Baire</u> if and only if the union of α nowhere-dense closed sets contains no non-empty open set.

For $\alpha = \aleph_0$ the above definition becomes simply the familiar definition of Baire space. For $\alpha > \aleph_0$ the property of being α-Baire has become important because of its connection with Martin's axiom. In fact the following topological version of Martin's axiom is known to be equivalent to the usual versions and has been used in a number of applications to topological questions, e.g. the Souslin hypothesis, the normal Moore space problem, etc. ([2], [9], [10]). Martin's axiom (together with $2^{\aleph_0}$ = anything reasonable) is relatively consistent with the usual axioms of set theory ([4], [8]).

Topological Version of Martin's Axiom. Let (X,σ) be a compact Hausdorff space in which any pairwise disjoint collection of open sets is countable. Then (X,σ) is α-Baire for all $\alpha < 2^{\aleph_0}$.

Thus Martin's axiom (together with $\aleph_1 < 2^{\aleph_0}$) provides examples of reasonable spaces which are α-Baire for $\alpha > \aleph_0$. In addition we see that any non-standard characterization of α-Baire spaces immediately provides a non-standard version of Martin's axiom.

Definition. Let D be an ultrafilter. (X,σ) is D-Baire if and only if for all $K \in {}^D C_\sigma$,

$$\exists V \in \sigma(\emptyset \neq V \subseteq K) \Rightarrow \exists V \in {}^D\sigma(\emptyset \neq V \subseteq K).$$

Note that the condition $\exists V \in \sigma(\emptyset \neq V \subseteq K)$ is external--it says that the set K in ${}^*\mathfrak{M}$ contains all the standard points of the non-empty open set V in $\mathfrak{M}$. Observe also that if (X,σ) contains no isolated points and if ${}^D\mathfrak{M}$ is an enlargement of $\mathfrak{M}$, then (X,σ) is not D-Baire. This follows from the fact that there is a *finite element of ${}^D C_\sigma$ which contains all the standard points.

Theorem 2.1. Let (X,σ) be a topological space, α an infinite cardinal, and $|I| = \alpha$.

(1) If D is an ultrafilter on I, then X is α-Baire $\Longrightarrow$ X is D-Baire.

(2) If D is a regular ultrafilter on I, then X is D-Baire $\Longrightarrow$ X is α-Baire.

Proof of (1). Let $K \in {}^D C_\sigma$, thus $K = \{K_i\}_{i\in I}/D$ where $K_i \in C_\sigma$. Suppose $V \in \sigma$, $\emptyset \neq V \subseteq K$. Let $H = \{i \in I \mid \not\exists\, \mathcal{V} \in \sigma(\emptyset \neq \mathcal{V} \subseteq K_i)\}$ and suppose $H \in D$. Let $x \in V$. Since $V \subseteq K$ we have $x \in K = \{K_i\}_{i\in I}/D$, hence $\{i \mid x \in K_i\} \in D$. Thus $\{i \mid x \in K_i\} \cap H \in D$, in particular $x \in \bigcup_{i\in H} K_i$. This shows that $V \subseteq \bigcup_{i\in H} K_i$ where

$|H| \leq \alpha$. Since (X,σ) is α-Baire there must be some $i \in H$ such that K_i contains a non-empty open set. But this contradicts the definition of H.

The above contradiction shows $H \notin D$, hence $I - H \in D$. Define $W = \{W_i\}_{i \in I}/D$ as follows. For $i \in I - H$ choose $W_i \in \sigma$ such that $\emptyset \neq W_i \subseteq K_i$ and for $i \in H$ let $W_i = \emptyset$. Then $\{i \mid \emptyset \neq W_i$ and $W_i \in \sigma$ and $W_i \subseteq K_i\} \in D$ so $\emptyset \neq W \in {}^D\sigma$ and $W \subseteq K$. This completes the proof of (1).

Proof of (2). Let D be a regular ultrafilter on I, $I = S_\omega(J)$, with $|I| = |J| = \alpha$. Suppose (X,σ) is D-Baire. Let $\{K_j\}_{j \in J}$ be a collection of α closed sets such that for some open set V, $\emptyset \neq V \subseteq \bigcup_{j \in J} K_j$. For $i \in I$, $i = \{j_1, \cdots, j_n\}$, let $L_i = K_{j_1} \cup K_{j_2} \cup \cdots \cup K_{j_n}$ and consider the set $L = \{L_i\}_{i \in I}/D \in {}^D C_\sigma$. If $x \in V$ then $x \in \bigcup_{j \in J} K_j$ so $x \in K_g$ for some $g \in J$. But then for $i \in \hat{g}$, $K_g \subseteq L_i$ so $x \in L_i$. Thus $\{i \mid x \in L_i\} \in D$, consequently $x \in L$. This shows $V \subseteq L$.

Since (X,σ) is D-Baire there is a $W \in {}^D\sigma$ such that $\emptyset \neq W \subseteq L$. If we write $W = \{W_i\}_{i \in I}/D$ then $\{i \mid \emptyset \neq W_i\} \cap \{i \mid W_i \subseteq L_i\} \cap \{i \mid W_i \in \sigma\} \in D$. Thus we may choose an $h \in I$ such that $\emptyset \neq W_h \in \sigma$ and $W_h \subseteq L_h$. But $L_h = K_{j_1} \cup K_{j_2} \cup \cdots \cup K_{j_n}$ for some $j_1, j_2, \cdots, j_n \in J$. If $W_h \nsubseteq K_{j_1}$ then $W' = W_h \cap (X - K_{j_1}) \subseteq K_{j_2} \cup K_{j_3} \cup \cdots \cup K_{j_n}$ where $\emptyset \neq W' \in \sigma$. By a repetition of this argument either $W' \subseteq K_{j_2}$ or $K_{j_3} \cup K_{j_4} \cup \cdots \cup K_{j_n}$ contains a non-empty open set. By continuing this reasoning we conclude that for some m, $1 \leq m \leq n$, K_m contains a non-empty open set. This proves that (X,σ) is α-Baire which completes the proof of theorem 2.1.

Since every non-principle ultrafilter on a countable set is isomorphic to a regular ultrafilter we have the following:

Corollary. Let D be a non-principle ultrafilter on ω. (X,σ) is a Baire space if and only if (X,σ) is D-Baire.

The above corollary may be used to give typical non-standard proofs of results which depend on the Baire category theorem. For example:

Uniform Boundedness Theorem. Let $\{T_\eta\}_{\eta \in J}$ be a collection of linear operators on a Banach space X such that for each $x \in X$ $\sup_{\eta \in J} \|T_\eta(x)\| = m(x) < \infty$. Then there exists an m such that $\|T_\eta\| \leq m$ for all $\eta \in J$.

Proof. By the Baire category theorem, X is a Baire space so by the previous corollary it is D-Baire for any ultrafilter D on ω. In ${}^D\mathcal{M}$ let ν be an infinite integer so $X \subseteq K = \{y \in {}^DX \mid T_\eta(y) \leq \nu$ for all $\eta \in {}^*J\}$ where K is a * closed set in ${}^DX$. Since X is D-Baire there is a * open set $S \subseteq K$ and $y \in {}^DX$ such that $S = \{z \in {}^DX \mid \|z - y\| < \eta\}$. Hence for $\|z\| \leq 1$ and $\eta \in {}^*J$, $\|T_\eta(z)\| \leq \frac{2}{\eta}\ {}^*m(y) + \nu$. Transferring to $\mathcal{M}$, $\exists\, m$ such that $\|T_\eta\| \leq m$ for all $\eta \in J$ which completes the proof.

We conclude this section with the observation that theorem 2.1, (2) is not necessarily true for arbitrary ultrafilters D. For example if (X,σ) is the real numbers with the usual topology and D is a β-complete ultrafilter on α, $2^{\aleph_0} \leq \beta \leq \alpha$, then (X,σ) is D-Baire but not α-Baire. Indeed if $\{x_i\}_{i \in \alpha}/D \in {}^DX$ then $\bigcup_{r \in X} \{i \mid x_i = r\} = \alpha \in D$ and since D is β-complete, $\{i \mid x_i = r\} \in D$ for some $r \in R$. Thus ${}^DX = X$ and therefore ${}^DS = S$ for any subset S of X, in particular if S is any open set. This shows that (X,σ) is D-Baire. Of course (X,σ) is not α-Baire since it can be written as the union of $2^{\aleph_0}$ singleton sets which are nowhere dense since there are no isolated points in X.

III. EQUIVALENCE CONDITIONS BETWEEN TOPOLOGICAL SPACES

In the previous section we showed how the property of being a Baire space could be captured by reference to any non-trivial countably indexed ultrapower of of the space. In this section we shall examine whether or not countably indexed ultrapowers may be used to characterize certain basic notions of equivalence between topological spaces which depend on the cardinality $\aleph_0$.

Once again the type of non-standard model used is critical. For example in enlargements the monads completely determine the topology, i.e. σ and τ are the same topology on X if and only if for any $x \in X$, $\mu_\sigma(x) = \mu_\tau(x)$. At the other extreme, if D is a countably complete ultrafilter and σ and τ are any two first countable Hausdorff topologies on a set X then they have the same monads, in fact for any $x \in X$, $\mu_\sigma(x) = \mu_\tau(x) = \{x\}$. To verify this let $x \in X$ and let $\mathcal{V}_1 \supseteq \mathcal{V}_2 \supseteq \mathcal{V}_3 \supseteq \cdots$ be a basis for the neighborhoods of x. If $\{x_i\}_{i\in I}/D \in {}^D\mu_\sigma(x)$, then $\bigcap_{n\in\omega} \{i \mid x_i \in \mathcal{V}_n\} = \{i \mid x_i \in \bigcap_{n\in\omega} \mathcal{V}_i\} \in D$. But $\bigcap_{n\in\omega} \mathcal{V}_n = \{x\}$ so $\{x_i\}_{i\in I}/D = x$, thus ${}^D\mu_\sigma(x) = \{x\}$ and similarly for ${}^D\mu_\tau(x)$.

Let σ and τ be two topologies on the same set X.

Definition. (X,σ) and (X,τ) are countably equivalent if and only if for every countable subset A of X, $c\ell_\sigma(A) = c\ell_\tau(A)$.

Definition. (X,σ) and (X,τ) are sequentially equivalent if and only if for every sequence $\{x_i\}_{i\in\omega}$ of elements of X and every $x \in X$,

$\lim_{i\to\infty} x_i = x$ (in σ) $\Longleftrightarrow \lim_{i\to\infty} x_i = x$ (in τ).

Let D be an ultrafilter on a set I.

Definition. (X,σ) and (X,τ) are D-equivalent if and only if for every $x \in X$, ${}^D\mu_\sigma(x) = {}^D\mu_\tau(x)$.

Theorem 3.1. (X,σ) and (X,τ) are countably equivalent if and only if they are D-equivalent for all ultrafilters D on ω.

Proof. Suppose that (X,σ) and (X,τ) are countably equivalent. Let D be an ultrafilter on ω and let $x \in X$, $y \in {}^D\mu_\sigma(x)$. Since $y \in {}^DX$ we may write $y = \{y_i\}_{i\in\omega}/D$ with $y_i \in X$ for all i. Suppose that $y \notin {}^D\mu_\tau(x)$. Then there is a $V \in \tau$ such that $x \in V$ and $y \notin {}^DV$. Let $A = \{i \mid y_i \in V\}$, thus $A \notin D$.

Now consider the set $S = \{y_i \mid i \notin A\} = \{y_i \mid y_i \notin V\}$. If W is a σ-open set containing x, then $\{i \mid y_i \in W\} \cap \omega - A \in D$ since $y \in {}^D\mu_\sigma(x)$ and $\omega - A \in D$. In particular, there is a $k \notin A$ such that $y_k \in W$, which shows that $W \cap S \neq \emptyset$. Therefore $x \in c\ell_\sigma(S)$. Since (X,σ) and (X,τ) are countably equivalent we have $x \in c\ell_\tau(S)$. However $x \in V \in \tau$, so there must be some $y_i \in S$ such that $y_i \in V$. But this contradicts the definition of S. Hence $y \in {}^D\mu_\tau(x)$ and ${}^D\mu_\sigma(x) = {}^D\mu_\tau(x)$.

Now suppose ${}^D\mu_\sigma(x) = {}^D\mu_\tau(x)$ for all $D \in \omega$ and all $x \in X$. Let S be a countable subset of X, $S = \{s_1, s_2, s_3, \cdots\}$, and let $x \in c\ell_\sigma(S)$. For every σ-open set $\mathcal{V}$ containing x let $I_\mathcal{V} = \{i \mid s_i \in \mathcal{V}\}$. The set $F = \{I_\mathcal{V} \mid x \in \mathcal{V} \in \sigma\}$ has the finite intersection property so we may let D be an ultrafilter on ω which includes F. If $s = \{s_i\}_{i\in\omega}/D$ then $s \in {}^D\mu_\sigma(x)$ since for any σ-open set $\mathcal{V}$ containing x, $\{i \mid s_i \in \mathcal{V}\} = I_\mathcal{V} \in D$. But ${}^D\mu_\sigma(x) = {}^D\mu_\tau(x)$, so $s \in {}^D\mu_\tau(x)$. Thus if W is a τ-open set containing x then $\{i \mid s_i \in W\} \in D$, in particular $W \cap S = \emptyset$. Hence $x \in c\ell_\tau(S)$ which completes the proof.

We now turn to the notion of sequential equivalence. In one direction there is an immediate connection with countably-indexed ultrapowers (actually with any non-standard extension in which ${}^*\omega \neq \omega$).

<u>Theorem 3.2</u>. Suppose for some non-principle ultrafilter D on ω that ${}^D\mu_\sigma(x) = {}^D\mu_\tau(x)$ for all $x \in X$. Then (X,σ) and (X,τ) are sequentially equivalent.

<u>Proof</u>. Let $\{x_i\}_{i\in\omega}$ be a sequence of elements of X and $x \in X$. Then $\lim_{i\to\infty} x_i = x$ (in σ) $\Longleftrightarrow x_\nu \in {}^D\mu_\sigma(x)$ for all $\nu \in {}^D\omega - \omega \Longleftrightarrow x_\nu \in {}^D\mu_\tau(x)$ for all $\nu \in {}^D\omega - \omega \Longleftrightarrow \lim_{i\to\infty} x_i = x$ (in τ).

Of course countable equivalence implies sequential equivalence. The converse is true for first countable spaces but may be false otherwise. In fact we will show that the converse to theorem 3.2 is false.

<u>Example</u>: Two spaces (X,σ) and (X,τ) which are sequentially equivalent but not D-equivalent for any non-principle ultrafilter D on ω.

Let $D \in Npu(\omega)$. We first construct two spaces (X_D, σ_D) and (X_D, τ_D) which are sequentially equivalent but not D-equivalent. Let $X_D = \{\hat{i} \mid i \in \omega\} \cup \{D\} \subseteq \beta\omega$ where $\beta\omega$ is the Stone-Céch compactification of ω and $\hat{i}$ is the principle ultrafilter determined by i. Let σ_D be the discrete topology on X_D and let τ_D be the topology induced on X_D by $\beta\omega$. Thus the basic τ_D-open sets of X_D which contain D are of the form $N_X = \{F \in \mathcal{V}(\omega) \mid X \in F\}$ where $X \in D$. Now let $y = \{\hat{i}\}_{i\in\omega}/D$. If $D \in N_X \in \tau_D$, then $\{i \mid \hat{i} \in N_X\} = \{i \mid X \in \hat{i}\} = \{i \mid i \in X\} = X \in D$ so $y \in {}^D\mu_{\tau_D}(D)$. However ${}^D\mu_{\sigma_D}(D) = \{D\}$ so $y \notin {}^D\mu_{\sigma_D}(x)$. Therefore ${}^D\mu_{\sigma_D} \neq {}^D\mu_{\tau_D}(D)$

which shows (X_D,σ_D) and (X_D,τ_D) are not D-equivalent.

To show these spaces are sequentially equivalent clearly we need only show that no sequence of elements of $\{\hat{i} \mid i \in \omega\}$ converges in (X_D,τ_D) to D. Suppose $\lim_{n\to\infty} \hat{a}_n = D$ (in τ_D) where $a_n \in \omega$ for all $n \in \omega$. Clearly we may suppose $\{a_n\}_{n\in\omega}$ is unbounded. Let N_X be a τ_D-basic neighborhood of D. Then for some $m \in \omega$, $E_m = \{\hat{a}_n \mid n > m\} \subseteq N_X$ so $E_m' = \{a_n \mid n > m\} \subseteq X$. E_m' is infinite so we may write $E_m' = Y_1 \cup Y_2$ where Y_1 and Y_2 are infinite disjoint sets. Let $X = X_1 \cup X_2$ where $X_1 \cap X_2 = \emptyset$ and $Y_1 \subseteq X_1$, $Y_2 \subseteq X_2$. Then either $X_1 \in D$ or $X_2 \in D$, say $X_1 \in D$. But $Y_2 \cap X_1 = \emptyset$, hence $S = \{\hat{a}_k \mid a_k \in Y_2\} \cap N_{X_1} = \emptyset$. Since S is infinite, this contradicts $\lim_{n\to\infty} \hat{a}_n = D$. This shows that (X_D,σ_D) and (X_D,τ_D) are sequentially equivalent.

Finally, for each (X_D,τ_D) we choose an isomorphic copy (X_D',τ_D') in such a way that $\{X_D' \mid D \in Npu(\omega)\}$ is pairwise disjoint. Let $X = \bigcup\{X_D' \mid D \in Npu(\omega)\}$, σ be the discrete topology on X, and τ the topology generated by $\bigcup\{\tau_D' \mid D \in Npu(\omega)\}$. Then (X,σ) and (X,τ) are sequentially equivalent since no non-trivial sequences converge in either topology. However they are not D-equivalent for any $D \in Npu(\omega)$ since ${}^D\mu_\sigma(D) = \{D\} \neq {}^D\mu_{\tau_D}(D) = {}^D\mu_\tau(D)$.

We conclude this section by observing that for a sequence $\{x_i\}_{i\in\omega}$ in a first countable space there is a connection between having a convergent subsequence and belonging to a monad in a suitable countably indexed ultrapower. The previous example shows that this connection may be absent in spaces which are not first countable. Recall that an ultrafilter D on ω is a P-point if given any sequence $S_1 \supseteq S_2 \supseteq S_3 \supseteq \cdots$ of elements of D there is an $S \in D$ such that $S \cap (\omega - S_n)$ is finite for all n (cf. [7]).

<u>Theorem 3.3</u>. Let (X,σ) be a first countable topological space, D a P-point on ω, $x \in X$, and $\{y_i\}_{i\in\omega}$ a sequence of elements of X. Then $\{y_i\}_{i\in\omega}/D \in {}^D\mu_\sigma(x)$ if and only if there is a subsequence $\{y_{i_k}\}_{k\in\omega}$ of $\{y_i\}_{i\in\omega}$ such that $\lim_{k\to\infty} y_{i_k} = x$ and $\{i_1, i_2, \cdots\} \in D$.

Proof. One direction of the theorem is immediate. For the other suppose $\{y_i\}_{i\in\omega}/D \in {}^D\mu_\sigma(x)$. Let $W_1 \supseteq W_2 \supseteq W_3 \supseteq \cdots$ be a basis for the neighborhoods of X. Let $S_n = \{i \mid y_i \in W_n\}$ for each $n \in \omega$, thus $S_1 \supseteq S_2 \supseteq S_3 \supseteq \cdots$ with $S_n \in D$ for all n. Since D is a P-point we may choose $S \in D$ such that $S \cap (\omega - S_n)$ is finite for all n. Consider the sequence $\{y_{i_k}\}_{k\in\omega}$ where $S = \{i_1, i_2, i_3, \cdots\}$ with $i_1 < i_2 < i_3 < \cdots$. Given any $n \in \omega$, $i_k \in S_n$ except for finitely many k so $y_{i_k} \in W_n$ except for finitely many k. Thus $\lim_{k\to\infty} y_{i_k} = x$ which completes the proof.

REFERENCES

[1] A.R. Bernstein, "A new kind of compactness for topological spaces," Fund. Math., vol. 66(1970), 185-193.

[2] I. Juhász, "Cardinal functions in topology," Mathematical Center Tract 34, Math. Centrum Amsterdam, 1971.

[3] M. Machover and J. Hirshfeld, "Lectures on Non-Standard Analysis," Springer-Verlag, Berlin, 1969.

[4] D. Martin and R.M. Solovay, "Internal Cohen extensions," Ann. Math. Logic, vol. 2(1970), 143-178.

[5] W.A.J. Luxemburg (editor), Applications of Model Theory to Algebra, Analysis, and Probability, Holt, Rinehart and Winston, New York, 1969.

[6] A. Robinson, Non-Standard Analysis, North Holland, Amsterdam, 1966.

[7] W. Rudin, "Homogeneity problems in the theory of Céch compactifications," Duke Math. J., vol. 23, 409-420.

[8] R.M. Solovay and S. Tennenbaum, "Iterated Cohen extensions and Souslin's problem," Ann. of Math., vol. 94(1971), 201-245.

[9] F.D. Tall, "Souslin's conjecture revisited," Proc. Bolyai Janós Math. Soc. Colloquium on Topology, 1972, Keszthely, Hungary, to appear.

[10] ———, "The countable chain condition vs. separability--applications of Martin's Axiom, preprint. (See Notices of A.M.S., vol. 19(October,1972) A-725.)

[11] F. Wattenberg, "Nonstandard topology and extensions of monad systems to infinite points," J. Sym. Logic, vol. 36(1971), 463-476.

[12] ———, "Two topologies with the same monads," these proceedings.

<u>ADDENDUM</u>: In his retiring presidential address delivered January 1973 at the annual meeting of the Association for Symbolic Logic in Dallas, Texas, Abraham Robinson listed 12 open metamathematical problems. Problem number 9 is "to provide a metamathematical framework for Baire's theorem, more particularly for the metric case and its applications." Section 2 of this paper hence may be considered, at least in part, a solution to this problem.

ENLARGEMENTS CONTAIN VARIOUS KINDS OF COMPLETIONS

Harry Gonshor

Rutgers University, New Brunswick, New Jersey

I. Introduction. It is known that enlargements of models often contain as subquotients, extensions of the models that are of importance classically. For example, as in [8], distributions may be regarded as equivalence classes of internal functions. As another example, in [6] and [7] completions of uniform spaces are studied from this point of view. The aim of this paper is to discuss various examples of this phenomenon arising in nonstandard analysis. We shall see that not only completions in the topological sense but also rings of quotients and projective covers can be obtained from enlargements.

Projective covers are discussed in more detail in [5]. The talk at the symposium dealt primarily with completions of Boolean algebras. This is discussed in section 5 beginning on page 7. (Section 5 begins with simple observations on other examples of injective hulls before turning to this main example.)

For background in non-standard analysis the reader is referred to [6], [7], or [8]. All enlargements considered will be higher order non-standard models.

II. The real numbers. We begin by rapidly surveying results which are essentially known but are not explicitly stated in the literature in the form which we desire.

First, suppose that we begin with an enlargement J^* of the integers J. We remark that there is a simple way of extracting the rationals from J^*. Clearly J^* itself does not contain rationals, e.g. $2x = 1$ has no solution in J, hence it has no solution in J^*. On the other hand, since J has arbitrarily large primes, J^* has infinite primes. If p is an infinite prime $\frac{J^*}{(p)}$ is a field containing the rationals.

As an amusing application of number theory, fields of the form $\frac{J^*}{(p)}$ can be obtained having various special properties. For example, consider the relation $R(p,q)$ in J defined by p is either -1 or a positive prime and q is a positive prime with p a quadratic residue of q. It is an interesting exercise in number theory that this relation is concurrent. (The proof uses the quadratic reciprocity law including the theorems on the quadratic character of -1 and 2, the Chinese remainder theorem, and Dirichlet's theorem for primes in an arithmetic progression.)

Hence there exists a prime q^* in J^* such that $R(p,q^*)$ for all primes $p \in J$ and $R(-1,q^*)$. Such a q^* is necessarily in $J^* - J$. Hence in $\frac{J^*}{(q^*)}$ every rational has a square root.

By starting with an enlargement of the rationals, it is well known that the reals can be obtained as the quotient of the ring of finite elements by the ideal of infinitesimal elements. Without assuming the existence of the reals in advance this can be used as an alternative method for constructing the reals. This appears to have the disadvantage that the machinery from model theory is required. However, if the machinery is going to be developed anyway for many other purposes too, then we may just as well obtain the reals as one of the by-products of this development.

III. The Stone-Cech compactification. We consider a completely regular space X with enlargement X^*. We shall obtain a direct construction of the Stone-Cech compactification. In [6] and [7] this is done from the point of view of uniform structures. (In [6] X^* is assumed to be more than just an enlargement.) Every bounded continuous function f on X has an extension to X^* usually denoted by the same letter f. Furthermore, since the sentence $(\forall x)(f(x) \leq m)$ is preserved by transfer from X to X^*, f is bounded on X^*.

We now define an equivalence relation on X^*.

Let F be the class of bounded continuous functions on X. Then $x \sim y$ iff ${}^0f(x) = {}^0f(y)$ for all $f \in F$. [${}^0x$ is the standard part of x. This exists for all finite x.] $\sim$ is clearly an equivalence relation such that two points in X are equivalent only if they are identical. Let $\bar{x}$ be the class containing x. Every $f \in F$ induces a function $\bar{f}$ on the equivalence classes--namely $\bar{f}(\bar{x}) = {}^0f(x)$ for any $x \in \bar{x}$. We claim that $\beta X = X^*/\sim$ with the weak topology induced by the $\bar{f}$: $f \in F$ is the Stone-Cech compactification of X. Since X is completely regular, X may be regarded as a subspace of βX with $\bar{f}$ an extension of X. We need only prove that X is dense in βX and that βX is compact.

Let $\overline{x} \in \beta X$ and let $\overline{x} \in U$ where $U = \{\overline{y}: |\overline{f}_i(\overline{y}) - \overline{f}_i(\overline{x})| < \varepsilon$ for all $i: 1 \leq i \leq n.\}$ Consider the subset A of X where $A = \{y: |f_i(y) - {}^{O}f_i(x)| < \varepsilon$ for all $i: 1 \leq i \leq n.\}$ $x \in A^*$, hence A^* and therefore A is non-empty. Furthermore, $A \subset U$. Thus $U \cap X \neq \phi$ i.e. X is dense in βX.

Since βX is completely regular it suffices to prove that every ultrafilter on X converges in βX. Any ultrafilter F has a monad of the form $\mu_d(a)$. (See [6], page 46). Let U and A be as in the previous paragraph with x replaced by a. Then $A \in F$ since $a \in A^*$. Since $A \subset U$ this completes the proof.

A different kind of compactification result is obtained in [11]. See also [12].

IV. The second conjugate space of a Banach space. The technique used here is similar in many ways to that of part III. See also [9] and [10].

Let B be an infinite dimensional complex Banach space and B* an enlargement. Let $\gamma B = (x \in B^*: f(x)$ is finite for all $f \in B')$. For $x, y \in \gamma B$ define $x \sim y$ iff ${}^{O}f(x) = {}^{O}f(y)$ for all $f \in B'$. This is an equivalence relation and by the Hahn-Banach Theorem two points in B are equivalent only if they are identical. Define $\overline{f}(\overline{x}) = {}^{O}f(x)$. Then $\overline{f}$ may be regarded as an extension of f. Let $I = (x \in B^*: {}^{O}f(X) = 0$ for all $f \in B')$. Then $x \sim y$ iff $x - y \in I$. γB is a subspace of B* and I is a subspace of γB. (Note that all spaces considered are taken over the complex numbers even though B* has the structure of a space over an enlargement of the complex numbers.)

For $x \in \gamma B$ define $\hat{x} \in B''$ by $\hat{x}(f) = {}^{O}f(x)$. Then $\hat{x}$ is linear. Furthermore $x \to \hat{x}$ is a linear map with kernel I. Since $B \subset \gamma B$, the map $x \to \hat{x}$ may be regarded as an extension of the usual map of B into B".

Now suppose $T \in B''$. Let R be the relation: $(f,x) \in R$ iff $f \in B''$, $x \in B$ and $f(x) = T(f)$. It is well-known that this relation is concurrent. Hence there exists $x \in B^*: (\forall f \in B')(f(x) = T(f)$. Necessarily $x \in \gamma B$ and $\hat{x} = T$. Note that for such an x the ^{O} operation is not even needed.

We have shown that $\frac{\gamma B}{I}$ is the second conjugate space of B".

A natural question arises. How does $\|\hat{x}\|$ as a standard norm of B" compare with $\|x\|$ where the norm in the second case is the model theoretic extension of the usual norm. Now the inequality $f(x) \leq \|f\| \|x\|$ is valid for all $x \in \gamma B$. Hence $|\hat{x}(f)| = |{}^0 f(x)| \leq \|f\| \|x\|$. Hence $\|\hat{x}\| \leq \|x\|$. On the other hand let R be the relation consisting of pairs $(<f,n>,x)$ where $f \in B'$, n is a positive integer, $x \in B$, $f(x) = 0$, and $\|x\| \geq n$. Since B is infinite dimensional it is easily seen that this is concurrent. Hence there exists on $x \in B^*$ such that $(\forall f \in B')$ $[f(x) = 0]$ and $\|x\|$ is infinite. Thus the reverse inequality fails in the strongest possible way.

Finally let us note what this means in terms of uniform structures. Consider the uniformity on B generated by the functions in B'. It is easily seen from [6] that the pre-near-standard points (in [7] they are called approachable points) are precisely the points in γB and the uniformity monads are the equivalence classes defined earlier. Hence by [7] $\frac{\gamma B}{I}$ and therefore B" is the completion of B in this uniformity.

We suggest to readers referring to [6] that theorem 3.15.3 on page 81 can be proved easily using enlargements only and the technique applies to any family of semimetrics generated by a class of functions, in particular, to the case that concerns us. [7] might therefore be preferable as a reference.

V. Rings of quotients and injective hulls. The first example we considered in part II was that of a ring of quotients. We now consider rings of quotients of rings of continuous functions as studied in [1]. Let R(X) be the ring of real valued continuous functions on the compact Hausdorff space X. Then X extends to X* and at the same time the set of continuous functions R(X) extends to the set of * continuous functions (R(X))*. They are functions from X* into R*. It follows from an elementary application of Urysohn's lemma that for every n-tuple of points $p_1, p_2, \ldots, p_n \in X$ and every n-tuple of real numbers $r_1, r_2, \ldots, r_n$ there is a continuous function on X such that $(\forall i \leq n)[f(p_i) = r_i]$. It follows using the technique of enlargements that an arbitrary function from X into R can be extended to a * continuous function from X* into R*.

This seems paradoxical at first, e.g. let I be the unit interval and f the function which is 0 on the rationals and 1 on the irrationals. However, the explanation is clear. Although the ε,δ definition must still be satisfied, this can legally be done by an infinitesimal δ even if ε is not infinitesimal.

In particular, every continuous function on a dense open subset of X into R can be extended to a * continuous function. Hence by [1] the ring of quotients of $R(X)$ may be regarded as a subquotient of $[R(X)]^*$.

If the real numbers are replaced by the complex numbers and we consider the injective hull in the category of C^* algebras essentially the same result is obtained. From [4] it suffices to consider extensions from Borel functions on X.

The same technique can be applied to the category of Boolean algebras by using the Stone space X. The Boolean algebra B consists of the clopen sets in X and the injective hull of the Borel sets modulo sets of first category.

Let X^* be an enlargement of X and B^* the corresponding enlargement of B. Let C be an arbitrary subset of X and f its characteristic function. By the earlier technique it follows that there exists a * clopen subset D of X^* whose characteristic function g agrees with f on X, i.e. $D \cap X = C$. Hence <u>every</u> subset of X is a restriction of a * clopen subset of X^*! Thus the power set of X is a quotient of B^*. In particular, the injective hull of B is a subquotient of B^*.

We next study Boolean algebras directly without using the Stone space. Let B be an infinite Boolean algebra with B^* an enlargement. For $x \in B^*$ define $U(x) = (y \in B\colon y \geq x)$ and $L(X) = (y \in B\colon y \leq x)$. [The pair $\{L(x), U(x)\}$ is analogous to a Dedekind cut; however, since B is not linearly ordered there are important differences.] $U(x) \cap L(x) = \phi$ if $x \in B^* - B$.

<u>Definition 1</u>: $x \in \gamma B$ if $\bigcap_{\substack{y \in U(x) \\ z \in L(x)}} (y - z) = 0$ where $\bigcap$ is understood to be in B. It is clear from the definition that $B \subset \gamma B$.

<u>Definition 2</u>: $I = (x \in \gamma B\colon L(x) = \{0\})$.

<u>Theorem 4.1</u> γB is a subalgebra of B^*, I is an ideal in γB, and $\gamma B/I$ is the completion of B.

Proof: Suppose that $x, y \in \gamma B$. Then $L(x) \cup L(y) \subset L(x \cup y)$ and $U(x) \cup U(y) \subset U(x \cup y)$. [We are using the usual notation $A \circ B = (a \circ b\colon a \in A,\ b \in B)$ where $\circ$ is an arbitrary binary operation in the system considered.] We must show that $\bigcap[U(x \cup y) - L(x \cup y)] = 0$. By the above inclusions it suffices to show that $\bigcap\{[U(x) \cup U(y)] - [L(x) \cup L(y)]\} = 0$. Now in any Boolean algebra the inequality $(a \cup b) - (c \cup d) \leq (a - c) \cup (b - d)$ holds for arbitrary a, b, c, and d. Thus it suffices to show that $\bigcap\{[U(x) - L(x)] \cup [U(y) - L(y)]\} = 0$. Since $\bigcap[U(x) - L(x)] = \bigcap[U(y) - L(y)] = 0$ this follows from one of the distributive laws. We have shown that $x \cup y \in \gamma B$.

Now suppose that $x \in \gamma B$. It is easily seen that $U(x') = [L(x)]'$ and $L(x') = [U(x)]'$. From the elementary result that $a - b = a \cap b' = b' - a'$ it follows that $U(x') - L(x') = [L(x)]' - [U(x)]' = U(x) - L(x)$. Hence $\bigcap[U(x') - L(x')] = \bigcap[U(x) - L(x)] = 0$. Therefore $x' \in \gamma B$. Thus γB is a subalgebra of B^*.

It is clear that if $x \in I$ and $y \leq x$ then $y \in I$. Next assume that $x, y \in I$. Since $I \subset \gamma B$ we have $\bigcap U(x) = \bigcap U(y) = 0$. By an argument similar to the one which showed that γB is closed under union only easier it follows that $\bigcap U(x \cup y) = 0$. Hence $L(x \cup y) = \{0\}$.

Note: We shall see later that $(x\colon L(x) = 0)$ is not necessarily closed under union in many Boolean algebras! The above result concerns elements which are also in γB.

Since $B \subset \gamma B$ we have the natural composition $B \xrightarrow{i} \gamma B \xrightarrow{\varphi} \frac{\gamma B}{I}$, with composite f.

Let $b \in B$. Then $b \in L(b)$. If $b \neq 0$ then $L(b) \neq \{0\}$ hence $b \notin I$. Therefore $f(b) \neq 0$. So f is monic. We identify B and $f(B)$.

Let $\bar{x} \in \frac{\gamma B}{I}$ and x a representative of $\bar{x}$ in γB. We claim that $\bar{x} = \bigcup L(x)$ where $\bigcup$ is taken in $\frac{\gamma B}{I}$. It is clear that $\bar{x}$ is an upper bound to $L(x)$. Suppose $y \in \gamma B$ and $\bar{y} \in \frac{\gamma B}{I}$ is an upper bound to $L(x)$. We now show that $L(xy') = 0$. Let $c \in L(xy')$. Then $c \leq xy' \leq x$. Hence $c \in L(x)$. Since $\bar{y}$ is an upper bound to $L(x)$, $\bar{y} \geq c$. Again $c \leq xy' \leq y'$. Hence $c \leq \overline{y'} = (\bar{y})'$. Therefore $c \leq (\bar{y})' \cap \bar{y} = 0$. We have shown that $xy' \in I$. It follows that $\bar{x}(\bar{y})' = 0$, i.e. $\bar{x} \leq \bar{y}$. Hence x is an l.u.b. to $L(\bar{x})$.

Note that so far we used only the fact that B^* is an algebra containing B, e.g. all the previous results are trivially satisfied if $B^* = B$. The final step uses the fact that B^* is an enlargement.

It remains to prove that every subset of B has an l.u.b. in $\frac{\gamma B}{I}$. Let $D \subset B$ and let E be the set of upper bounds to D. We show that $\bigcap(E - D) = 0$. Let $0 \neq x \leq e - d$ for all $e \in E$ and $d \in D$. In particular $x \leq d'$ hence $d \leq x'$ thus $x' \in E$. Hence $x \leq x' - 0$. It follows that $x = 0$ and we obtain a contradiction. Therefore $\bigcap(E - D) = 0$.

We assume that $E \cap D = \phi$. Otherwise the result is trivial. Consider the following relation R: $-(x,y) \in R$ iff $(x \in D \cap x \leq y) \vee (x \in E \cap x \geq y)$. This relation is concurrent. In fact if $x_1, x_2, \ldots, x_m \in D$ and $x_{m+1}, x_{m+2}, \ldots, x_{m+n} \in E$, then $y = x_1 \cup x_2 \cup \ldots \cup x_m$ satisfies $x_i R y$ for all i. Hence there exists a $y \in B^*$ such that $x \in D \Rightarrow x \leq y$ and $x \in E \Rightarrow x \geq y$. Hence $E - D \subset \bigcap[U(y) - L(y)]$. Since $\bigcap(E - D) = 0$ then a-fortiori $\bigcap[U(y) - L(y)] = 0$, i.e. $y \in \gamma B$.

It is clear that $\overline{y}$ is an upper bound in D. (We have shown earlier that $\overline{y}$ is the l.u.b. of $L(y)$. Unfortunately this is not what we need now though it seems to come close in some sense.) Let $z \in \gamma B$ and $\overline{z} \in \frac{\gamma B}{I}$ be an upper bound to D. Let $c \in L(yz')$. $c \leq yz' \leq z'$. Hence $z \leq c'$. It follows that $\overline{z} \leq c'$ and since $\overline{z}$ is an upper bound to D that c' is an upper bound to D. $c \in L(y)$. Although we cannot conclude that $c \in D$ we know at least that c is a lower bound to E. Therefore every element in $E - D$ is bounded below by $c - c' = c$. Since $\bigcap(E - D) = 0$, it follows that $c = 0$. We can now continue as in the previous proof to show that $\overline{y} \leq \overline{z}$ and therefore that $\overline{y}$ is a l.u.b. to D. Q.E.D.

Now that we have obtained the completion of a Boolean algebra as a subquotient of B^* in the form $\frac{\gamma B}{I}$ several natural questions arise which fortunately have easy answers. When does $I = \{0\}$ and when does $\gamma B = B^*$?

<u>Theorem 4.2</u> $I \neq \{0\}$.

<u>Proof:</u> We first note the convenient fact that $x \in I$ iff $\bigcap U(x) = 0$. This is trivial to prove. Since B is infinite it has a non-principal ultrafilter F. Define the relation R as follows: F is the domain and $\langle x,y \rangle \in R$ iff $x \geq y$

and $y \neq 0$. This is clearly concurrent. Thus there exists a $y \in B^*$ such that $y \neq 0$ and $x \geq y$ for all $x \in F$. Then $F \subset U(y)$. Since $\bigcap F = 0$, certainly $\bigcap U(y) = 0$. Hence $y \in I$. Q.E.D.

We have seen that our completions are not obtained as subalgebras of B^*. The next theorem deals with the two extremes that can occur with respect to B^*.

Theorem 4.3

(1) $B^* = \gamma B$ iff the algebra is atomic.

(2) $(\exists x \in B^*)[U(x) = \{1\}$ and $L(x) = \{0\}]$ iff the algebra is atomless.

Proof: For the sake of efficiency we shall take advantage of the overlap of the proofs of $\Rightarrow$ in (1) and $\Leftarrow$ in (2), and vice-versa.

First, suppose that a is an atom of B. Then every $x \in B$ satisfies $a \leq x$ or $x \leq a'$. Since B^* is an elementary extension of B, $(\forall x)(a \leq x$ or $x \leq a')$ is valid in B^*, i.e. $(\forall x)[a \in L(x)$ or $a' \in U(x)]$. This already proves $\Rightarrow$ in (2). If $a \leq x$, then $1 - a \in U(x) - L(x)$ and if $x \leq a'$, then $a' - 0 \in U(x) - L(x)$. Hence in either case $a' \in U(x) - L(x)$. Therefore $U(x) - L(x)$ contains all elements of the form a' where a is an atom. In an atomic algebra this intersection is 0 hence a-fortiori $\bigcap[U(x) - L(x)] = 0$. This proves $\Leftarrow$ in (1).

Let S be the set of atomless elements of B. We consider the following relation R in B: $-\langle x,y\rangle \in R$ iff $x \in S \wedge \neg(x \leq y) \wedge \neg(y \leq x')$. We shall prove that this relation is concurrent. First S is the domain of R since $\langle x,y\rangle \in R$ for any $y \neq 0$ such that $y < x$. Such a y exists for any $x \in S$. Now suppose $a_1, a_2, \ldots, a_m \in S$ and $b_1, b_2, \ldots, b_n \in B - \{0\}$. We show by double induction that there is an $x \in B$ such that $(\forall i)(x \cap a_i \neq 0)$ and $(\forall i)(b_i \not\leq x)$. Let $m = n = 1$. If $b_1 \not\leq a$, then $x = a_1$ works. If $b_1 \leq a_1$ then $a_1 - b_1$ works unless $b_1 = a_1$. In the latter case any x satisfying $0 < x < a$ works. Such an x exists since a is atomless and a-fortiori non-atomic. Note that in all cases the x found satisfies $x \leq a$.

Suppose that the result is valid for $m = 1$ and $n = k$. Let $a \in S$; $b_1, b_2, \ldots, b_{k+1} \in B - \{0\}$ and x satisfy $x \cap a \neq 0$ and $(\forall i \leq k)(b_i \not\leq x)$.

By replacing x by $x \cap a$ if necessary we can always assume that $x \leq a$. Hence x is atomless and thus the first case can be applied to x and b_{k+1} to obtain $y \leq x$ such that $y \neq 0$ and $b_{k+1} \not\leq y$. Then $y \leq a$ and a-fortiori $(\forall i \leq k)$ $(b_i \not\leq y)$. This proves the result for $m = 1$ and $n = k + 1$. (It is clear that all elements x obtained by this procedure satisfy $x \leq a$ so that alternatively the latter condition could be included in the inductive hypothesis. Then $x \cap a \neq 0$ could be replaced by $x \neq 0$.)

Finally, suppose that $a_1, a_2, \ldots, a_{m+1} \varepsilon S$; $b_1, b_2, \ldots, b_n \varepsilon B - \{0\}$ and x satisfies $(\forall i \leq m)(x \cap a_i \neq 0)$ and $(\forall i \leq n)(b_i \not\leq x)$. Let $b_i - x = c_i$. Then $c_i \neq 0$. We now apply the previous case to a_{m+1} and $\{c_1, c_2, \ldots, c_n\}$ to obtain y such that $y \cap a_{m+1} \neq 0$ and $(\forall i \leq n)(c_i \not\leq y)$. Then $x \cup y$ satisfies $(\forall i \leq m + 1)[(x \cup y) \cap a_i \neq 0]$ and $(\forall i \leq n)(b_i \not\leq x \cup y)$.

The concurrency of the relation R now follows immediately. Given a finite subset of S: $-\{x_1, x_2, \ldots, x_n\}$ we apply the previous result to the case where $\{a_1, a_2, \ldots, a_n\}$ and $\{b_1, b_2, \ldots, b_n\}$ both equal $\{x_1, x_2, \ldots, x_n\}$.

It now follows that there exist a $y \varepsilon B^*$ such that $(\forall x \varepsilon S)$ $[\neg(x \leq y) \wedge \neg(y \leq x')]$. Fix y and let x be arbitrary in S. If $a \varepsilon L(y)$ then $a \cap x \varepsilon L(y)$. Assume that $a \cap x \neq 0$. Since x is atomless so is $a \cap x$ hence $\neg[a \cap x \varepsilon L(y)]$ and we have a contradiction. Thus $a \cap x = 0$ and $a \leq x'$. Again if $b \varepsilon U(y)$ then $[(b' \cap x)' = b \cup x' \varepsilon U(y)]$. Suppose that $b' \cap x \neq 0$. Then $b' \cap x$ is atomless and $\neg[(b' \cap x)' \varepsilon U(y)]$. By contradiction, $b' \cap x = 0$ hence $x \leq b$. Summarizing, we have $a \varepsilon L(y) \Rightarrow a \leq x'$ and $b \varepsilon U(y) \Rightarrow x \leq b$. Hence an arbitrary element $b - a$ in $U(y) - L(y)$ satisfies $b - a \geq x - x' = x$, i.e. x is a lower bound to $U(y) - L(y)$. Hence $y \not\varepsilon \gamma B$ if S is non-empty. This finally proves $\Rightarrow$ in (1). On the other hand, if the algebra is atomless, then $1 \varepsilon S$, hence 1 is a lower bound to $U(y) - L(y)$. Therefore $L(y) = \{0\}$ and $U(y) = \{1\}$. This proves $\Leftarrow$ in (2). Q.E.D.

Roughly speaking we have shown that in passing from B to B^*, in the atomic case we are simply filling in gaps (with duplication) but in the atomless case we include elements that are entirely "way out", i.e. incomparable with everything in b except 0 or 1. In the general case the proof leads to elements y which are as "way out" as possible, namely, the set of lower bounds to $U(y) - L(y)$ is precisely S. (We remind the reader that complements of atoms are necessarily in $U(y) - L(y)$, thus the set of lower bounds does not contain atoms and is therefore by transitivity a subset of S. As an immediate consequence of Theorem 4.3 we obtain examples of elements a and b such that $L(a) = \{0\}$, $L(b) = \{0\}$, and $L(a \quad b) \neq \{0\}$. In fact let a satisfy $L(a) = \{0\}$ and $U(a) = \{1\}$. Then $L(a') = \{0\}$. $L(a \cup a') = L(1) = B$.

VI. Remarks on Projective Covers. Since this example is studied in detail in [5], we shall limit ourselves here to indicating how the projective cover of a compact Hausdorff space is obtained without giving any proofs. For background we refer the reader to [3]. Let X be such a space and X^* an enlargement. Let γX be the set of all $x \in X^*$ which are not in the closure of two disjoint open sets. (e.g. $X \subset \gamma X$ iff X is extremely disconnected.) Let $F(x)$ be the class of regular open sets containing x and define: $x \sim y$ iff $F(x) = F(y)$. Then the quotient set δX with a suitable topology is the projective cover. The map from δX onto X is nothing but the map which takes each element into the standard part of one of its representatives.

Note. In order to keep our statements concise we did not distinguish between a subset A of X and the corresponding subset A^* of X^*, the distinction being hopefully clear from the context.

REFERENCES

[1] N. Fine, L. Gillman, and J. Lambek, Rings of quotients of rings of functions, (McGill Univ. Press), Montreal, 1965.

[2] L. Gillman and M. Jerson, Rings of continuous functions, 1960.

[3] A. M. Gleason, Projective topological spaces, Illinois J. Math., 12 (1958), 482-489.

[4] H. Gonshor, Injective hulls of C* algebras II, Proc. Amer. Math. Soc., 24 (1970), 486-491.

[5] H. Gonshor, Projective covers as subquotients of enlargements, accepted by Israel J. Math.

[6] W. A. J. Luxemburg, A general theory of monads, Applications of model theory to algebra, analysis and probability, 18-86.

[7] M. Machover, Lectures on non-standard analysis, Lecture notes in mathematics no. 94, Springer, 1969.

[8] A. Robinson, Non-standard Analysis, (Studies in Logic and the Foundations of Mathematics), Amsterdam, North Holland, 1966.

Rutgers University, The State University of New Jersey
New Brunswick, New Jersey

Additional references - (These have come to the attention of the author after the paper was written.)

[9] W.A.J. Luxemburg, On some concurrent binary relations occurring in analysis, contributions to non-standard analysis, edited by W.A.J. Luxemburg and A. Robinson, North Holland, 1972, 85-100.

[10] W.A.J. Luxemburg, Ultrapowers of normed linear spaces, Notices Am. Math. Soc. 11 (2) No. 73 64T-137.

[11] A. Robinson, Compactification of groups and rings, and non-standard analysis, J. Symb. Log., 34, 576-588.

[12] K. D. Stroyan, Additional remarks on the theory of monads, contributions to non-standard analysis, edited by W. A. J. Luxemburg and A. Robinson, North Holland, 1972.

SEMI-REFLEXIVITY OF THE NONSTANDARD HULLS OF A LOCALLY CONVEX SPACE

C. Ward Henson
L. C. Moore, Jr.
Duke University

In this paper we extend to locally convex spaces some results for normed spaces proved in Section 8 of [1]. It is shown there (assuming sufficient saturation) that the nonstandard hull $(\hat{E}, \hat{\rho})$ of a normed space (E, ρ) is reflexive if and only if the dual space of $(\hat{E}, \hat{\rho})$ is the nonstandard hull of (E', ρ'). If E is a real vector space, this is in turn equivalent to the following standard condition:

(#) for some $r \in R$, $0 < r < 1$, and some positive integer n there do not exist finite sequences $\{x_1, x_2, \ldots, x_n\}$ in E and $\{y_1, y_2, \ldots, y_n\}$ in E' which satisfy

$$\rho(x_i) = 1, \qquad \rho'(y_j) = 1 \qquad i, j = 1, 2, \ldots, n,$$

$$r < \langle x_i, y_j \rangle \text{ if } 1 \leq j \leq i \leq n, \text{ and}$$

$$0 = \langle x_i, y_j \rangle \text{ if } 1 \leq i < j \leq n.$$

If E is a complex normed space, the reflexivity of $(\hat{E}, \hat{\rho})$ is equivalent to the condition obtained from (#) by replacing $\langle x_i, y_j \rangle$ everywhere by $\mathrm{Re}\langle x_i, y_j \rangle$.

In the present paper these results are extended to locally convex spaces (E, θ). Specifically, let $\langle E, F \rangle$ be paired vector spaces and let θ be a locally convex vector topology on E which is admissible for the pairing (in the sense that θ has a local base at 0 which consists of absolutely convex $\sigma(E, F)$ - closed sets.) Let $^*\mathfrak{M}$ be an enlargement of some set-theoretical structure $\mathfrak{M}$ which contains E and F. Let $(\hat{E}, \hat{\theta})$ be the nonstandard hull of (E, θ) constructed using $^*\mathfrak{M}$. (See [1].)

In Section 1 is introduced the space $\hat{F}$ of all $\hat{\theta}$-continuous linear functionals on $\hat{E}$ which are representable by elements of $^*F$. (In the language of [1], $\hat{F}$ is the quotient space M_θ / m_θ.) There is a natural pairing between $\hat{E}$ and $\hat{F}$, relative to which $\hat{\theta}$ is an admissible topology. If, for example, (E, θ) is normable with norm ρ and F is the dual space (E', ρ'), then $\hat{F}$ is just the nonstandard hull $(\widehat{E'}, \widehat{\rho'})$. However if (E, θ) is not normable, then $\hat{F}$ is not a nonstandard hull in any natural way.

In Section 2 it is shown that if $^*\mathfrak{M}$ is κ-saturated and if S is any subspace of $\hat{E}$ which has Hamel dimension less than κ, then for each $\hat{\theta}$-continuous linear functional φ on $\hat{E}$ there exists $\psi \in \hat{F}$ such that $\varphi = \psi$ on S. In particular it follows that if $^*\mathfrak{M}$ is $\aleph_1$-saturated, then a subset B of $\hat{E}$ is $\hat{\theta}$-bounded if and only if B is $\sigma(\hat{E}, \hat{F})$-bounded. In Section 3 it is shown (assuming $^*\mathfrak{M}$ to be sufficiently saturated) that $(\hat{E}, \hat{\theta})$ is semi-reflexive if and only if $\hat{F}$ is $\beta(\hat{E}', \hat{E})$-dense in $\hat{E}'$. (Here $\hat{E}'$ is the dual space $(\hat{E}, \hat{\theta})'$.) In addition, the semi-reflexivity of $(\hat{E}, \hat{\theta})$ is shown to be equivalent to a standard condition on (E, θ) which is analogous to (#) above.

Preliminaries. Throughout this paper K will denote the field of real or complex numbers, E and F will denote vector spaces over K which are paired by a bilinear form $\langle \ldots, \ldots \rangle$. (We require that each non-zero element of E determine a non-zero linear functional on F, via the pairing, and vice versa.) Also θ will denote a Hausdorff locally convex vector topology on E which has a local base at 0 consisting of absolutely convex, $\sigma(E, F)$-closed sets. The weak topology on E defined by F is denoted by $\sigma(E, F)$ and the strong topology on E defined by F is denoted by $\beta(E, F)$.

The basic nonstandard theory used here is developed in [1]; a few definitions and details will be reproduced here for convenience. Let $\mathfrak{M}$ be a set-theoretical structure which contains E, F and K. Recall that an enlargement $^*\mathfrak{M}$

of $\mathfrak{M}$ is K-saturated [7] (where K is an infinite cardinal number) if whenever X is in $\mathfrak{M}$ and $\mathcal{C}$ is a collection of internal subsets of $^*X$ which has the finite intersection property and has cardinality less than K, then there exists $p \in {}^*X$ such that $p \in A$ for each $A \in \mathcal{C}$. $^*\mathfrak{M}$ is $\aleph_0$-enlarging [1] if for each X in $\mathfrak{M}$ and each collection $\mathcal{C}$ of internal subsets of $^*X$ such that $\mathcal{C}$ has the finite intersection property and has only finitely many nonstandard elements, then there exists $p \in {}^*X$ such that $p \in A$ for each $A \in \mathcal{C}$.

Let $^*\mathfrak{M}$ be an enlargement of $\mathfrak{M}$ and let $\mathcal{U}$ be a local base at 0 for θ. An element p of $^*E$ is <u>θ-finite</u> if for each $U \in \mathcal{U}$ there is a standard integer n such that $p \in n{}^*U$. Also, p is <u>θ-pre-nearstandard</u> if for each $U \in \mathcal{U}$ there exists $x \in E$ such that $p \in x + {}^*U$. The set of θ-finite elements of $^*E$ is denoted by $\text{fin}_\theta({}^*E)$, while the set of θ-pre-nearstandard elements of $^*E$ is denoted by $\text{pns}_\theta({}^*E)$. The θ-monad of 0 is $\mu_\theta(0)$, and these subsets of $^*E$ are related by:

$$\mu_\theta(0) \subseteq \text{pns}_\theta({}^*E) \subseteq \text{fin}_\theta({}^*E).$$

The collection $\{{}^*U \cap \text{fin}_\theta({}^*E) \mid U \in \mathcal{U}\}$ is a filter base for a locally convex vector topology $\tilde{\theta}$ on $\text{fin}_\theta({}^*E)$ under which $\text{pns}_\theta({}^*E)$ and $\mu_\theta(0)$ are closed sets. The nonstandard hull of (E, θ), with respect to the enlargement $^*\mathfrak{M}$, is the Hausdorff quotient space

$$(\hat{E}, \hat{\theta}) = (\text{fin}_\theta({}^*E), \tilde{\theta}) / \mu_\theta(0).$$

Let $\pi: \text{fin}_\theta({}^*E) \longrightarrow \hat{E}$ be the natural quotient map. The map taking x to $\pi({}^*x)$ is a topological vector space isomorphism of (E, θ) into $(\hat{E}, \hat{\theta})$.

If S is a subset of $^*F$ define

$$S^f = \{p \in {}^*E \mid \langle p, q\rangle \text{ is finite for all } q \in S\},$$

$$S^o = \{p \in {}^*E \mid |\langle p, q\rangle| \leq 1 \text{ for all } q \in S\}, \text{ and}$$

$$S^i = \{p \in {}^*E \mid \langle p, q\rangle \text{ is infinitesimal for all } q \in S\}.$$

If T is a subset of $^*E$, then T^f, T^o, and T^i are the analogous subsets of $^*F$.

Define $K(\theta)$ to be the smallest infinite cardinal number K such that there is a local base $\mathcal{U}$ at 0 for θ with the cardinality of $\mathcal{U}$ equal to κ. Let $\kappa(\theta)^+$ be the smallest cardinal number strictly greater than $\kappa(\theta)$. If $^*\mathfrak{M}$ is $\kappa(\theta)^+$-saturated, then $(\hat{E}, \hat{\theta})$ is complete. Finally we make the following notational conventions. The set of positive integers is denoted by N. If $a, b \in {}^*K$ and $a-b$ is infinitesimal, then we write $a =_1 b$. If $a \in {}^*K$ is near-standard to x, write $x = st(a)$. If S is a subset of $\mathfrak{M}$, $^*[S] = \{ ^*x \mid x \in S \}$.

<u>Section</u> 1. <u>The space</u> $\hat{F}$. Let $\mathcal{A}$ be the collection of subsets of F defined by $\mathcal{A} = \{ U^o \mid U$ is a θ-neighborhood of $0 \}$. Since θ is an admissible topology for the pairing between E and F, θ is identical with the topology of uniform convergence on members of $\mathcal{A}$. Equivalently, the collection $\{ A^o \mid A \in \mathcal{A} \}$ is a local base at 0 for θ. We define a subset M_θ of $^*F$ by

$$M_\theta = \cup \{ ^*A \mid A \in \mathcal{A} \},$$

noting that $M_\theta \supseteq {}^*[F]$. As is proved in Section 5 of [1] $fin_\theta({}^*E) = (M_\theta)^f$ and $\mu_\theta(0) = (M_\theta)^i$. Clearly $M_\theta \subseteq (fin_\theta({}^*E))^f$. Moreover, if $^*\mathfrak{M}$ is $\kappa(\theta)^+$-saturated, then

$$M_\theta = (fin_\theta({}^*E))^f = \mu_\theta(0)^i = \mu_\theta(0)^f.$$

(The proof of Theorem 5.9 in [1] will go through easily in this more general setting.) Also define a subset m_θ of $^*F$ by $m_\theta = (fin_\theta({}^*E))^i$. Then $m_\theta \subseteq M_\theta$ and both m_θ and M_θ are vector spaces over K.

Define $\hat{F}$ to be the quotient space M_θ/m_θ as a vector space over K and let $\pi' : M_\theta \longrightarrow \hat{F}$ be the quotient map. Since $^*[F] \subseteq M_\theta$ and $^*[E] \subseteq fin_\theta({}^*E)$, the mapping $y \longmapsto \pi'({}^*y)$ is a vector space isomorphism of F into $\hat{F}$.

Note that $\langle p, q \rangle$ is finite for each $p \in fin_\theta({}^*E)$ and $q \in M_\theta$, since $M_\theta = fin_\theta({}^*E)^f$. Also, if $p, p' \in fin_\theta({}^*E)$, $q, q' \in M_\theta$ and $p - p' \in \mu_\theta(0)$, $q - q' \in m_\theta$, then $\langle p, q \rangle =_1 \langle p', q' \rangle$. Therefore a bilinear form $\langle \ldots, \ldots \rangle$ between $\hat{E}$ and $\hat{F}$ may be defined by $\langle x, y \rangle = st \langle p, q \rangle$ where $x = \pi(p)$ and $y = \pi'(q)$.

Lemma 1. The bilinear form defined above is a pairing between $\hat{E}$ and $\hat{F}$, relative to which $\hat{\theta}$ is an admissible topology.

Proof. If $x = \pi(p)$ is a non-zero element of $\hat{E}$, then $p \in fin_\theta(*E) \sim \mu_\theta(0)$. Therefore $p \in (M_\theta)^f \sim (M_\theta)^i$, so that for some $q \in M_\theta$, $\langle p, q\rangle$ is not infinitesimal. Hence $\langle x,y\rangle \neq 0$ if $y = \pi'(q)$.

If $y = \pi'(q)$ is any non-zero element of $\hat{F}$, then $q \in M_\theta \sim m_\theta \subseteq fin_\theta(*E)^f \sim fin_\theta(*E)^i$. Hence there exists $p \in fin_\theta(*E)$ so that $\langle p,q\rangle$ is not infinitesimal. If $x = \pi(p)$, then $\langle x,y\rangle \neq 0$. This shows that $\hat{E}$ and $\hat{F}$ are paired by the given bilinear form.

The topology $\hat{\theta}$ has a local base at 0 which consists of sets $\hat{U}$ of the form $\hat{U} = \pi(*U \cap fin_\theta(*E))$, where U is an absolutely convex, $\sigma(E,F)$-closed θ-neighborhood of 0 in E. Then $U^{oo} = U$, so that $(*U)^{oo} = *U$. It follows that $\hat{U}$ is the polar in $\hat{E}$ of $\pi'(*U^o)$, so that $\hat{U}$ is absolutely convex and $\sigma(\hat{E}, \hat{F})$-closed. This shows that $\hat{\theta}$ is admissible.

Note that Lemma 1 implies that $\hat{F}$ is canonically isomorphic to a vector subspace of the dual space $(\hat{E}, \hat{\theta})'$. In the remainder of this paper we will identify $\hat{F}$ with this space.

Suppose (E, θ) is normable, with norm ρ. Then $M_\theta = fin_{\rho'}(*F)$ and $m_\theta = \mu_{\rho'}(0)$, where ρ' is the dual norm to ρ on F. Thus $\hat{F}$ is the nonstandard hull of (F, ρ'). It is proved in Theorem 5.12 of [1] that this is the only case in which $\hat{F}$ is canonically isomorphic to a nonstandard hull of F. Note that part of Theorem 8.5 of [1] can be stated as follows (where $F = (E, \theta)'$): if (E, θ) is normable, then $(\hat{E}, \hat{\theta})$ is reflexive if and only if $\hat{F} = (\hat{E}, \hat{\theta})'$.

In general, even in the normable case, the space $\hat{F}$ depends on F (as well as on (E, θ) and on $*\mathcal{M}$). For example, let E be the sequence space ℓ_∞ of all bounded sequences and let θ be the topology generated by the supremum norm. Let F_1 be the dual space of (E, θ) and let F_2 be the sequence space ℓ_1, considered as a subspace of F_1 in the usual way. Then θ is an admissible topology for the pairings $\langle E, F_1\rangle$ and $\langle E, F_2\rangle$. It can easily be shown that $\hat{F}_2$ is a proper subspace of $\hat{F}_1$, and that $\hat{F}_2$ is $\beta(\hat{F}_1, \hat{E})$-closed in $\hat{F}_1$.

Section 2. Representation of elements of $\hat{E}'$ by elements of $\hat{F}$.

As above, let $\langle E, F\rangle$ be paired spaces and let θ be a Hausdorff locally convex topology on E which is admissible for the pairing. Also let $*\mathfrak{M}$ be an enlargement of $\mathfrak{M}$. Note that if $\mathcal{U}$ is a local base at 0 for θ, then $\{\hat{U} = \pi(*U \cap \text{fin}_\theta(*E)) \mid U \in \mathcal{U}\}$ is a local base at 0 for $\hat{\theta}$. The following results indicate the extent to which the elements of $\hat{F}$ represent the $\hat{\theta}$-continuous linear functionals on $\hat{E}$.

Lemma 2. Let S be a finite dimensional subspace of $\hat{E}$ and let $\varphi \in (\hat{E}, \hat{\theta})'$. If U is an absolutely convex θ-closed neighborhood of 0 in E such that $\varphi \in \hat{U}^{\circ}$ (in $(\hat{E}, \hat{\theta})'$), then there exists $q \in *U^{\circ}$ such that $\pi'(q) = \varphi$ on S.

Proof. Let S, φ, and U be as above and let T be the subspace of $\hat{E}$

$$T = \{x \in \hat{E} \mid nx \in \hat{U} \text{ for all } n \in N\}.$$

Denote by ψ the quotient map of $\hat{E}$ onto $\hat{E}/T$. Then a basis $\{x_1, x_2, \ldots, x_n\}$ of S may be selected so that for some $r \leq n$, $\{\psi(x_1), \ldots, \psi(x_r)\}$ is a basis for $\psi(S)$ and $x_{r+1}, \ldots, x_n$ are in T.

Pick $p_1, \ldots, p_r$ in $\text{fin}_\theta(*E)$ such that $\pi(p_i) = x_i$ for $i = 1, \ldots, r$. By Theorem 1.8 of [1], $\{p_1, \ldots, p_r\}$ are $*$-independent in $*E$ (i.e., if $\sum_{i=1}^{r} \lambda_i p_i = 0$ for coefficients $\lambda_1, \lambda_2, \ldots, \lambda_r$ in $*K$, then $\lambda_1 = \lambda_2 = \ldots = \lambda_r = 0$). Let $H = \{\sum_{i=1}^{r} \lambda_i p_i \mid \lambda_i \in *K,\ i = 1, 2, \ldots, r\}$ be the $*$-linear subspace of $*E$ spanned over $*K$ by $\{p_1, p_2, \ldots, p_r\}$. Then a $*$-linear functional q_o may be defined on H by

$$\langle \sum_{i=1}^{r} \lambda_i p_i, q_o\rangle = \sum_{i=1}^{r} \lambda_i \langle x_i, \varphi\rangle.$$

Now let ρ be the Minkowski functional on E generated by U. If $\sum_{i=1}^{r} \lambda_i p_i \in H$ ($\lambda_i \in *K$, $i = 1, \ldots, r$) and $*\rho\,(\sum_{i=1}^{r} \lambda_i p_i) \leq 1$, then λ_i is finite for $i = 1, \dots, r$, as we prove next. Otherwise we may assume $|\lambda_1| \geq |\lambda_2| \geq \dots \geq |\lambda_r|$ and λ_1 is infinite. Then ${}^*\rho(\sum_{i=1}^{r}(\lambda_i/\lambda_1)p_i) = (1/\lambda_1)\,{}^*\rho(\sum_{i=1}^{r}\lambda_i p_i) =_1 0$ and so $\pi(\sum_{i=1}^{r}(\lambda_i/\lambda_1)p_i) \in T$. But again by Theorem 1.8 of [1]

$$\pi(\sum_{i=1}^{r}(\lambda_i/\lambda_1)p_i = \sum_{i=1}^{r} st(\lambda_i/\lambda_1)x_i.$$

Since $\{\psi(x_1), \dots, \psi(x_r)\}$ is independent in $\hat{E}/T$, it follows that $st(\lambda_i/\lambda_1) = 0$ for $i = 1, \dots, r$, which is impossible for $i = 1$. Thus λ_i is finite for $i = 1, \dots, r$. It follows that

$$st\langle \sum_{i=1}^{r} \lambda_i p_i, q_o \rangle = \sum_{i=1}^{r} st(\lambda_i)\, st\langle p_i, q_o \rangle$$

$$= \sum_{i=1}^{r} st(\lambda_i)\langle x_i, \varphi \rangle$$

$$= \langle \sum_{i=1}^{r} st(\lambda_i)x_i, \varphi \rangle.$$

But since $\sum_{i=1}^{r} \lambda_i p_i \in {}^*U$ and $\pi(\sum_{i=1}^{r} \lambda_i p_i) = \sum_{i=1}^{r} st(\lambda_i)x_i \in \hat{U}$, it follows that $st|\langle \sum_{i=1}^{r} \lambda_i p_i, q_o \rangle| \leq 1$. Thus for some $\eta =_1 0$ we have that if ${}^*\rho(p) \leq 1$ and $p \in H$, then $|\langle p, q_o \rangle| \leq 1 + \eta$.

Passing the Hahn-Banach Theorem to ${}^*\mathfrak{M}$ shows that for some $\tilde{q} \in {}^*F$

(i) $\tilde{q} \in (1 + \eta)\,{}^*U^o$

(ii) $\tilde{q} = q_o$ on S.

Now $1/(1+\eta) =_1 1$ so $q = (1/(1+\eta))\tilde{q} \in {}^*U^o$ and $\pi'(q) = \pi'(\tilde{q})$. Then for $i = 1, 2, \dots, r$

$$\langle x_i, \pi'(q) \rangle = st\langle p_i, \tilde{q} \rangle = \langle x_i, \varphi \rangle .$$

Finally since $q \in {}^*U^o$, $\pi'(q) = 0$ on T. Thus $0 = \langle x_i, \pi'(q) \rangle = \langle x_i, \varphi \rangle$ for $r+1 \leq i \leq n$. Hence $\varphi = \pi'(q)$ on S.

Theorem 1 (Retraction Theorem). _Assume ${}^*\mathfrak{M}$ is κ-saturated, where κ is an uncountable cardinal. Let S be a subspace of $\hat{E}$ which has Hamel dimension less than κ. For each $\varphi \in (\hat{E}, \hat{\theta})'$ there exists ψ in $\hat{F}$ such that $\varphi = \psi$ on S. In particular, if $\varphi \in \hat{U}^o = \pi({}^*U \cap fin_\theta({}^*E))^o$, where U is a θ-closed,_

<u>absolutely convex, θ-neighborhood of</u> 0, <u>then there exists</u> $q \in {}^*U^o$ <u>such that</u> $\varphi = \pi'(q)$ <u>on</u> S.

Proof. Pick $\{ p_i \mid i \in I \} \subseteq \text{fin}_\theta({}^*E)$ so that $\{x_i \mid i \in I\} = \{\pi(p_i) \mid i \in I\}$ is a Hamel basis for S. For each $n \in N$ and each finite subset J of I, let

$$A(n,J) = \{ q \mid q \in {}^*U^o \text{ and } |\langle p_i, q\rangle - \langle x_i, \varphi\rangle| < 1/n \text{ for } i \in J \}.$$

Lemma 2 shows that this family of internal sets has the finite intersection property. Since there are fewer than κ such sets, $\cap \{ A(n,J) \mid n \in N,\ J \subseteq I,\ J \text{ finite} \}$ is not empty. Suppose q is in this intersection. Then $\langle p_i, q\rangle =_1 \langle x_i, \varphi\rangle$ for each $i \in I$; that is, $\langle x_i, \pi(q)\rangle = \langle x_i, \varphi\rangle$ for each $i \in I$. Hence $\varphi = \pi'(q)$ on S. Moreover $q \in {}^*U^o$, which completes the proof.

<u>Corollary</u> 1. <u>Assume</u> ${}^*\mathfrak{M}$ <u>is</u> $\aleph_1$-<u>saturated</u>. <u>Then a subset of</u> $\hat{E}$ <u>is</u> $\hat{\theta}$-<u>bounded if and only if it is</u> $\sigma(\hat{E},\hat{F})$-<u>bounded</u>.

Proof. Since $\hat{F}$ is contained in $(\hat{E},\hat{\theta})'$, every $\hat{\theta}$-bounded subset of $\hat{E}$ is $\sigma(\hat{E},\hat{F})$-bounded.

Assume $B \subseteq \hat{E}$ and B is $\sigma(\hat{E},\hat{F})$-bounded. Suppose for some $\varphi \in (\hat{E},\hat{\theta})'$, φ is not bounded on B. Then there exists a sequence $\{x_n\} \subseteq B$ such that $|\langle x_n, \varphi\rangle| \uparrow \infty$. By the theorem above there exists $\pi'(q) \in \hat{F}$ such that $\langle x_n, \varphi\rangle = \langle x_n, \pi'(q)\rangle$ for $n = 1,2,\ldots$. Hence B is not $\sigma(\hat{E},\hat{F})$-bounded, which is a contradiction. Thus B is $\sigma(\hat{E},(\hat{E},\hat{\theta})')$-bounded and so is $\hat{\theta}$-bounded.

<u>Section</u> 3. <u>Semi-reflexivity of</u> $(\hat{E},\hat{\theta})$. Let $\mathcal{U}$ be a local base at 0 for θ consisting of closed absolutely convex sets. Then as noted above $\{ \hat{U} \mid U \in \mathcal{U} \}$ is a local base at 0 for $\hat{\theta}$ consisting of absolutely convex sets. Thus if B is a $\hat{\theta}$-bounded subset of $\hat{E}$, then there is a mapping $\alpha : \mathcal{U} \to R^+$ such that $B \subseteq \cap\{ \alpha(U)\hat{U} \mid U \in \mathcal{U} \}$. On the other hand if α is any mapping of $\mathcal{U} \to R^+$ then $\cap\{ \alpha(U)\hat{U} \mid U \in \mathcal{U} \}$ is a $\hat{\theta}$-bounded subset of $\hat{E}$.

Further if $\alpha : \mathcal{U} \to R^+$ then

$$\cap \{ [\alpha(U) + 1/n] \hat{U} \mid U \in \mathcal{U} \text{ and } n \in N \}$$

is a $\hat{\theta}$-closed, absolutely convex, bounded subset of $\hat{E}$. By the remarks above every $\hat{\theta}$-bounded subset of $\hat{E}$ is contained in a bounded set of this type.

Theorem 2. *Assume $\mathcal{M}$ is $\kappa(\theta)^+$-saturated and $\mathcal{U}$ is a local base at 0 for θ consisting of θ-closed, absolutely convex sets. Let $\alpha : \mathcal{U} \to R^+$ and let B be the $\hat{\theta}$-bounded set*

$$B = \cap \{ [\alpha(U) + 1/n] \hat{U} \mid U \in \mathcal{U} \text{ and } n \in N \}.$$

If φ belongs to the $\beta((\hat{E}, \hat{\theta})', \hat{E})$-closure of $\hat{F}$ in $(\hat{E}, \hat{\theta})'$, then there exists $b \in B$ such that $|\langle x, \varphi \rangle| \le \langle b, \varphi \rangle$ for all $x \in B$.

Proof. Since φ is bounded on B we may set $\beta = \sup \{ |\langle x, \varphi \rangle| : x \in B \}$. Since B is absolutely convex, we may select $x_n \in B$ with $|\langle x_n, \varphi \rangle - \beta| < 1/n$ for each $n \in N$. Since φ belongs to the $\beta(\hat{E}', \hat{E})$-closure of $\hat{F}$ in $\hat{E}'$, for each $k \in N$ we may select $y_k \in \hat{F}$ such that $|\langle x, \varphi - y_k \rangle| < 1/k$ for all $x \in B$.

Pick $p_n \in \mathrm{fin}_\theta({}^*E)$ such that $\pi(p_n) = x_n$ for $n \in N$ and pick $q_k \in M_\theta$ such that $\pi'(q_k) = y_k$ for $k \in N$. Now for each $n \in N$ and $U \in \mathcal{U}$, $\pi(p_n) \in [\alpha(U) + 1/n] \hat{U}$. Hence there exists $p \in [\alpha(U) + 1/n] {}^*U$ (p depending on U and n) such that $p_n - p \in \mu_\theta(0)$. Thus

$$p_n \in \mu_\theta(0) + [\alpha(U) + 1/n] {}^*U \subseteq [\alpha(U) + 2/n] {}^*U.$$

Now since ${}^*\mathcal{M}$ is $\aleph_1$-saturated we may extend the mappings $n \to p_n$ and $k \to q_k$ of $N \to {}^*E$ and $N \to {}^*F$ to internal mappings of ${}^*N \to {}^*E$ and ${}^*N \to {}^*F$ respectively. By our construction, for each $n \in N$ and $k \in N$ we have $|\langle x_n, y_k \rangle - \beta| < 1/n + 1/k$ so $|\langle p_n, q_k \rangle - \beta| < 1/n + 1/k$. Since ${}^*\mathcal{M}$ is $\kappa(\theta)^+$-saturated and $\aleph_0 \cdot \mathrm{card}(\mathcal{U}) < \kappa(\theta)^+$, there must exist $\omega \in {}^*N \sim N$ such that

(i) if $n \in {}^*N$, $n \le \omega$ and $U \in \mathcal{U}$ then $p_n \in (\alpha(U) + 2/n) {}^*U$ and

(ii) if $n, k \in {}^*N$ and $n, k \le \omega$ then $|\langle p_n, q_k \rangle - \beta| < 1/n + 1/k$.

Then $p_\omega \in (\alpha(U) + 2/\omega)*U$ for each $U \in \mathcal{U}$, so $p_\omega \in \text{fin}_\theta(*E)$ and $\pi(p_\omega) \in B$. If $k \in N$ then $|<p_\omega, q_k> - \beta| < 1/\omega + 1/k$, so $|<\pi(p_\omega), y_k> - \beta| \leq 1/k$ for each $k \in N$. Thus $|<\pi(p_\omega), \varphi> - \beta| \leq |<\pi(p_\omega), \varphi - y_k>| + |<\pi(p_\omega), y_k> - \beta| \leq 2/k$ for each $k \in N$. Hence $<\pi(p_\omega), \varphi> = \beta$ and the proof is complete.

The following theorem for real locally convex spaces is contained in [2], Theorem 2.

Theorem 3. *Let (E, θ) be a complete locally convex space over R and let B be a θ-bounded, $\sigma(E, E')$-closed subset of E. Then B is $\sigma(E, E')$-compact if and only if each linear functional $\varphi \in E'$ achieves its supremum on B.*

If E is complex and $\varphi \in E'$, let Re φ denote the real part of φ. It is easy to see that Theorem 3 implies the following result for complex locally convex spaces.

Theorem 3'. *Let (E, θ) be a complete locally convex space over C and let B be a θ-bounded $\sigma(E, E')$-closed subset of E. Then B is $\sigma(E, E')$-compact if and only if for each $\varphi \in E'$ the linear functional Re φ achieves its supremum on B.*

Theorem 4. *Assume $*\mathfrak{m}$ is $\kappa(\theta)^+$-saturated. Then $(\hat{E}, \hat{\theta})$ is semi-reflexive if and only if $\hat{F}$ is $\beta(\hat{E}', \hat{E})$-dense in $\hat{E}'$.*

Proof. Assume $(\hat{E}, \hat{\theta})$ is semi-reflexive. If $\hat{F}$ is not $\beta(\hat{E}', \hat{E})$-dense in $\hat{E}'$, then by the Hahn-Banach Theorem there exists a non-trivial $\beta(\hat{E}', \hat{E})$-continuous linear functional ψ on $\hat{E}'$ which is zero on $\hat{F}$. But ψ is represented by some element $\pi(p)$ of $\hat{E}$. Since ψ is zero on $\hat{F}$ it follows that $p \in M^i_\theta = \mu_\theta(0)$. Thus $\pi(p) = 0$ which is impossible.

Now suppose $\hat{F}$ is $\beta(\hat{E}', \hat{E})$-dense in $\hat{E}'$. Let $\alpha : \mathcal{U} \to R^+$ and let B be the $\hat{\theta}$-bounded set

$$B = \cap \{ [\alpha(U) + 1/n] \hat{U} \mid U \in \mathcal{U} \text{ and } n \in N \}.$$

Since $(\hat{E}, \hat{\theta})$ is complete, by Theorem 3 (or Theorem 3') and Theorem 2, B is $\sigma(\hat{E}, \hat{E}')$-compact. Since every $\hat{\theta}$-bounded subset of $\hat{E}$ is contained in a bounded set of this type, every $\hat{\theta}$-bounded set is $\sigma(\hat{E}, \hat{E}')$-relatively compact. Hence $(\hat{E}, \hat{\theta})$ is semi-reflexive.

If (E, θ) is normable and $(\hat{E}, \hat{\theta})$ is reflexive, then $\hat{F} = \hat{E}'$ since $\hat{F}$ is a nonstandard hull with respect to the dual norm topology and thus it is $\beta(\hat{E}', \hat{E})$-complete. In general it is an open question whether the semi-reflexivity of $(\hat{E}, \hat{\theta})$ implies $\hat{F} = \hat{E}'$, even if it is assumed that $F = (E, \theta)'$.

Finally we show, assuming $^*\mathfrak{M}$ is $\kappa(\theta)^+$-saturated, that the semi-reflexivity of $(\hat{E}, \hat{\theta})$ is equivalent to a standard condition on (E, θ). We recall another theorem of James [2].

Theorem 5. Let (E, θ) be a complete locally convex space over R and let B be a bounded $\sigma(E, E')$-closed subset of E. Then B is $\sigma(E, E')$-compact if and only if there does not exist a positive number r, a sequence $\{x_k\}$ in B, and an equicontinuous sequence $\{y_n\}$ in E' such that $\langle x_k, y_n \rangle > r$ if $n \le k$ and $\langle x_k, y_n \rangle = 0$ if $n > k$.

Again Theorem 5 has a complex version.

Theorem 5'. Let (E, θ) be a complete locally convex space over C and let B be a bounded $\sigma(E, E')$-closed subset of E. Then B is $\sigma(E, E')$-compact if and only if there does not exist a positive number r, a sequence $\{x_k\}$ in B, and an equicontinuous sequence $\{y_n\}$ in E' such that $\langle x_k, \mathrm{Re}\, y_n \rangle > r$ if $n \le k$ and $\langle x_k, \mathrm{Re}\, y_n \rangle = 0$ if $n > k$.

Theorem 6. Assume $^*\mathfrak{M}$ is $\kappa(\theta)^+$-saturated. Let (E, θ) be a locally convex space over R and let $\mathcal{U}$ be a local base at 0 for θ. Then $(\hat{E}, \hat{\theta})$ is semi-reflexive if and only if for each $V \in \mathcal{U}$, for each function $\alpha : \mathcal{U} \to R^+$, and each $r \in R^+$, the following statement holds:

(##) for some finite subset $\mathcal{F}$ of $\mathcal{U}$ and some $n \in N$ there do not exist

$x_i \in E$, $i = 1,2,\ldots, n$, <u>and</u> $y_j \in V^o$, $j = 1,2,\ldots, n$, <u>such that:</u>

$x_i \in \alpha(U)\, U$ <u>for each</u> $U \in \mathcal{F}$ <u>and</u> $i = 1,2,\ldots, n$,

$\langle x_i, y_j \rangle > r$ <u>if</u> $j \leq i \leq n$, <u>and</u>

$|\langle x_i, y_j \rangle| \leq r/n$ <u>if</u> $i < j \leq n$.

Proof. Assume $(\hat{E}, \hat{\theta})$ is semi-reflexive. Suppose that for some $V \in \mathcal{U}$, $\alpha : \mathcal{U} \longrightarrow R^+$, and $r > 0$ condition (##) fails. $\mathcal{U}$ is contained in a *-finite subset $\mathcal{B}$ of $*\mathcal{U}$. Passing the negation of (##) to $*\mathfrak{M}$, it follows that for any $\omega \in *N \sim N$, there exist $p_1, p_2, \ldots, p_\omega$ in *E and $q_1, q_2, \ldots, q_\omega$ in $*V^o$ such that

$$p_i \in \alpha(U)\, *U \text{ for each } U \in \mathcal{B},$$

$$\langle p_i, q_j \rangle > r \text{ if } j \leq i \leq \omega, \text{ and}$$

$$|\langle p_i, q_j \rangle| \leq r/\omega \text{ if } i < j \leq \omega .$$

It follows that for each i, $p_i \in \mathrm{fin}_\theta(*E)$ and $\pi(p_i)$ belongs to the bounded set $B = \cap \{ \alpha(U)\, \hat{U} \mid U \in \mathcal{U} \}$. Now $\pi'(q_j) \in \pi'(*V^o) \subseteq \hat{V}^o$ for all $j \in N$. Moreover $r/2 < r \leq \langle \pi(p_i), \pi'(q_j) \rangle$ if $j \leq i$, and $0 = \langle \pi(p_i), \pi'(q_j) \rangle$ if $i < j$. Since $(\hat{E}, \hat{\theta})$ is complete, it follows by Theorem 5 that B is not $\sigma(\hat{E}, \hat{E}')$-relatively compact. Thus $(\hat{E}, \hat{\theta})$ is not semi-reflexive, which is a contradiction.

Now assume $(\hat{E}, \hat{\theta})$ is not semireflexive. Then there is a $\sigma(\hat{E}, \hat{E}')$-closed, bounded subset of $\hat{E}$ which is not $\sigma(\hat{E}, \hat{E}')$-compact. By Theorem 5 there exists $V \in \mathcal{U}$, $r > 0$, $\{z_i \mid i \in N\} \subseteq B$, and $\{w_j \mid j \in N\} \subseteq \hat{V}^o$ such that $r < \langle z_i, w_j \rangle$ if $j \leq i$, and $0 = \langle z_i, w_j \rangle$ if $i < j$. Since B is $\hat{\theta}$-bounded, for each $U \in \mathcal{U}$ there exists $\alpha(U) \in R^+$ such that $B \subseteq \alpha(U)\, \hat{U}$.

Let $\mathcal{F}$ be a finite subset of $\mathcal{U}$ and $n \in N$. Then by Lemma 2 there exist $q_1, q_2, \ldots, q_n \in *V^o$ such that $r < \langle z_i, \pi'(q_j) \rangle$ if $j \leq i \leq n$, and $0 = \langle z_i, \pi'(q_j) \rangle$ if $i < j \leq n$. Pick $p_i \in \cap \{\alpha(U)\, *U \mid U \in \mathcal{F}\}$ with $\pi(p_i) = z_i$ for $i = 1,2,\ldots, n$. Then $r < \langle p_i, q_j \rangle$ if $j \leq i \leq n$, and $|\langle p_i, q_j \rangle| \leq r/n$ if $i < j \leq n$. Pulling this back to $\mathfrak{M}$, it follows that for this

V, α, and r, condition (##) fails. Thus if condition (##) holds for each V, α, and r, then $(\hat{E}, \hat{\theta})$ is semi-reflexive. This completes the proof.

Using Theorem 5', the following analogous result for complex spaces can be proved.

Theorem 6'. Assume ${}^*\mathfrak{M}$ is $\kappa(\theta)^+$-saturated. Let (E, θ) be a locally convex space over C and let $\mathcal{U}$ be a local base at 0 for θ. Then $(\hat{E}, \hat{\theta})$ is semi-reflexive if and only if for each $V \in \mathcal{U}$, for each map $\alpha: \mathcal{U} \to R^+$ and each $r \in R^+$ the following statement holds.

(##') for some finite subset $\mathcal{F}$ of $\mathcal{U}$ and some $n \in N$, there do not exist $x_i \in E$, $i = 1,2,\ldots, n$, and $y_j \in V^o$, $j = 1,2,\ldots, n$, such that

$$x_i \in \alpha(U)\, U \text{ for each } U \in \mathcal{F} \text{ and } i = 1,2,\ldots, n,$$

$$\langle x_i, \operatorname{Re} y_j \rangle > r \quad \text{if } j \leq i \leq n, \quad \text{and}$$

$$|\langle x_i, \operatorname{Re} y_j \rangle| \leq r/n \quad \text{if } i < j \leq n.$$

Remark. Since completing this paper the authors have proved the following result: for each Banach space (E, ρ), the nonstandard hull $(\hat{E}, \hat{\rho})$ is reflexive if and only if (E, ρ) is super-reflexive in the sense of James ([3], [4], [5], and [6]). ((E, ρ) is said to be super-reflexive if no non-reflexive Banach space is finitely represented in (E, ρ).) The equivalence of the super-reflexivity of (E, ρ) and the geometric condition (#) on (E, ρ), which follows from this result and Theorem 8.5 of [1], was proved by James in [3].

References

1. Henson, C. Ward and L. C. Moore, Jr., The nonstandard theory of topological vector spaces, Trans. Amer. Math. Soc. 172 (1972), 405-435.

2. James, Robert C., Weak compactness and reflexivity, Israel J. Math. 2 (1964), 101-119.

3. ______________, Some self-dual properties of normed linear spaces, Symposium on Infinite Dimensional Topology, Annals of Math. Studies 69(1972), 159-175.

4. ______________, Super-reflexive spaces with bases, Pacific J. Math. 41(1972), 409-419.

5. ______________, Super-reflexive Banach spaces, Canad. J. Math. 24(1972), 896-904.

6. ______________, and J. J. Schäffer, Super-reflexivity and the girth of spheres, Israel J. Math. 11(1972), 398-404.

7. Luxemburg, W. A. J., A general theory of monads, in Applications of Model Theory (W. A. J. Luxemburg, editor), Holt, Rinehart and Winston (New York, 1969), 18-86.

INVARIANCE OF THE NONSTANDARD HULLS OF A UNIFORM SPACE

C. Ward Henson
L. C. Moore, Jr.
Duke University

For each uniform space $(X, \mathcal{U})$ and each enlargement $^*\mathfrak{M}$ of a set-theoretical structure which contains $(X, \mathcal{U})$, let $(\hat{X}, \hat{\mathcal{U}})$ denote the nonstandard hull of $(X, \mathcal{U})$ constructed using the set $\mathrm{fin}_{\mathcal{U}}(^*X)$ of $\mathcal{U}$-finite points, as defined in [1]. The nonstandard hulls $(\hat{X}, \hat{\mathcal{U}})$ are said to be invariant if they are all equal to the completion of $(X, \mathcal{U})$ and are therefore independent of $^*\mathfrak{M}$. We show in Section 1 that this is the only reasonable concept of invariance or stability for such nonstandard hulls. Indeed, if the nonstandard hulls of $(X, \mathcal{U})$ are not invariant in this sense, then for each cardinal number κ there is an enlargement $^*\mathfrak{M}$ such that the associated space $(\hat{X}, \hat{\mathcal{U}})$ has cardinality greater than κ. Similar results hold for the classes of nonstandard hulls defined in [4] and are discussed at the end of this paper.

A necessary condition for invariance of the nonstandard hulls $(\hat{X}, \hat{\mathcal{U}})$ is the following: if Y is a subset of X and every uniformly continuous, real valued function on $(X, \mathcal{U})$ is bounded on Y, then Y is totally bounded. This condition is also sufficient for invariance if $\mathcal{U}$ is the uniformity defined by some metric on X. However it is not a sufficient condition for invariance in general.

In Section 2 we consider the metrizable topological spaces (X, τ) for which there is a metric d which defines τ and such that (X, d) has invariant nonstandard hulls (as a uniform space). Under certain assumptions, which are satisfied when (X, τ) is a connected space, the existence of such a metric d implies that (X, τ) is separable. Conversely, if (X, τ) is separable and metrizable, then a metric with the given properties must exist. If, in addition, (X, τ) is topologically complete, then the metric d which defines τ may be chosen so that (X, d) has invariant nonstandard hulls and is a complete metric space.

Preliminaries. Let X be a set and let $\mathcal{U}$ be a filter on $X \times X$ which is a uniformity on X such that the corresponding uniform topology on X is Hausdorff. For each $A \in \mathcal{U}$ and each integer $n \geq 1$, define A^n recursively by the equations $A^1 = A$ and $A^{n+1} = \{ (x,z) \mid \text{for some } y, (x,y) \in A^n \text{ and } (y,z) \in A \}$. That is, $(x,y) \in A^n$ if and only if there is a sequence $x_0, \ldots, x_n$ in X such that $x_0 = x$, $x_n = y$ and $(x_i, x_{i+1}) \in A$ for each $i = 0, \ldots, n-1$. Also, for each $A \in \mathcal{U}$ and each $x \in X$, let $A(x)$ denote the set $\{ y \mid (y,x) \in A \}$.

Let $^*\mathcal{M}$ denote an enlargement of some set-theoretical structure which contains $(X, \mathcal{U})$. We will make use of the nonstandard approach to uniform spaces developed in [4]. In particular, recall that the monad of the filter $\mathcal{U}$ (that is, the intersection of all sets of the form $^*A$ where A is in $\mathcal{U}$) is an equivalence relation on $^*X$. For each $p \in {}^*X$ we denote by $\mu(p)$ the equivalence class of p under this equivalence relation.

Let $\tilde{\mathcal{U}}$ be the uniformity on $^*X$ which is described in [1]: $\tilde{\mathcal{U}}$ is the filter on $^*X \times {}^*X$ which is generated by the collection $\{ {}^*A \mid A \in \mathcal{U} \}$. Let $X_0 = \{ \mu(p) \mid p \in {}^*X \}$ and let $\mathcal{U}_0$ be the quotient uniformity on X_0 obtained from $\tilde{\mathcal{U}}$. Therefore $\mathcal{U}_0$ has a base consisting of all sets of the form

$$\{ (\mu(p), \mu(q)) \mid (p,q) \in {}^*A \}$$

where A ranges over $\mathcal{U}$. Denote by π the natural quotient map of $(^*X, \tilde{\mathcal{U}})$ onto $(X_0, \mathcal{U}_0)$ defined by $\pi(p) = \mu(p)$ for $p \in {}^*X$. As is observed in [1], the space $(X_0, \mathcal{U}_0)$ is obtained from $(^*X, \tilde{\mathcal{U}})$ by simply identifying any two points in $^*X$ which have exactly the same neighborhoods in the $\tilde{\mathcal{U}}$-topology.

Recall that $p \in {}^*X$ is said to be $\mathcal{U}$-finite [1] if for each $A \in \mathcal{U}$ there is a finite sequence $p_0, \ldots, p_n$ in ${}^*X$ such that $p_0 = p$, p_n is a standard point ${}^*x$ for some $x \in X$ and for each $i = 0, \ldots, n-1$ the pairs (p_i, p_{i+1}) and (p_{i+1}, p_i) are in ${}^*A$. Since $\mathcal{U}$ has a base of symmetric sets, p is $\mathcal{U}$-finite if and only if for each $A \in \mathcal{U}$ there exist $n \geq 1$ and $x \in X$ such that $p \in {}^*(A^n(x))$. The set of $\mathcal{U}$-finite points is denoted by $\text{fin}_{\mathcal{U}}({}^*X)$.

The nonstandard hulls we discuss are all uniform subspaces of $(X_0, \mathcal{U}_0)$ of the form $\pi(F)$, where F is a certain set of "finite" elements of ${}^*X$. In particular, $(\hat{X}, \hat{\mathcal{U}})$ is just $\pi(\text{fin}_{\mathcal{U}}({}^*X))$. For simplicity, $(\hat{X}, \hat{\mathcal{U}})$ will be referred to simply as the nonstandard hull of $(X, \mathcal{U})$ constructed using ${}^*\mathfrak{M}$.

Section 1. Throughout this section $(X, \mathcal{U})$ will denote a Hausdorff uniform space and ${}^*\mathfrak{M}$ will denote an enlargement of some set-theoretical structure $\mathfrak{M}$ which contains $(X, \mathcal{U})$.

If $\mathcal{F}$ is a Cauchy filter on X relative to $(X, \mathcal{U})$, then the filter monad $\mu(\mathcal{F})$ is contained in $\text{pns}_{\mathcal{U}}({}^*X)$. Conversely, if p is in $\text{pns}_{\mathcal{U}}({}^*X)$, then the ultrafilter $\text{Fil}(p) = \{Y \mid Y \subseteq X \text{ and } p \in {}^*Y\}$ is a Cauchy filter on X. (See Sections 3.11 and 3.12 of [4].) A property of filters on X which corresponds in the same way to elements of $\text{fin}_{\mathcal{U}}({}^*X)$ is described next.

Definition 1.1. *A filter $\mathcal{F}$ on X is $\mathcal{U}$-finite if for each $A \in \mathcal{U}$ there exist $x \in X$ and $n \geq 1$ such that $A^n(x) \in \mathcal{F}$.*

Observe that if $\mathcal{F}$ is a $\mathcal{U}$-finite filter on X, then $\mu(\mathcal{F})$ is contained in $\text{fin}_{\mathcal{U}}({}^*X)$. If $p \in \text{fin}_{\mathcal{U}}({}^*X)$, then the ultrafilter Fil(p) is $\mathcal{U}$-finite, also. This observation, together with the remarks above, lead immediately to the following result.

Theorem 1.2. $\text{fin}_{\mathcal{U}}({}^*X) = \text{pns}_{\mathcal{U}}({}^*X)$ *if and only if every $\mathcal{U}$-finite ultrafilter on*

X *is a* $\mathcal{U}$ *-Cauchy filter.*

The subspace $\pi(\mathrm{pns}_{\mathcal{U}}(^*X))$ of $(X_0, \mathcal{U}_0)$ is naturally isomorphic to the completion of $(X, \mathcal{U})$. Therefore, if $\mathrm{pns}_{\mathcal{U}}(^*X) = \mathrm{fin}_{\mathcal{U}}(^*X)$, then the nonstandard hull of $(X, \mathcal{U})$ is also isomorphic to this completion. Theorem 1.2 shows that the equality $\mathrm{pns}_{\mathcal{U}}(^*X) = \mathrm{fin}_{\mathcal{U}}(^*X)$ is independent of the particular enlargement $^*\mathfrak{M}$ being used.

Definition 1.3. The nonstandard hulls of $(X, \mathcal{U})$ *are invariant if* $\mathrm{pns}_{\mathcal{U}}(^*X) = \mathrm{fin}_{\mathcal{U}}(^*X)$ *for some (equivalently, every) enlargement* $^*\mathfrak{M}$.

Theorem 1.4. If $p \in {}^*X$ *but* $p \notin \mathrm{pns}_{\mathcal{U}}(^*X)$ *and if* $^*\mathfrak{M}$ *is* κ *-saturated, then the subset* $\pi(\mu(\mathrm{Fil}(p)))$ *of* X_0 *has cardinality greater than or equal to* κ .

Proof. Assume that $^*\mathfrak{M}$ is κ -saturated and that p is an element of $^*X$ but not of $\mathrm{pns}_{\mathcal{U}}(^*X)$. Let $\mathcal{F}$ be the ultrafilter $\mathrm{Fil}(p)$. If $\pi(\mu(\mathcal{F}))$ has fewer than κ elements, then there is a subset V of $^*X$ which satisfies $\mathrm{card}(V) < \kappa$ and

$$\mu(\mathcal{F}) \subseteq \cup \{\mu(q) \mid q \in V\} .$$

Since $^*\mathfrak{M}$ is κ -saturated, there must be a $*$-finite (and hence internal) subset W of $^*X$ which has V as a subset. (See the proof of Lemma 2.1 in [2] .)

Let A be an arbitrary element of $\mathcal{U}$. For each $q \in {}^*X$, $\mu(q) \subseteq {}^*A(q)$. Therefore

$$\mu(\mathcal{F}) \subseteq \cup \{^*A(q) \mid q \in V\} \subseteq \cup \{^*A(q) \mid q \in W\} .$$

Now let Z be an element of $^*\mathcal{F}$ which satisfies $Z \subseteq \mu(\mathcal{F})$. Then Z is contained in the set $\cup \{^*A(q) \mid q \in W\}$, which is therefore an element of $^*\mathcal{F}$. Since W is $*$-finite and the function taking q to $^*A(q)$ is internal, there must exist $q \in W$ which satisfies $^*A(q) \in {}^*\mathcal{F}$. Passing the existence of such a q back to $\mathfrak{M}$, it follows that for some $x \in X$ the set $A(x)$ is in $\mathcal{F}$. Since A is any element of $\mathcal{U}$, this shows that $\mathcal{F}$ is a Cauchy filter.

But $p \in \mu(\mathcal{F})$ and hence $p \in pns_{\mathcal{U}}(*X)$, which is a contradiction.

Corollary 1.5. *If the nonstandard hulls of $(X, \mathcal{U})$ are not invariant, then for each cardinal κ there is an enlargement $*\mathfrak{M}$ so that the nonstandard hull $\pi(fin_{\mathcal{U}}(*X))$ has cardinality greater than or equal to κ.*

Proof. Let $*\mathfrak{M}$ be a κ-saturated enlargement. Since the nonstandard hulls of $(X, \mathcal{U})$ are not invariant, there exists p in $fin_{\mathcal{U}}(*X) \sim pns_{\mathcal{U}}(*X)$. Since p is $\mathcal{U}$-finite, the filter monad $\mu(Fil(p))$ is contained in $fin_{\mathcal{U}}(*X)$. By Theorem 1.4, the subset $\pi(\mu(Fil(p)))$ of $\pi(fin_{\mathcal{U}}(*X))$ has at least κ-elements.

Corollary 1.5 shows that the concept given in Definition 1.3 is the only reasonable concept of invariance or stability for the nonstandard hulls of $(X, \mathcal{U})$.

Recall that a subset Y of X is *finitely chainable* (relative to $(X, \mathcal{U})$) [1] if for each $A \in \mathcal{U}$ there exist $y_1, \ldots, y_n$ in Y and $k \geq 1$ such that

$$Y \subseteq \cup \{A^k(y_i) \mid i = 1, \ldots, n\}.$$

By Theorem 3.3 of [1], Y is finitely chainable if and only if every uniformly continuous, real valued function on $(X, \mathcal{U})$ is bounded on Y.

Theorem 1.6. *If the nonstandard hulls of $(X, \mathcal{U})$ are invariant, then every finitely chainable subset of X is totally bounded.*

Proof. If the nonstandard hulls of $(X, \mathcal{U})$ are invariant, then $pns_{\mathcal{U}}(*X) = fin_{\mathcal{U}}(*X)$. If Y is a finitely chainable subset of X, then $*Y \subseteq fin_{\mathcal{U}}(*X)$ by Theorem 3.2 of [1]. That is, $*Y \subseteq pns_{\mathcal{U}}(*X)$, from which it follows that Y is totally bounded.

The converse of Theorem 1.6 is not true in general (although it is true for metric spaces; see Theorem 2.1 below.) In [3] is given an example of a locally convex topological vector space (E, θ) in which every θ-bounded set is relatively compact (hence θ-totally bounded) but such that *not* every θ-finite element of

*E is θ-pre-nearstandard. (See Section 3 of [2] for definition of these terms.) Let $\mathcal{U}$ be the unique translation-invariant uniformity on E which defines the topology θ. Then by Theorem 3.3 of [2], $\text{fin}_{\mathcal{U}}(^*E) = \text{fin}_{\theta}(^*E)$. Since $\text{pns}_{\mathcal{U}}(^*E) = \text{pns}_{\theta}(^*E)$ by definition, it follows that $\text{fin}_{\mathcal{U}}(^*E) \neq \text{pns}_{\mathcal{U}}(^*E)$. Thus the nonstandard hulls of $(E, \mathcal{U})$ are not invariant. Now it is easy to prove that a subset Y of E is θ-bounded if and only if every uniformly continuous, real valued function on $(E, \mathcal{U})$ is bounded on Y. Hence the finitely chainable subsets of E are totally bounded, proving that the uniform space $(E, \mathcal{U})$ is a counter-example to the converse of Theorem 1.6.

Section 2. Next we consider uniform spaces which have a countable base, or equivalently, metric spaces. If d is a metric on X and d defines the uniformity $\mathcal{U}$ on X, then the properties of $(X, \mathcal{U})$ dealt with in this paper can all be translated in a straightforward way to refer to the metric space (X,d). For example, $p \in {}^*X$ is $\mathcal{U}$-pre-nearstandard if and only if for each $\delta > 0$ there exists $x \in X$ such that $^*d(p, {}^*x) < \delta$. Also, p is $\mathcal{U}$-finite if and only if for each $\delta > 0$ there is a finite sequence $p_0, \ldots, p_n$ in *X such that $p_0 = p$, $p_n = {}^*x$ for some $x \in X$ and $^*d(p_i, p_{i+1}) < \delta$ for $i = 0, \ldots, n-1$. In order to emphasize the role of the metric in this setting we will write $\text{pns}_d(^*X)$ for $\text{pns}_{\mathcal{U}}(^*x)$ and $\text{fin}_d(^*X)$ for $\text{fin}_{\mathcal{U}}(^*X)$.

Theorem 2.1. The nonstandard hulls of a metric space (X,d) are invariant if and only if every finitely chainable subset of X is totally bounded.

Proof. One implication of this theorem is contained in Theorem 1.6. For the converse, suppose that every finitely chainable subset of X is totally bounded (relative to the metric d). If the nonstandard hulls of (X,d) are not invariant, then there is an element p of $\text{fin}_d(^*X)$ which is not in $\text{pns}_d(^*X)$.

For each $n \geq 1$ let $A_n = \{ (x,y) \mid d(x,y) \leq 1/n \}$. Since $p \in \text{fin}_d(^*X)$, for each $n \geq 1$ there exist $x_n \in X$ and $k_n \geq 1$ so that

$$p \in {}^*(A_n^{k_n}(x_n)).$$

Also, since $p \notin \mathrm{pns}_d({}^*X)$ there exists $r \geq 1$ such that

$$p \notin {}^*(A_r(x)) \quad \text{for every } x \in X.$$

We will obtain a sequence $\{y_n\}$ in X which satisfies

$$\text{(a)} \qquad y_{n+1} \in A_j^{k_j}(x_j) \sim A_r(y_j)$$

for every pair j,n of integers which satisfy $1 \leq j \leq n$. This sequence is obtained inductively, as follows. Let y_1 be any element of X. Suppose that $y_1, \ldots, y_k$ have been chosen so that (a) is satisfied whenever $1 \leq j \leq n < k$. By assumption,

$$p \in {}^*(A_j^{k_j}(x_j)) \sim {}^*(A_r(y_j))$$

for $1 \leq j \leq k$. Pulling the existence of such a p back to $\mathfrak{M}$ shows that y_{n+1} can be chosen so that (a) is satisfied when $n = k$ and $1 \leq j \leq n$. This shows that the infinite sequence $\{y_k\}$ exists with the desired properties.

Now consider the set $Y = \{y_1, y_2, \ldots\} \subseteq X$. If $1 \leq i < j$ then $y_j \notin A_r(y_i)$; that is, $d(y_i, y_j) > 1/r$, and hence Y is not totally bounded. On the other hand, for each $n \geq 1$

$$\{y_{n+1}, y_{n+2}, \ldots\} \subseteq A_n^{k_n}(x_n).$$

Since the sets $\{A_n\}$ form a base for the d-uniformity, it follows that Y is finitely chainable. These properties of Y contradict the original assumption, and hence the nonstandard hulls of (X,d) must be invariant.

We now consider the following natural questions:

(I) If (X, τ) is a metrizable topological space, can τ be defined by a metric d such that (X,d) has invariant nonstandard hulls?

(II) If (X, τ) is metrizable and topologically complete, can τ be defined by a metric d such that (X,d) has invariant nonstandard hulls and is complete? (A metrizable space (X, τ) is <u>topologically complete</u> if τ is definable by a complete metric.)

If (X, τ) is separable and metrizable, then τ is definable by a metric d such that the completion of (X,d) is compact, as is well-known. In that case

$$*X = \text{pns}_d(*X) = \text{fin}_d(*X)$$

so that (X,d) does have invariant nonstandard hulls. Therefore the answer to (I) is positive when (X, τ) is separable. The answer to (II) is also positive in that case, as the next theorem asserts.

Theorem 2.2. _If_ (X, τ) _is a separable, topologically complete, metrizable space, then there is a metric_ d _which defines_ τ _and such that_ (X,d) _is complete and has invariant nonstandard hulls._

Proof. Since (X, τ) is topologically complete, there is a metric d_1 which defines τ and such that (X, d_1) is complete. Also, since (X, τ) is separable there is a dense subset $\{x_i \mid i \geq 1\}$ of X. For each $m,n \geq 1$ define

$$S_{m,n} = \{ x \mid d_1(x, x_i) \leq 1/n \quad \text{for some } i = 1,\ldots, m \}.$$

Note that $S_{1,n} \subseteq S_{2,n} \subseteq \cdots \subseteq S_{m,n}$. Also since $\{x_i\}$ is dense in (X, d_1), for each given $n \geq 1$

$$X = \bigcup\{S_{m,n} \mid m \geq 1\}.$$

These observations justify defining functions f_n on X (for $n \geq 1$) by

$$f_n(x) = \sum_{m=1}^{\infty} d_1(x, S_{m,n}).$$

For each $x \in X$ and $n \geq 1$ there exists $r \geq 1$ such that

$$f_n(y) = \sum_{m=1}^{r} d_1(y, S_{m,n})$$

for all y in the open neighborhood $\{y \mid d_1(x,y) < 1/2n\}$ of x. (Just take r so that $x \in S_{r,2n}$.) Therefore, each function f_n is continuous on (X, d_1).

If p is a τ-nearstandard element of $*X$, then $*f_n(p)$ is a finite element of $*R$ for each (standard) $n \geq 1$, by the continuity of the functions f_n. Conversely, suppose $p \in *X$ is not τ-nearstandard. Since (X, d_1) is complete,

$pns_{d_1}(*X) = ns_\tau(*X)$. Therefore there exists a (standard) integer $n \geq 1$ such that $*d_1(p, *x) > 1/n$ for every $x \in X$. Let $m \geq 1$ be any standard integer and let $q \in *S_{m,2n}$. Then

$$*d_1(p,q) \geq *d_1(p, *x_i) - *d_1(q, *x_i)$$

holds for $1 \leq i \leq m$. Hence $*d_1(p,q) \geq 1/2n$ for every such q. That is, $*d_1(p, *S_{m,2n}) \geq 1/2n$ for every standard $m \geq 1$. From this it follows that $*f_{2n}(p)$ must be infinite. Therefore we have shown that p is τ-nearstandard if and only if $*f_n(p)$ is finite for every standard $n \geq 1$.

Now let d be any metric on X which defines the topology τ, satisfies $d_1(x,y) \leq d(x,y)$ for all $x,y \in X$ and relative to which the functions f_n are all uniformly continuous. For example, we may take

$$d(x,y) = d_1(x,y) + \sum_{n=1}^{\infty} \frac{|f_n(x) - f_n(y)|}{2^n(|f_n(x) - f_n(y)| + 1)}.$$

It is immediate that (X,d) is a complete metric space. It remains only to show that $fin_d(*X) = pns_d(*X)$ ($= ns_\tau(*X)$ since (X,d) is complete.) But if $p \in fin_d(*X)$, then each $*f_n(p)$ is finite, by Theorem 1.4 of [1], since the functions f_n are all uniformly continuous. By the argument above, this implies that p is τ-nearstandard, and the proof is complete.

A partition $\{X_i \mid i \in I\}$ of the metric space (X,d) is <u>uniformly open</u> if there exists $\delta > 0$ such that for each $i \in I$

$$x \in X_i \text{ and } d(x,y) < \delta \text{ imply } y \in X_i.$$

The following theorem states that if (X,d) has invariant nonstandard hulls and is not separable, then there is a uniformly open partition $\{X_i\}$ of (X,d) such that the number of sets X_i is uncountable. As a consequence, if (X,τ) is a connected space (or, more generally, if X cannot be partitioned into uncountably many non-empty, open sets) then the answer to questions (I) and (II) for (X,τ) is positive <u>if and only if</u> (X,τ) is separable. In contrast, it should be noted that questions

(I) and (II) have a positive answer when (X, τ) is a discrete space of any cardinality.

Theorem 2.3. If (X,d) is a metric space with invariant nonstandard hulls and every uniformly open partition of (X,d) is countable, then (X,d) is separable.

Proof. Assume that (X,d) satisfies the stated conditions. We will show first that there is a sequence $\{f_n\}$ of uniformly continuous, real valued functions on (X,d) such that for each $p \in X$, $p \in \mathrm{fin}_d(*X)$ if and only if $*f_n(p)$ is finite for every standard $n \geq 1$. Let $A_n = \{(x,y) \mid d(x,y) \leq 1/n\}$ and let $\equiv_n$ be the equivalence relation on X defined by

$$x \equiv_n y \text{ if and only if } (x,y) \in (A_n)^k \text{ for some } k \geq 1.$$

(That is, $x \equiv_n y$ if and only if there is a finite sequence $x_0, \ldots, x_k$ in X such that $x_0 = x$, $x_k = y$ and $d(x_i, x_{i+1}) \leq 1/n$ for $i = 0, \ldots, k-1$. The equivalence classes for $\equiv_n$ form a uniformly open partition of (X,d) and therefore there are only countably many such classes, say $Y_{1,n}, Y_{2,n}, \ldots$ (perhaps a finite list). Define a function $g_{0,n}$ on X by

$$g_{0,n}(x) = k \text{ if and only if } x \in Y_{k,n}.$$

Also let $y_{k,n}$ be a fixed element of $Y_{k,n}$. By Lemma 2.1 of [1], there is a semi-metric $d_{k,n}$ defined on the equivalence class $Y_{k,n}$ and satisfying the following conditions:

(i) the uniformity defined on $Y_{k,n}$ by $d_{k,n}$ is weaker than the uniformity defined on $Y_{k,n}$ by d, and

(ii) for each $p,q \in *(Y_{k,n})$, $*d_{k,n}(p,q)$ is finite if and only if $(p,q) \in *(A_n)^r$ for some standard $r \geq 1$.

Define a function $g_{k,n}$ on X by

$$g_{k,n}(x) = \begin{cases} 0 & \text{if } x \notin Y_{k,n} \\ d_{k,n}(x,y_{k,n}) & \text{if } x \in Y_{k,n} \end{cases}.$$

Evidently the functions $g_{k,n}$ are all uniformly continuous on (X,d). Now suppose $p \in *X$ and every $*g_{k,n}(p)$ is finite. Then $k = *g_{0,n}(p)$ is a standard integer and $p \in *(Y_{k,n})$. Also, $*g_{k,n}(p) = *d_{k,n}(p, *y_{k,n})$ is finite, so that for some standard $r \geq 1$, $p \in *(A_n^r(y_{k,n}))$. Therefore, if $*g_{k,n}(p)$ is finite for every n and every appropriate k, then p is in $\text{fin}_d(*X)$. The converse of this statement is also true, by Theorem 1.4 of [1]. That is, we may arrange the functions $\{g_{k,n}\}$ into a sequence $\{f_n\}$ which has the desired properties.

Now suppose that (X,d) is not separable. It will be shown that there is an infinite set $W \subseteq X$ such that (i) each function f_n is bounded on W and (ii) there exists $\delta > 0$ such that $d(x,y) \geq \delta$ for each distinct $x,y \in W$. The non-separability of (X,d) implies that there exist $\delta > 0$ and an uncountable set $Y \subseteq X$ such that $d(x,y) \geq \delta$ for each distinct $x,y \in Y$. Given any uncountable set $S \subseteq X$ and $n \geq 1$, one of the sets $\{x \mid x \in S \text{ and } |f_n(x)| \leq m\}$, for $m = 1,2,\ldots,$ must be uncountable. Therefore there is a sequence $Y_1 \supseteq Y_2 \supseteq \cdots$ of uncountable subsets of Y such that f_n is bounded on Y_n for each $n \geq 1$. A sequence $\{x_n\}$ may be selected so that for each $n \geq 1$, x_n is in Y_n and is distinct from $x_1, \ldots, x_{n-1}$. It follows that the set $W = \{x_1, x_2, \ldots\}$ has the desired properties.

Since W is infinite, there is an element p in $*W$ which is not standard. For each $n \geq 1$, f_n is bounded on W and therefore $*f_n(p)$ is finite. By the construction of $\{f_n\}$ this implies $p \in \text{fin}_d(*X)$. Hence p is in $\text{pns}_d(*X)$, since the nonstandard hulls of (X,d) are assumed to be invariant. Thus there is an $x \in X$ which satisfies $*d(p, *x) < \delta/2$. Since p is not standard, there must be infinitely many elements y of W which satisfy $d(y,x) < \delta/2$. But this implies that $d(y_1, y_2) < \delta$ for some two distinct elements y_1, y_2 of W. This contradicts the fact that W is a subset of Y, and completes the proof.

Section 3. In this section we consider briefly the nonstandard hulls described by Luxemburg in [4]. These may be constructed as follows: let $\mathcal{C}$ be any set of uniformly continuous, real valued functions on the uniform space $(X, \mathcal{U})$. For each enlargement $*\mathcal{M}$ of a set-theoretical structure $\mathcal{M}$ which contains $(X, \mathcal{U})$, define

$$\text{fin}_{\mathcal{C}}(*X) = \{p \mid *f(p) \text{ is finite for every } f \in \mathcal{C}\}.$$

The $\mathcal{C}$-nonstandard hull of $(X, \mathcal{U})$ is defined to be the uniform space $\pi(\text{fin}_{\mathcal{C}}(*X))$, as a subspace of $(X_0, \mathcal{U}_0)$. (The fact that this construction gives exactly the same nonstandard hulls as [4] is discussed in [1], which contains a detailed comparison of the various $\mathcal{C}$-nonstandard hulls and the nonstandard hull $\pi(\text{fin}_{\mathcal{U}}(*X))$.)

Theorem 3.1. The following conditions are equivalent:

(i) For some $*\mathcal{M}$, $\text{fin}_{\mathcal{C}}(*X) = \text{pns}_{\mathcal{U}}(*X)$,

(ii) for every $*\mathcal{M}$, $\text{fin}_{\mathcal{C}}(*X) = \text{pns}_{\mathcal{U}}(*X)$,

(iii) there is an upper bound on the cardinality of the $\mathcal{C}$-nonstandard hulls, as $*\mathcal{M}$ varies, and

(iv) if $\mathcal{F}$ is an ultrafilter on X and for each $f \in \mathcal{C}$ there is an element of $\mathcal{F}$ on which f is bounded, then $\mathcal{F}$ is a Cauchy filter.

Proof. If $\mathcal{F}$ is a filter on X and if each function f in $\mathcal{C}$ is bounded on some element of $\mathcal{F}$ (perhaps depending on f) then the filter monad $\mu(\mathcal{F})$ is obviously contained in $\text{fin}_{\mathcal{C}}(*X)$. Conversely, if p is in $\text{fin}_{\mathcal{C}}(*X)$, then each function in $\mathcal{C}$ is bounded on some element of the ultrafilter $\text{Fil}(p)$. The equivalence of conditions (i), (ii) and (iv) follows immediately from these observations, as in the proof of Theorem 1.2.

Evidently (ii) implies (iii), since $\pi(\text{pns}_{\mathcal{U}}(*X))$ is the completion of $(X, \mathcal{U})$ for each enlargement $*\mathcal{M}$. Conversely, suppose (i) is false and let $*\mathcal{M}$ be a κ-saturated enlargement of $\mathcal{M}$, which contains $(X, \mathcal{U})$. If $p \in \text{fin}_{\mathcal{C}}(*X) \sim \text{pns}_{\mathcal{U}}(*X)$, then $\mu(\text{Fil}(p))$ is a subset of $\text{fin}_{\mathcal{C}}(*X)$, by the observations above. By Theorem 1.4, this subset $\pi(\mu(\text{Fil}(p)))$ of the $\mathcal{C}$-nonstandard hull has at least κ elements. This shows that (iii) implies

(i) and completes the proof.

When the equivalent conditions in Theorem 3.1 hold we say that the $\mathcal{C}$-nonstandard hulls of $(X, \mathcal{U})$ are invariant.

Theorem 3.2. If the $\mathcal{C}$-nonstandard hulls of $(X, \mathcal{U})$ are invariant, then any subset of X, on which every function in $\mathcal{C}$ is bounded, must be totally bounded.

Proof. If the $\mathcal{C}$-nonstandard hulls are invariant, and the functions in $\mathcal{C}$ are bounded on Y, then $*Y \subseteq fin_{\mathcal{C}}(*X) = pns_{\mathcal{U}}(*X)$. From this it follows that Y is totally bounded.

The converse to Theorem 3.2 is true when $\mathcal{C}$ is a countable set of functions, as is proved next, but not in general. (See the example after Theorem 1.6, taking $\mathcal{C}$ to be the set of all uniformly continuous, real valued functions.)

Theorem 3.3. If $\mathcal{C}$ is countable, then the $\mathcal{C}$-nonstandard hulls of $(X, \mathcal{U})$ are invariant if and only if every subset of X, on which every function in $\mathcal{C}$ is bounded, is totally bounded.

Proof. By Theorem 3.2 we need only prove one implication. Assume that if every function in $\mathcal{C}$ is bounded on a subset Y of X, then Y is totally bounded. If the $\mathcal{C}$-nonstandard hulls of $(X, \mathcal{U})$ are not invariant, then there is an element p of $fin_{\mathcal{C}}(*X)$ which is not in $pns_{\mathcal{U}}(*X)$. Let $\mathcal{C} = \{ f_n \mid n \geq 1 \}$. Since p is in $fin_{\mathcal{C}}(*X)$, there are standard integers $k_1, k_2, \ldots$ such that $| *f_n(p)| \leq k_n$ for each $n \geq 1$. Since p is not in $pns_{\mathcal{U}}(*X)$, there exists $A \in \mathcal{U}$ such that $(p, *x) \notin *A$ for every $x \in X$. Proceeding as in the proof of Theorem 2.1, we may obtain a sequence $\{x_n \mid n \geq 1\}$ in X such that $| f_n(x_i)| \leq k_n$ for all $1 \leq i \leq n$ and $(x_i, x_j) \notin A$ for all $1 \leq i < j$. But then $Y = \{x_1, x_2, \ldots\}$ is not totally bounded, while every function in $\mathcal{C}$ is bounded on Y. This contradiction completes the proof.

Example. Let (X,d) be a metric space, with $\mathcal{U}$ the uniformity defined by d, and let $\mathcal{C}$ consist of the single uniformly continuous function $f(x) = d(x, x_0)$, where

x_0 is a fixed element of X. Note that f is bounded on a subset Y of X if and only if Y has finite d-diameter. Also, $\mathrm{fin}_{\mathcal{C}}(^*X)$ is just the principal galaxy of *X in Robinson's sense. Theorem 3.3 implies that the principal galaxy is equal to $\mathrm{pns}_d(^*X)$ if and only if all subsets of X with finite d-diameter are totally bounded.

References

1. Henson, C. Ward, The nonstandard hulls of a uniform space, Pacific J. Math., 43(1972), 115-137.

2. ______________ and L. C. Moore, Jr., The nonstandard theory of topological vector spaces, Trans. Amer. Math. Soc. 172 (1972), 405-435.

3. ______________, Invariance of the nonstandard hulls of locally convex spaces, Duke Math. J., to appear.

4. Luxemburg, W. A. J., A general theory of monads, in Applications of Model Theory (W. A. J. Luxemburg, editor), Holt, Rinehart and Winston (New York, 1969), 18-86.

MODELS OF ARITHMETIC AND THE SEMI-RING OF RECURSIVE FUNCTIONS

Joram Hirschfeld
Yale University - Tel Aviv University

S. Feferman, D. Scott and S. Tennenbaum proved that every non trivial homomorphic image of the semi-ring R of recursive functions fails to be a model of arithmetic [1]. The aim of this paper is to show that every countable model of full arithmetic can be embedded in such a homomorphic image. To prove this we modify a theorem by H. Friedman [2] to obtain a sufficient condition for a model of arithmetic to be embeddable in a model of a fragment of arithmetic. We then introduce recursive ultrapowers - homomorphic images of R which are models of that fragment. Finally, given a model we show how to construct a recursive ultrapower which satisfies the condition of Friedman's theorem.

We deal with the complete theory T(N) of the natural numbers in the language with symbols $0,1,+,\cdot,<$. Bounded quantifiers $\exists x<y$ and $\forall x<y$ are introduced as usual. We denote by $\overline{x},\overline{y}$ etc. strings of variables (not necessarily of fixed length) and use also $\exists\overline{x}$ for $\exists x_1,\cdots\exists x_n$ and $\exists\overline{x}<y$ for $\exists x_1<y\cdots\exists x_n<y$. A formula is called bounded if all its quantifiers are bounded. Σ_1 formulas are obtained from bounded ones by adding (at most) a block of existential quantifiers at the front. A similar addition of universal quantifiers leads us from the bounded formulas to the Π_1 formulas and from the Σ_1 formulas to the Π_2 ones. From now on we denote by T the Π_2 theorems which are true in the natural numbers. This is the fragment of arithmetic which was mentioned in the introduction.
Every Σ_1 formula describes in N a r.e. predicate and vice versa. If the formula describes a function it does so in every model of T and thus recursive functions of N extend to every model of T (which contains N as an initial segment but generally also infinite elements). Every bounded formula (but not only those) describes in N a recursive predicate. We note that at least for bounded formulas $\phi(\overline{x},y)$

$$T \vdash \forall \bar{x}(\phi(\bar{x},0) \wedge \forall y[\phi(\bar{x},y) \rightarrow \phi(\bar{x},y+1)] \rightarrow \forall y \phi(\bar{x},y)).$$

0.1 In particular, if $M \models T$ and $N \subsetneq M$ then N is not definable in M by any bounded formula with parameters in M.

Finally we state Kleene's enumeration theorem in terms of Σ_1 formulas:

0.2 For any given n there is a Σ_1 formula $F^n(z,x_1 \cdots x_n)$ such that to any n-place Σ_1 formula $e(x_1 \cdots x_n)$ there corresponds an element $i \in N$ for which

$$\forall x_1 \cdots \forall x_n [F^n(i,x_1 \cdots x_n) \leftrightarrow e(x_1 \cdots x_n)]$$

This equivalence holds also in all models of T as it has a Π_2 normal form.

Theorem 1 is a version of H. Friedman's theorem which applies to models of T.

Theorem 1:

Let M and M' be models of T such that M is countable and such that the following holds:

1.1 For every Σ_1 formula $e(x,\bar{y})$ and parameters $\bar{a} \in M$ there is a bounded formula $\phi(x,\bar{z})$ and parameters $\bar{b} \in M'$ for which

$$\{i \in N \mid M \models e(i,\bar{a})\} = \{i \in N \mid M' \models \phi(i,\bar{b})\}.$$

Then for every element $c \in M'-N$ there is a monomorphism of M into the initial segment determined in M' by c.

Proof:

We order M as a sequence $a_1 \cdots a_n \cdots$ $n < \omega$ and construct by induction on n a map $f: M \rightarrow M'$ such that:

1.2 For every $\sum_1$ formula $e(\overline{x},\overline{y})$, if $M \models \exists \overline{x} e(\overline{x},\overline{a})$ then $M' \models \exists \overline{x}<c\ e(\overline{x},\overline{f}(a))$.

This assumption holds for $m = 0$: if $M \models \exists \overline{x} e(\overline{x})$ then $T \vdash \exists \overline{x} e(\overline{x})$ so that $N \models \exists \overline{x} e(\overline{x})$. Thus $N \models e(\overline{k})$ for some $\overline{k} \in N$ and again $T \vdash e(\overline{k})$ and $M' \models e(\overline{k})$. As c is bigger than all elements of N we get $M' \models \exists \overline{x}<c\ e(\overline{x})$. (Note that this is the only place that we use the fact that T is the Π_2 part of full arithmetic and not of some weaker arithmetical theory).

Assume now that $f(a_i) = b_i$, $i = 1 \cdots n$ and that the induction assumption 1.2 holds for these elements. With F^{n+2} as in 0.2 we put

$$A = \{i \in N \mid M \models \exists z F^{n+2}(i,\overline{a},a_{n+1},z)\}$$

By assumption there is a bounded formula $\phi(y)$ (with parameters in M') for which

$$A = \{i \in N \mid M' \models \phi(i)\}\ .$$

$F^{n+2}(y,\overline{x},x_{n+1},z)$ is $\sum_1$ and is therefore of the form $\exists \overline{v} F'(y,\overline{x},x_{n+1},z,\overline{v})$ where F' is bounded. Thus the following predicate is bounded:

1.3 $\psi(j) \equiv \exists y<c\ \forall i<j[\phi(i) \rightarrow \exists z<c \exists \overline{v}<c F'(i,\overline{b},y,z,\overline{v})]$.

We show that $M' \models \psi(j)$ for all $j \in N$. Let $i_1 \cdots i_r$ be the elements less then j for which ϕ holds in M'. By the choice of ϕ

$$M \models \exists y \left(\bigwedge_{k=1}^{r} \exists z \exists \overline{v} F'(i_k,\overline{a},y,z,\overline{v}) \right)$$

as it holds with $y = a_{n+1}$.

By induction assumption

$$M' \models \exists y<c(\bigwedge_{k=1}^{r} \exists z<c \exists \bar{v}<c F'(i_k,\bar{b},y,z,\bar{v}))$$

so that $\psi(j)$ holds.

Thus by 0.1 $\psi(j_o)$ holds also for some infinite j_o. For such a j_o we choose an element $y = b_{n+1}$ whose existence is asserted in 1.3 and put $b_{n+1} = f(a_{n+1})$.

Finally we check that 1.2 holds with this choice. Let $e(\bar{x},y,\bar{u})$ be a $\sum_1$ formula such that $M \models \exists \bar{u} e(\bar{a},a_{n+1},\bar{u})$. We note first that

1.4 $\quad T \vdash \forall \bar{x},y[\exists \bar{u} e(\bar{x},y,\bar{u}) \leftrightarrow \exists z \exists \bar{u}<z e(\bar{x},y,\bar{u})]$

Put $e'(\bar{x},y,z) \equiv \exists \bar{u}<z e(\bar{x},y,\bar{u})$. Then $M \models \exists z e'(\bar{a},a_{n+1},z)$. By 0.1 there is an $i \in N$ such that

1.5 $\quad T \vdash \forall \bar{x},y,z[F^{n+2}(i,\bar{x},y,z) \leftrightarrow e'(\bar{x},y,z)]$.

For this i we have $M' \models \phi(i)$ and by the choice of b_{n+1} $M' \models \exists z<c F^{n+2}(i,\bar{b},b_{n+1},z)$.

From 1.4 and 1.5 we conclude now that $M' \models \exists \bar{u}<c e(\bar{b},b_{n+1},\bar{u})$.

Thus f extends to all of M and by the way it was constructed it is clearly a monomorphism.

q.e.d.

Let R be the semi-ring of recursive functions. A maximal collection of recursive sets with the finite intersection property is called a <u>recursive ultrafilter</u>. If F is such an ultrafilter we define the recursive ultrapower $R/_F$ whose elements are the classes defined by the equivalence relation

$$f \equiv g \quad \text{iff} \quad \{x | f(x) = g(x)\} \in F$$

It is easy to see that this is really an equivalence relation and that the following are well defined for equivalence classes:

$$[f] + [g] = [f+g], \quad [f][g] = [fg] \quad \text{and } [f] < [g] \text{ iff } \{x|\ f(x)<g(x)\} \in F$$

Thus $R/_F$ is a homomorphic image of R.

We omit the proof of the following theorem which is straightforward:

Theorem 2:

Let F be a recursive ultrafilter.

i) If $\phi(x_1 \cdots x_n)$ is a bounded formula then
$R/_F \models \phi([f_1]\cdots[f_n])$ iff $\{x|\ N \models \phi(f_1(x),\cdots,f_n(x))\} \in F$.

ii) $R/_F \models T$.

Our main result is:

Theorem 3:

Let M be a countable model of T. Then there is a recursive ultrafilter F such that M can be embedded in $R/_F$.

Proof:

Let $A_1 \cdots A_j \cdots_{\ j<\omega}$ be all the subsets of N which are defined in M by a $\sum_1$ formula in the sense of 1.1. To apply theorem 1 we must construct an ultrafilter F such that in $R/_F$ all these sets are defined by bounded formulas. We define

$$B_{ij} = \begin{cases} \{x|\quad P_i \mid (x)_j\} & i \in A_j \\ \\ \{x|\quad P_i \nmid (x)_j\} & i \notin A_j \qquad i,j < \omega \end{cases}$$

where $a|b$ $(a\nmid b)$ means that a divides b (a does not divide b), P_i is the i-th prime and $(x)_j$ is the highest power k for which $P_j^k|x$.

Let $G = \{B_{ij} \mid i,j<\omega\}$

Every set in G is recursive and G has the finite intersection property, therefore G extends to a recursive ultrafilter F.

Let f_j be the recursive function $(x)_j$. We claim

$$3.1 \qquad A_j = \{i \in N \mid R/_F \models P_i \mid [f_j]\}$$

The formula $x|y$ is bounded so that by theorem 2 for every i

$$R/_F \models P_i \mid [f_j] \quad \text{iff} \quad \{x \mid P_i \mid (x)_j\} \in F.$$

This is true iff $B_{ij} = \{x \mid P_i \mid (x)_j\}$ i.e; iff $i \in A_j$. To apply theorem 1 it remains to show that we can replace the formula $P_x|y$ in 3.1 by a bounded formula with parameters in $R/_F$.

Let $e(z,x) = \exists \overline{v} e'(z,x,\overline{v})$ be the $\sum_1$ formula $z = P_x$, so that e' is bounded. For $i \in N$ it is easy to see that b is the i-th prime iff

$$R/_F \models \exists \overline{v}<[f_j] e'(b,i,\overline{v})$$

hence $A_j = \{i \in N \mid \exists z,y<[f_j](z\cdot y = [f_j] \wedge \exists \overline{v}<[f_j] e'(z,i,\overline{v}))\}$.

This completes the proof in view of theorems 1 and 2.

q.e.d.

REFERENCES

1. S. Feferman, D. Scott and S. Tennenbaum: Models of arithmetic through function rings. Notices of the A.M.S. 173 (1959) 556-31.

2. H. Friedman: Countable models of set theory. To appear in the Proceeding of the Cambridge symposium (1971).

NEAR PERIODS AND BOHR COMPACTIFICATIONS

A. E. Hurd

University of Victoria

Kugler developed a non-standard approach to almost periodic functions on the real line [2, 3] and on arbitrary groups [4]. The important result in [4] was a non-standard characterization of a certain kind of Bohr compactification. However the standard Bohr compactification (relative to all almost periodic functions) was not discussed. In this note we introduce a general class of subalgebras of the almost periodic functions, including those considered by Kugler, and also the almost periodic functions themselves, for which an associated compactification result holds.

I. Let G be a group. We denote by B the Banach algebra of bounded complex-valued functions on G with pointwise operations and the sup-norm. The subalgebras of left (right) almost periodic functions and the (two-sided) almost periodic functions will be denoted by LAP (RAP) and AP respectively. We will work in a fixed enlargement of the set-theoretic structure based on $G \cup C$ (where C denotes the complex numbers). If $a,b \in {}^*C$ are infinitesimally close, we write $a \simeq b$.

Recall that the set of <u>near periods</u> of $f \in B$ is the set

$$E(f) = \{t \in {}^*G : f(xty) \simeq f(xy) \quad \text{for all } x,y \in {}^*G\} .$$

$E(f)$ is a normal (external) subgroup of ${}^*G$. It is easy to see that

$$E(f) = \cap\, {}^*E(\epsilon;f) \qquad (\epsilon > 0 \text{ and standard})$$

where

$$E(\epsilon;f) = \{t \in G : |f(xty) - f(x,y)| < \epsilon \quad \text{for all } x,y \in G\}$$

is the set of <u>translation elements</u> of f corresponding to ϵ. We will also write

$$E(\epsilon_1,\ldots,\epsilon_n;f_1,\ldots,f_n) = \bigcap_{i=1}^{n} E(\epsilon_i;f_i) .$$

If E is an arbitrary subset of $^*G$ (not necessarily internal) consider the subalgebra of B defined by

$$\mathcal{B}(E) = \{f \in B : E(f) \supset E\} .$$

It is not hard to see that $\mathcal{B}(E)$ is a closed, translation invariant subalgebra of B. $\mathcal{B}(E)$ could be quite large; for example, if $E = \{e\}$, where e is the identity in G, then $\mathcal{B}(E) = B$. If $\mathcal{B}(E) \subset AP$ we will call E a <u>Bohr set</u> and $\mathcal{B}(E)$ a <u>Bohr subalgebra</u>.

II. We now set out to construct a compact topological group b_EG in which G is densely imbedded and for which $\mathcal{B}(E)$ is isomorphic, under the naturally associated homomorphism, to the algebra of all continuous functions on b_EG. Every continuous function on b_EG is uniformly continuous which shows that there must be some restriction on $\mathcal{B}(E)$ in order for the construction to work. In fact our construction will yield the desired result only for Bohr subalgebras.

There is a natural candidate for b_EG. Let

$$\hat{E} = \cap\, E(f) \qquad (f \in \mathcal{B}(E)) .$$

Clearly $\hat{E} \supset E$ and $\hat{E}$ is a normal (external) subgroup of $^*G$. Now define

$$b_EG = {}^*G/\hat{E} .$$

Let $\nu : {}^*G \to b_EG$ denote the projection homomorphism and $\nu_0 : G \to b_EG$ denote its (external) restriction to G. Elements in b_EG will be distinguished by primes, with $x' = \nu(x)$.

If $f \in \mathcal{B}(E)$ then it extends to $^*G$ and, being bounded, has a standard part ${}^of$. It is easy to see that ${}^of$ is constant on cosets and hence induces a function f' on b_EG. Let $\mathcal{B}'(E) = \{f' : f \in \mathcal{B}(E)\}$.

We now topologize b_EG with the "weak" topology induced by $\mathcal{B}'(E)$. A basis for the neighbourhood system at e' is given by sets of the form

$$E'(\epsilon_1,\ldots,\epsilon_n ; f_1',\ldots,f_n') = \bigcap_{i=1}^{n} E'(\epsilon_i;f_i')$$

where

$$E'(\epsilon;f') = \{t' \in b_E G : |f'(x't'y') - f'(x'y')| < \epsilon \text{ for all } x',y' \in b_E G\} .$$

The monad of e' in $*b_E G$ is the set

$$\mu' = \{t' \in *b_E G : f'(x't'y') \simeq f'(x'y') \text{ for all } f' \in B'(E) \text{ and } x',y' \in *b_E G\} .$$

Clearly $\mu'\mu' \subseteq \mu'$ and $(\mu')^{-1} \subseteq \mu'$ so this defines a group topology on $b_E G$ [6].

We will now show that one should not expect good results with this topology unless $B(E)$ is a Bohr subalgebra.

THEOREM 1. If $b_E G$ is compact in the "weak" topology, then $B(E)$ is a Bohr subalgebra.

Proof. Suppose that $b_E G$ is compact and let $f \in B(E)$. Then for any standard $\epsilon > 0$, the set $E'(\epsilon;f')$ is an open neighbourhood of e' , and $b_E G = \cup\, x'E'(\epsilon;f')$ $(x' \in b_E G)$. By compactness there is a finite subcover; i.e. there exist elements $x_i' \in b_E G$ $(i-1,\ldots,n)$ so that $b_E G = \bigcup_{i=1}^{n} x_i'E'(\epsilon;f')$.

Then $G^* = \bigcup_{i=1}^{n} \nu^{-1}(x_i'E'(\epsilon;f')) = \bigcup_{i=1}^{n} x_i{*E}(\epsilon;f)$ since $\nu^{-1}(x_i'E'(\epsilon;f')) = x_i{*E}(\epsilon;f)$.

Thus the statement "There exists a finite sequence $x_i \in *G$ such that $*G = \bigcup_{i=1}^{n} x_i{*E}(\epsilon;f)$" is *-true and so by transfer there exist sequence $x_i \in G$ $(i=1,\ldots,n)$ such that $G = \bigcup_{i=1}^{n} x_i E(\epsilon;f)$ so $f \in$ RAP. The same proof shows that $f \in$ LAP . Q.E.D.

The converse and more will be proved with the help of the following standard lemma which is a generalization of a classical result on AP functions on the line [1; chap. 1, §1,11°].

LEMMA 1. If the functions $f_i (i=1,\ldots,N)$ are in LAP (RAP), then the sets $E(\epsilon_1,\ldots,\epsilon_N;f_1,\ldots,f_N)$ are left (right) relatively dense.

Proof. By induction. We will only prove the result in the "right" case. For $k = 1$ this is Theorem 3 in [4]. Assume the result for k functions and let $\epsilon_1,\dots,\epsilon_{k+1}$ be arbitrarily prescribed positive real numbers. By the induction hypothesis there exist elements $r_i\,(i=1,\dots,n)$ and $s_j\,(j=1,\dots,m)$ so that

$$\bigcup_{i=1}^{n} r_i E\left(\frac{\epsilon_1}{2},\dots,\frac{\epsilon_k}{2};f_1,\dots,f_k\right) = \bigcup_{j=1}^{m} s_j E\left(\frac{\epsilon_{k+1}}{2};f_{k+1}\right) = G .$$

Put $K = \{x_i \in G : i=1,\dots,M = 2(n+m)\}$ where the x_i are the relabelled elements r_i and s_i and their inverses. Then $K = K^{-1}$ and

$$K\,E\left(\frac{\epsilon_1}{2},\dots,\frac{\epsilon_k}{2};f_1,\dots,f_k\right) = K\,E\left(\frac{\epsilon_{k+1}}{2};f_{k+1}\right) = G . \tag{1}$$

Let α be any fixed element in G. By (1) there exist elements

$$t_1(\alpha) \in E\left(\frac{\epsilon_1}{2},\dots,\frac{\epsilon_k}{2};f_1,\dots,f_k\right)$$

and

$$t_2(\alpha) \in E\left(\frac{\epsilon_{k+1}}{2};f_{k+1}\right)$$

with $t_i(\alpha) \in K\alpha\,(i=1,2)$ and so $t_1(\alpha)t_2(\alpha)^{-1} \in K^2$. Thus the set of elements $\{t_1(\alpha)t_2(\alpha)^{-1} : \alpha \in G\}$ is divided into at most M^2 equivalence classes, the elements in any equivalence class being equal to some element in K^2. Letting the elements $t_1^i(t_2^i)^{-1}$ $(i=1,\dots,M')$, with $M' \le M^2$, be representatives for these equivalence classes, we have $t_1(\alpha)(t_2(\alpha))^{-1} = t_1^i(t_2^i)^{-1}$ for some i depending on α. Then

$$t_1(\alpha)(t_1^i)^{-1} = t_2(\alpha)(t_2^i)^{-1} = t(\alpha)$$

and $t(\alpha) \in E(\epsilon_1,\dots,\epsilon_k;f_1,\dots,f_k) \cap E(\epsilon_{k+1};f_{k+1})$, since, if $t_1 \in E(\epsilon,f)$ and $t_2 \in E(\epsilon,f)$, then $t_1t_2^{-1} \in E(2\epsilon,f)$. Remembering that $t_1(\alpha) = t_1^i t(\alpha)$ is in $K\alpha$, we see that

$$\alpha \in Kt_1^i\, t(\alpha) \subseteq \overline{K}\, t(\alpha) \subseteq \overline{K}\, E(\epsilon_1,\dots,\epsilon_{k+1};f_1,\dots,f_{k+1})$$

where $\overline{K} = KT$ and $T = \{t_1^i : i=1,\dots,M'\}$. Since α was arbitrary,

$G = \overline{K}\, E(\epsilon_1,\ldots,\epsilon_{k+1};f_1,\ldots,f_{k+1})$ and $\overline{K}$ is a finite set,

i.e. $E(\epsilon_1,\ldots,\epsilon_{k+1};f_1,\ldots,f_{k+1})$ is right relatively dense. Q.E.D.

We can now go through the relevant proofs in [4], replacing neighbourhoods of the form $E(\epsilon;f)$ by ones of the form $E(\epsilon_1,\ldots,\epsilon_n;f_1,\ldots,f_n)$, and use Lemma 1 to establish the following result.

THEOREM 2. If E is a Bohr set, then $b_E G$ is compact in the "weak" topology, $\nu(G)$ is dense in $b_E G$, and the Bohr subalgebra $\mathcal{B}(E)$ is isometrically isomorphic to the algebra $CB(b_E G)$ of all bounded continuous functions on $b_E G$.

Using this result we see that a Bohr subalgebra acquires the characteristic structure of the group algebra of a compact topological group. In particular the approximation theorem and the L^2 uniqueness theorem (using Haar integral on $b_E G$) are available. Their meaning in this general context should be investigated.

III. The question now arises: what sets $E \subset {}^*G$ are Bohr sets, i.e. give rise to Bohr algebras. Clearly any set $E(g)$, where $g \in AP$, is a Bohr set since any f for which $E(f) \supset E(G)$ is also AP. More generally, call a set E (possibly external) *-relatively dense if, for every infinite integer N, there exist elements $r_i, s_i (i=1,\ldots,N)$ in $^*G$ so that $\bigcup_{i=1}^{N} r_i E = \bigcup_{i=1}^{N} E s_i = {}^*G$ (such a set is not in general an $E(g)$ for some $g \in AP$). Then any *-relatively dense set is a Bohr set (use Theorem 4 in [4]).

The complementary and perhaps more interesting question is: what algebras $A \subset AP$ are Bohr subalgebras. The natural candidate for the Bohr set of A is the set

$$E(A) = \cap\, E(f) \quad (f \in A)$$
$$= \cap\, {}^*E(\epsilon,f) \quad (\epsilon > 0 \text{ standard, } f \in A)\,.$$

Note that this set is *-relatively dense. For by concurrence we can find a *-finite family F of functions in $^*A$ which contain all functions in A and similarly a *-finite set P of nonstandard reals which contain all positive standard reals. Then

$$E(A) \supset \cap\, E(\epsilon,f) \quad (f \in F, \epsilon \in P)\,.$$

But since $A \subset \mathrm{AP}$, the set on the right is *-relatively dense by Lemma 1, and so $\mathrm{E}(A)$ is *-relatively dense and hence a Bohr set. Now define

$$\hat{A} = B(\mathrm{E}(A)) .$$

Clearly $\hat{A} \supset A$. We say that A is <u>full</u> if $\hat{A} = A$.

THEOREM 3. A is a Bohr subalgebra iff it is full.

<u>Proof</u>. If A is full it is clearly a Bohr subalgebra. Conversely, suppose that A is not full, i.e. $\hat{A}$ properly contains A. If $A = B(\mathrm{E})$ for some E then $\mathrm{E}(A) \supset \mathrm{E}$ and so $\hat{A} = A$ (contradiction). Q.E.D.

Thus only full subalgebras have an associated Bohr compactification. Happily, AP is full, for by the above, E(AP) is a Bohr set so $\widehat{\mathrm{AP}} \subset \mathrm{AP}$; i.e. $\widehat{\mathrm{AP}} = \mathrm{AP}$. In this case our compactification is identical with the usual Bohr compactification.

A somewhat interesting example of a subalgebra which is not full is afforded by the set LMP of limit periodic functions on the real line [2]. It is easy to see using Theorems 5.4 and 6.3 in [2] that E(LMP) = E(AP) so that $\widehat{\mathrm{LMP}} = \mathrm{AP}$. Thus LMP does not possess an associated Bohr compactification.

The results presented here are not unrelated to those of Robinson [7] to which the reader is referred.

The research presented in this paper was sponsored in part by National Research Council Grant No. A8198.

References

[1] A. S. Besicovitch, Almost periodic functions, Dover Publications, New York, c. 1954.

[2] L. D. Kugler, Non-standard analysis of almost periodic functions, Thesis, UCLA, 1966.

[3] ___________, Non-standard analysis of almost periodic functions, in [5], 150-166.

[4] ___________, Non-standard almost periodic functions on a group, Proc. A. M. S. 22 (1969), 527-533.

[5] Applications of model theory to algebra, analysis, and probability, Edited by W. A. J. Luxemburg, Holt, New York, 1969.

[6] R. Parikh, A nonstandard theory of topological groups, in [5], 279-284.

[7] A. Robinson, Compactification of groups and rings and nonstandard analysis, J. Symbolic Logic 34 (1969), 576-588.

MONOTONE COMPLETE FIELDS

H. Jerome Keisler
University of Wisconsin, Madison

Let F be an ordered field. F is said to be monotone complete if every bounded increasing function $h : F \to F$ has a limit $\lim_{x\to\infty} h(x)$ in F. The cofinality of F is the least cardinal of an unbounded subset of F. The field R of real numbers is the unique monotone complete field of cofinality ω. In non-standard analysis one uses a proper elementary extension R^* of R. R^* can never be complete; in this paper we show that R^* can be taken to be monotone complete.

Proposition 1. Let F be monotone complete of cofinality κ. Then

(a) Every bounded increasing sequence of length κ has a supremum in F.

(b) No infinite strictly increasing sequence of length $< \kappa$ has a supremum in F.

(c) There is no strictly increasing sequence of length $\geq \kappa^+$ in F.

Monotone completeness is related to a weaker notion of completeness introduced by Scott [5]. A function $h : F \to F$ is said to be Cauchy if for every $\epsilon > 0$ in F there exists b in F such that for all $x, y > b$, $|h(x) - h(y)| < \epsilon$. F is said to be Scott complete if every Cauchy function $h : F \to F$ has a limit $\lim_{x\to\infty} h(x)$ in F. Scott proved that every ordered field F has a unique Scott complete extension in which F is dense.

Proposition 2. F is monotone complete if and only if

(a) F is Scott complete.

(b) Every bounded increasing function $h : F \to F$ is Cauchy.

Corollary. The Scott completion of F is monotone complete if and only if (b) above holds.

Let R be the field of real numbers and let $\mathfrak{M}$ be any model of the form $\mathfrak{M} = \langle M, R, +, \cdot, \leq, S_0, S_1, \ldots \rangle$.

Theorem 3. $\mathfrak{M}$ has an elementary extension $\mathfrak{M}^* = \langle M^*, R^*, \ldots \rangle$ such that R^* is a non-archimedean monotone complete field.

The proof is similar to the construction used by Gaifman [1] and MacDowell-Specker [4] to study non-standard models of number theory.

Stronger forms of Theorem 3: If $\kappa > 2^\omega$ is a regular cardinal, $\mathfrak{M}$ has an elementary extension $\mathfrak{M}^*$ such that R^* is a monotone complete field of cofinality κ and cardinality κ. Every elementary extension $\mathfrak{M}_1$ of $\mathfrak{M}$ has an elementary extension $\mathfrak{M}^*$ such that R^* is monotone complete.

Theorem 4. Let κ be a regular cardinal. Then there exists a monotone complete field of cofinality κ and cardinality $> \kappa$ if and only if there exists a Kurepa tree of length κ.

The proof of Theorem 4 uses methods in [3].

Monotone complete fields have an application to the axioms for non-standard calculus in [2]. These axioms for a pair of structures (F^*, F) are as follows.

I. F is a real closed ordered field.

II. F is a subfield of F^*.

III. For each relation S on F there is a relation S^* on F^* such that

$$(F, S)_{S \subset \bigcup_n F^n} \prec (F^*, S^*)_{S \subset \bigcup_n F^n}.$$

IV. F^* has a positive infinitesimal relative to F.

V. Every finite $x^* \in F^*$ is infinitely close to some $x \in F$.

VI. F is archimedean.

Kunen asked whether the archimedean axiom VI can be eliminated. The following theorem answers his question. A cardinal κ is said to be <u>measurable</u> if there is a κ-additive non-trivial two-valued measure on the set of all subsets of κ.

Theorem 5. For any real closed ordered field F, the following are equivalent.

(a) There exists an F^* such that (F^*, F) satisfies axioms I - V.

(b) F is monotone complete and its cofinality is a measurable cardinal.

It follows that the archimedean axiom is needed if and only if there exist uncountable measurable cardinals.

REFERENCES

[1] Gaifman, H., Uniform extension operators for models. Sets, Models, and Recursion Theory, Ed. by J. M. Crossley, Amsterdam 1967, pp. 122-155.

[2] Keisler, H. J., Elementary Calculus; An Approach Using Infinitesimals. Bogden & Quigley, Inc., Tarrytown, N. Y., 1970.

[3] Keisler, H. J., Models with tree structures. To appear, Tarski Symposium.

[4] MacDowell, R., and Specker, E., Modelle der Arithmetik. Infinitistic Methods, New York-Oxford-London-Paris and Warsaw, 1961, pp. 257-263.

[5] Scott, D., On completing ordered fields. Applications of Model Theory to Algebra, Analysis, and Probability, Ed. by W. A. J. Luxemburg, pp. 274-278.

QUANTUM MECHANICS, QUANTUM FIELD THEORY AND HYPER-QUANTUM MECHANICS

Peter J. Kelemen

New York University, University Heights, Bronx, N.Y.

It can be said that quantum mechanics provides the foundations for atomic and nuclear physics while quantum field theory provides the foundations for elementary particle physics. Quantum mechanics very accurately predicts every phenomenon in its domain of validity. The physically interesting quantities can be calculated rigorously. These calculations are aided by several well founded approximation schemes. At the same time quantum field theoretic calculations of strong interaction phenomenon do not always agree with experiements. In quantum electrodynamics, where the agreement is extremely good, several mathematically not yet justified steps are employed in the calculations. These calculations are more difficult and cumbersome than the corresponding quantum mechanical ones and only heuristic approximation techniques are available to simplify them.

These differences are most interesting but hardly surprising. They arise because the number of degrees of freedom is finite in quantum mechanics and infinite in quantum field theory. This distinction is the source of a further difficulty. Schroedinger's quantum mechanics is equivalent to Heisenberg's quantum mechanics although they appear quite different. No such equivalence exists among the possible representations for quantum field theory.

To describe the motion of a particle when the quantum effects are not negligible, it is postulated that the position and the momentum of a particle are self-adjoint operators q and p, respectively, on a separable Hilbert space satisfying the commutation relations $\exp(icp)\exp(ibq) = \exp(icb)\exp(ibq)\exp(icp)$, where c and b are constants. The different formulations of quantum mechanics are equivalent because all irreducible representations of these commutation relations are unitarily equivalent.[1] In quantum field theory, there is an infinite number of unitarily inequivalent representations for the corresponding commutation relations.[2] Presumably, in quantum field theory one can calculate the physically interesting quantities only if the "right" representation is used. Since physicists presently have no intuitions for choosing the right representation the following prescription is followed:[3]

(i) Start with the representation used for free fields.

(ii) Introduce the interaction as a perturbation of the Hamiltonian.

(iii) Use several not necessarily rigorous steps (renormalization) to find the right representation.

(iv) Verify that the representation obtained is indeed the correct one.

Non-standard analysis offers a new approach to renormalization. It exploits the fact that the distinction between finite and infinite in non-standard analysis is replaced by the distinction between internal and external. If one selects the number of degrees of freedom a definite infinite integer, α, then all representations of the canonical commutation relations are unitarily equivalent in the non-standard sense. The standard quantum field theory is realized as the restriction of the representation of the canonical commutation

---

1. von Neumann, J.; Math. Ann. 104 570 (1931).

2. Garding, L. and Wightman, A.S.; Proc. Nat. Acad. Sci. 40 622 (1954)

3. Currently progress has been made in this direction. For two dimensional space-time and even degree polynomial interactions this program was carried out by J. Glimm, A. Jaffe and others.

relations to an external subspace of some Hilbert space[4] $F(\alpha)$. Thus the renormalization is replaced by "cutting out" of $F(\alpha)$ the proper external subspace.[5]

More explicitly, consider the case of a system with a finite number of degrees of freedom, n, corresponding to non-interacting particles. It is convenient and natural to realize the canonical commutation relations for such a system on the finite tensor product space $\bigotimes_{j=1}^{m} L_2^j(R)$ where $L_2(R)$ is the space of square integrable functions and $n \leq m$. The position operator q^k of the k-th degree of freedom is realized as the multiplication operator by the coordinate q on $L_2^k(R(q))$ tensored with the identity operator on $L_2^j(R)$ for $j \neq k$. The momentum operator p^k of the k-th degree of freedom is realized as the differentiation operator $-i \frac{\partial}{\partial q}$ on $L_2^k(R(q))$ tensored with the identity operator on $L_2^j(R)$ for $j \neq k$. Clearly[6] $[p^i, q^j] = -i\,\delta_{ij}$. The Hamiltonian, which is the generator of the time translation operator, is such that p^k and q^k evolve in time independently from p^j and q^j, $j \neq k$; expressing the fact that the particles move on straight lines without influencing each other. When an interaction is "turned on", i.e. when a potential term is added to the free Hamiltonian, the evolution of p^k and q^k may depend on p^j and q^j, $j \neq k$. Nevertheless one can still realize the canonical commutation relations on $\bigotimes_{j=1}^{m} L_2^j(R)$.

It follows from the transfer theorem that the above statements are also true in the non-standard sense for a system with a

---

4. $F(\alpha)$ is separable only in the non-standard sense.

5. This program was carried out for the model in ref. 3 in Kelemen, P.J. and Robinson, A.; J.M.P. 13 (Dec. 1972).

6. Earlier, the exponentiated form of this commutation relation was given, because q and p are unbounded operators and we wished to avoid the question of their domains.

finite or infinite number of degrees of freedom, α, on $\bigotimes_{j=1}^{m} L_2^j(R)$, $\alpha \leq m$. We refer to the resulting theory as hyper-quantum mechanics.

With standard mathematics, treating an infinite number of particles is extremely difficult. For free fields the canonical commutation relations can be realized on the infinite product space,[7] T_o, that is based on the eigenvector x_o corresponding to the lowest eigenvalue of the free Hamiltonian. In general the eigenvector x_1 corresponding to the lowest eigenvalue of a Hamiltonian that contains an interaction term is orthogonal to T_o. Therefore, T_o and the infinite tensor product space T_1 based on x_1 are perpendicular.[8] Hence to find T_1 one is forced to modify the Hamiltonian, i.e. introduce a cutoff, and trace the evolution of x_1 as the cutoff is removed. Since an infinite tensor product space is not the limit of finite tensor product spaces this is an extremely difficult and nonintuitive procedure.

In the non-standard approach one starts with hyper-quantum mechanics, and extracts from $F(\alpha)$ an external subspace. The restriction of hyper-quantum mechanics to this external sub-space is the standard field theory. Then the task is to show that for a given interaction the model constructed in this manner satisfies the Wightman axioms[9] of field theory.

To extract the external subspace one modifies the construction described in reference 5. One constructs the set of vectors

---

7. von Neumann.; Compos. Math. 6 1 (1939).

8. We are assuming that the interaction term used in the Hamiltonian allows the realization of the canonical commutation relations on an infinite tensor product space. A large class of representations of the canonical commutation relations cannot be realized on an infinite tensor product space.

9. Streater, R.F. and Wightman, A.S.; PCT, Spin & Statistics and All That, W.A. Benjamin, Inc., New York (1964).

$$V = \{\exp(it_1 H)\exp(ia_1 s(k_1))\exp(-it_1 H)\ldots\exp(ia_n s(k_n))\exp(-it_n H)x_1\}$$

where n takes on all values in N; $t_1,\ldots,t_n$, $a_1,\ldots,a_n$ are standard rationals; $k_1,\ldots,k_n$ εN; s(m) is either p^m or q^m; and x_1 is the eigenfunction corresponding to the lowest eigenvalue of the Hamiltonian H. The standard hull of the subspace spanned by finite length vectors of V is the external subspace we are looking for. One then needs to prove that for the given interaction, p^k and q^k are defined on a dense set of vectors and that exp(itH) as a function of t is a continuous one parameter unitary group on this space.

In the case of free fields, because of the form of the Hamiltonian the set of vectors V can be explicitly constructed. Hence it is not difficult to show that we have obtained T_0 in a roundabout way, and indeed recovered the standard quantum field theory for free fields. As of yet we have found no way to show in the interacting case that our construction leads to a standard field theory; but standard methods have been no more successful.

There are indications that hyper-quantum mechanics will provide the foundations for a theory of elementary particles. In that case the above construction of field theory is unnecessary. Thus, one should first perform explicit hyper-quantum mechanics calculations and compare them with experiments. We believe, that if these calculations are discouraging, one should then introduce simple external assumptions into the theory of hyper-quantum mechanics before rejecting it in favor of field theory.

The external assumptions we have in mind are of three types. The first one would reduce the restrictions that invariance forces on the theory. In the standard approach invariance under translation often reduces certain functions to constants. Boundary conditions at infinity then determine these constants. The external assumption would demand invariance only up to an infinitesimal for finite translations. For example: $^\circ f(x) \equiv {}^\circ\exp[-\alpha x^2] = {}^\circ\exp[-\alpha(x+c)^2]$ for all finite c and infinitesimal α; yet f(x) is not a constant.

The second would place less restriction on approximating a function with an infinite sum. We define a series $\Sigma f_k(x)$ to be R convergent to f(x) if given a standard $\varepsilon > 0$, there exists an infinite n_o such that $|f(x) - \sum_{k=1}^{n} f_k(x)| < \varepsilon$ for all $n > n_o$.

The third would allow one to expand the class of admissable functions. We define an internal function f(x) to be R continuous at b if given a standard $\varepsilon > 0$ there exists an infinitesimal δ such that $|f(x) - f(b)| < \varepsilon$ for all $0 < |x-b| < \delta$. The function

$$f(x) = \begin{cases} 0 & x = 0 \\ \alpha \sin(x^{-1}) & 0 < |x| < \alpha \\ \alpha^{-1} x & \alpha \leq |x| \end{cases}$$

for infinitesimal α is R continuous but neither Q nor S continuous at $x = 0$.

As outlined above, non-standard analysis provides a new method for constructing a standard field theory. It might provide, in hyper-quantum mechanics, the foundation for the theory of elementary particle physics. And most of all it allows one, by introducing external notions, to modify the mathematical content of a theory without changing its physical content. It is safe to conclude, that non-standard analysis is going to be an important and useful tool of physicists.

Acknowledgements

Many of the ideas expressed in this paper are joint work of the author with A. Robinson. The author has benefited from discussions with J. Glimm and C. Newman.

TOPOS-THEORETIC FACTORIZATION

OF NON-STANDARD EXTENSIONS

Anders Kock and Chr. Juul Mikkelsen

Aarhus Universitet

In the present paper, we attempt to give an analysis (in the form of a factorization theorem for certain functors) of the basic principle of higher order non-standard analysis; this basic principle is the contradiction that "higher order properties are preserved and yet not preserved" by non-standard extensions.

As is well known (see Robinson, [16]), any non-standard argument starts by embedding the structures under investigation in larger ones; in particular, if the argument is a higher order one, so that both a set A and (say) its power-set $\mathcal{P}A$ is considered, one has

$$A \subseteqq A^* \quad ; \quad \mathcal{P}A \subseteqq (\mathcal{P}A)^* \quad ;$$

and in general $(\mathcal{P}A)^*$ will not be the power set $\mathcal{P}(A^*)$, but rather a subset of it

$$(0.0) \qquad (\mathcal{P}A)^* \subseteqq \mathcal{P}(A^*),$$

namely that subset which consists of the internal subsets of A^*.

The $*$-operation (which in this note is denoted φ) thus does not preserve power-set formation, and more generally it does not "preserve truth of higher order logic sentences"; yet it does preserve such sentences provided every quantifier ranging over a class of higher order entities is understood to range over internal entities of that kind; for instance if $\mathbb{R}^*$ is a non-standard extension of the reals, the sentence

"Every bounded subset of real numbers has a sup"

is true for $\mathbb{R}$ and false for $\mathbb{R}^*$, but true for $\mathbb{R}^*$ if we replace the phrase "every bounded subset" by the phrase "every <u>internal</u> bounded subset". So without changing the elements of $\mathbb{R}^*$, a false higher-order sentence becomes true by a change in the ambient logic.

The category theoretic line on logic is that logic (including higher order logic) is built into the category $\mathcal{S}$ of sets (or into any topos $\underline{E}$, in fact). (This line of thought was initiated by Lawvere in 1963, [9]). In particular, one does not need a formalized language to describe the concepts of "first-order logic preserving functor" or "higher order logic preserving functor", (as we will see soon). It furthermore turns out that the idea of "changing the logic without changing the elements" can be made precise, again without introducing any syntactical notions: the toposes themselves play the role of logic, so "changing the logic" means "changing the topos". That this is possible is the content of the main theorem (the notions entering into the statement of the Theorem will be explained later in the paper):

THEOREM. Any first-order logic preserving functor between toposes $\varphi : \underline{E} \longrightarrow \underline{E}_0$ admits a factorization $\underline{E} \longrightarrow \underline{E}^* \longrightarrow \underline{E}_0$ where the first functor preserves higher-order logic and the second functor preserves elements (as well as first-order logic).

The theorem appears as Theorem 3.5 below.

The reader may have the following example in mind for φ; take a set I with a non-principal ultrafilter D on it; define the functor φ from the category $\mathcal{S}$ of sets to itself by the ultrapower

(0.1) $$\varphi(X) = \prod_I X/D$$

for any set $X \in |\mathcal{S}|$. (The factorization given in the theorem seems not in general to yield a known category in the middle, but for this special case, the middle category can be shown to be equivalent to the category $(\prod_I \mathcal{S})/D$).

The factorization was constructed and some of its properties conjectured by the first named author in [5]. A fuller account was given in our joint preliminary version [6] .

There are three sections. In the first, we recall notions from the theory of elementary toposes, and develop some of the intrinsic logic in a topos. In the second section we describe the notion of first-order-logic-preserving functor between toposes, and develop the notions: standard map, internal map, internal subobject, and related notions needed for defining the middle category $\underline{E}^*$ of the factorization. Finally, the third section contains the proof that $\underline{E}^*$ is a topos, (as well as properties of the functors in the factorization).

Let us remark that from closed category theory one knows that any closed functor admits a factorization into a "residual" closed functor followed by a "normal" closed functor, [4]. The middle category in this factorization will not in general be a topos. But it will sit as a subcategory of our middle topos (as "the category of internal maps").

1. The logic of an elementary topos $\underline{E}$

We consider a category $\underline{E}$ with finite inverse limits, in particular with a terminal object 1. An object Ω of $\underline{E}$ together with a map

$$1 \xrightarrow{\text{true}} \Omega$$

(necessarily monic) is said to be a <u>subobject classifier</u> [12],[13], if for any monic map $A' \overset{a}{\rightarrowtail} A$ in $\underline{E}$, there is precisely one map $A \longrightarrow \Omega$ (called the <u>characteristic map</u> ch(a) of a) such that

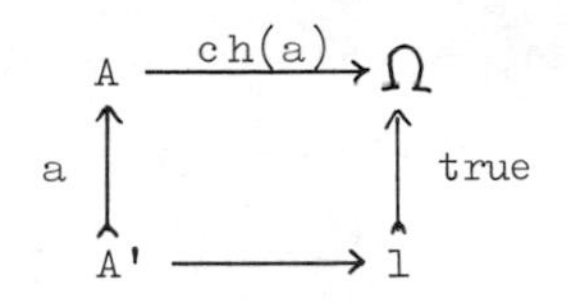

is a pull-back. A category with finite inverse limits and a subobject classifier is said to be an (elementary) topos if it further is cartesian closed. This means that there is a natural 1-1 correspondence between maps $A \times B \to C$ and maps $A \to C^B$ for a suitable object C^B (here also denoted $B \pitchfork C$).

The original definition of the concept of elementary topos (as given by Lawvere and Tierney [13], [3]) also required existence of finite colimits, but this can be shown to be redundant, [15].

Further, Cartesian closed-ness calls for existence of exponential objects B^A for any A, B; we show below that exponential objects of the form Ω^A will suffice to get all the others.

One should think of Ω^A as the power-set object of A in the category $\underline{E}$; this is clear when we look at $\underline{E} = \mathcal{S}$; then $\Omega = 2$ (any two-element set will do), and B^A is the set of mappings from A to B; so Ω^A is the set of mappings from A to 2, which clearly indexes the power set $\mathcal{P}A$ of A. When we illustrate in the category $\mathcal{S}$ the constructions to come, we shall simply identify Ω^A with $\mathcal{P}A$, thus saying, for instance

"let $A' \subseteqq A$ be an element of Ω^A".

From now on, $\underline{E}$ denotes an arbitrary topos, for instance the category $\mathcal{S}$ of sets.

For any object $A \in |\underline{E}|$, we have the diagonal

$$A \xrightarrow{\Delta_A} A \times A \tag{1.1}$$

which is monic since it has a right inverse: projection to the first (or second) factor (we compose maps from left to right). The subobject of $A \times A$

defined by the monic map (1.1) may be viewed as the extension of the equality predicate for "elements in A". It has a characteristic function

$$(1.2) \qquad A \times A \xrightarrow{\delta_A} \Omega,$$

denoted δ_A since it specializes to Kronecker's δ in the case $\underline{E} = \mathcal{S}$ (and $\Omega = \{1,0\} = \{\text{true, false}\}$). The exponential adjoint of δ_A is a map, "singleton-map",

$$(1.3) \qquad A \xrightarrow{\{\cdot\}_A} \Omega^A$$

denoted this way, because in the set case, it assigns to $a \in A$ the element $\{a\} \subseteq A$ of Ω^A.

These three maps appear in the work of Lawvere and Tierney under the same names. They also proved that (1.3) is monic (see e.g. [8] for a proof). Therefore we can take the characteristic function s_A of $\{\cdot\}_A$:

$$(1.4) \qquad \Omega^A \xrightarrow{s_A} \Omega$$

In the set-case, it takes $A' \subseteq A$ into "true" iff A' is a singleton $\{a\}$.

For any map $f: A \to B$, consider the exponential adjoint $\ulcorner f \urcorner$ of the composite

$$1 \times A \cong A \xrightarrow{f} B \;;$$

it is a map

$$1 \xrightarrow{\ulcorner f \urcorner} B^A ,$$

which Lawvere in [9] called "the name of f".

For any object A, the characteristic map of the maximal subobject of A (which is A itself) is denoted "true_A"; alternatively

$$\text{true}_A = A \longrightarrow 1 \xrightarrow{\text{true}} \Omega .$$

Thus we have "the name of true_A:

$$1 \xrightarrow{\ulcorner true_A \urcorner} \Omega^A$$

Recall that Y^X depends functorially on Y (and also contravariant functorially on X). If $g\colon Y \longrightarrow Z$, we denote the resulting map $Y^X \longrightarrow Z^X$ by g^X, or $1 \pitchfork g$. Similarly in the upper variable.

PROPOSITION 1.1. Let A and B be arbitrary objects in $\underline{E}$. Then there is a pull-back diagram of the following form

$$\begin{array}{ccc} B^A & \longrightarrow & (\Omega^B)^A \\ \downarrow & & \downarrow (s_B)^A \\ 1 & \xrightarrow[\ulcorner true_A \urcorner]{} & \Omega^A \end{array} \tag{1.5}$$

We shall not here give the full proof (which may be found in the preliminary version [6]), but rather argue heuristically for the case $\underline{E} = \mathcal{S}$. Clearly $(\Omega^B)^A$ may be identified with $\Omega^{A \times B}$, the power set of $A \times B$, which in turn is the set of relations R between the sets A and B. The upper map associates to an element $\ulcorner f \urcorner \in B^A$ the <u>graph</u> of the corresponding map $f\colon A \longrightarrow B$. To say that the diagram is a pull-back is to say that a relation $R \subseteq A \times B$ is the graph of some map $f\colon A \longrightarrow B$ if and only if

$$(s_B)^A(R) = A \tag{1.6}$$

(the maximal subset of A). But recalling that s_B is the characteristic map of the singleton-construction $\{\cdot\}_B$, it is not hard to see that

$$(s_B)^A(R) = \{a \in A \mid \{b \mid aRb\} \text{ is a singleton}\} ;$$

So (1.6) says that for <u>every</u> $a \in A$, $\{b \mid aRb\}$ is a singleton; so, for every $a \in A$, precisely one $b \in B$ stands in the relation R to it; so R is (the graph of) a map $A \longrightarrow B$.

COROLLARY 1.2. If a category has finite inverse limits, a subobject classifier Ω, and exponentiation of the form Ω^A for all A, then it is a topos.

Proof. If exponentials of the form Ω^A exist, then also $(\Omega^B)^A$, since this object can be taken to be $\Omega^{B\times A}$. Now one can construct B^A as the pull-back of $\ulcorner true_A \urcorner$ with $(s_B)^A$, as in (1.5). For finite colimits: see Mikkelsen [15]).

We next need topos-theoretic versions of the notion "for all $a \in A$", and "there exists an $a \in A$".

For any map $f: A \longrightarrow B$ in E, "pulling back along f" defines a monotone map

$$\mathcal{P}(B) \xrightarrow{f^{-1}} \mathcal{P}(A)$$

(here $\mathcal{P}(A)$ denotes the set of subobjects of A - a subobject of A being as usual an equivalence class of monic maps with codomain A). We have that $\mathcal{P}(A)$ is a partially ordered set; we denote its elements A', A", etc. Now it is a well-known fact from the theory of (elementary) toposes (see e.g. [8], p.32) that f^{-1} has a right adjoint $\forall_f$ in the sense that for any $B' \in \mathcal{P}(B)$, $A' \in \mathcal{P}(A)$

$$(1.7) \qquad f^{-1}(B') \subseteq A' \quad \text{iff} \quad B' \subseteq \forall_f(A') \; ;$$

$\forall_f$ itself is a monotone map $\mathcal{P}(A) \longrightarrow \mathcal{P}(B)$, "universal quantification along f" ; in the set case

$$\forall_f(A') = \{ b \mid f^{-1}(b) \subseteq A' \} ,$$

as the reader may easily check. The fact that "right adjoint for pulling back" is a kind of universal quantification was pointed out by Lawvere in 1965, [10].

Besides the universal quantification $\forall_f$ considered in (1.7), there is also an existential quantification $\exists_f$; for an $f: A \longrightarrow B$, $\exists_f$ is a monotone map $\mathcal{P}(A) \longrightarrow \mathcal{P}(B)$ satisfying the dual of (1.7): for $A' \in \mathcal{P}(A)$, $B' \in \mathcal{P}(B)$

$$(1.8) \qquad A' \subseteq f^{-1}(B') \text{ iff } \exists_f(A') \subseteq B'.$$

To construct $\exists_f(A')$, just take the mono part of an epi-mono factorization of the composite $A' \rightarrowtail A \xrightarrow{f} B$. Of course, in a <u>boolean</u> topos (like the category of sets) $\exists_f$ and $\forall_f$ can be constructed in terms of each other, by means of negation.

Besides the $\forall_f$ and $\exists_f$ as given above, we shall need their <u>intrinsic</u> versions. Note that $\mathcal{P}(A)$ is the power <u>set</u> (<u>set</u> of subobjects) of A - but that this also lives intrinsically in $\underline{E}$, namely as $A \pitchfork \Omega = \Omega^A$. The functor $\hom_{\underline{E}}(1,-)$ takes $A \pitchfork \Omega$ to $\mathcal{P}(A)$. The intrinsic version of $\exists_f$ and $\forall_f$ should be maps in $\underline{E}$:

$$(1.9) \qquad \exists(f), \forall(f): A \pitchfork \Omega \longrightarrow B \pitchfork \Omega$$

which by $\hom_{\underline{E}}(1,-)$ go to the previously considered $\exists_f, \forall_f$. Sometimes we write $\exists_f$ instead of $\exists(f)$. We shall need the construction of the intrinsic $\exists(f)$ in Proposition 2.4. We now recall its construction:

For any X,Y the exponential adjointness applied to the identity map of $X \pitchfork Y$ gives rise to an "evaluation" map

$$(X \pitchfork Y) \times X \xrightarrow{ev} Y.$$

In particular, we have

$$(A \pitchfork \Omega) \times A \xrightarrow{ev} \Omega \ ;$$

this map is characteristic map for a subobject

$$(1.10) \qquad \in_A \rightarrowtail (A \pitchfork \Omega) \times A \ ;$$

in the set case, $\in_A$ consists of pairs $\langle A' \subseteq A, a \rangle$ such that $a \in A'$. It is an intrinsic version of the $\in$ -relation from set theory.

Now, for $f: A \longrightarrow B$, we get the intrinsic $\exists(f): A \pitchfork \Omega \longrightarrow B \pitchfork \Omega$ as the exponential adjoint of a map $(A \pitchfork \Omega) \times B \longrightarrow \Omega$, which we in turn get as characteristic map of that subobject of $(A \pitchfork \Omega) \times B$ which is the image of

$$\in_A \subseteq (A \pitchfork \Omega) \times A$$

under $1 \times f: (A \pitchfork \Omega) \times A \longrightarrow (A \pitchfork \Omega) \times B$.

(This construction is due to Lawvere-Tierney).

A simple construction of $\forall(f)$ may be found in [7].

2. Exact functors which preserve Ω

In this section we shall study functors which preserve smaller or larger parts of the intrinsic logic.

The functors we consider are all left exact functors (that is, finite inverse limit preserving functors)

$$\varphi : \underline{E} \longrightarrow \underline{E}_o \tag{2.1}$$

between two toposes. For such a functor φ, since $\varphi(1) = 1$ (terminal objects are preserved), we will have a monic map

$$\varphi(\text{true}): 1 = \varphi(1) \longrightarrow \varphi(\Omega)$$

which will have a characteristic map

$$\gamma : \varphi(\Omega) \longrightarrow \Omega_o \tag{2.2}$$

where Ω_o is the subobject classifier of the topos $\underline{E}_o$.

Although it from some examples seems conceivable that one can give the factorization desired under weaker assumptions, we shall in general assume that φ preserves Ω in the sense that γ in (2.2) is invertible, and also we shall in general assume that φ is exact in the sense of [1] meaning that it is left exact and preserves epic maps (this does not imply that it preserves coequalizer diagrams).

Because φ is left exact, it will preserve monic maps and thus give rise to a monotone map

$$\varphi'_A: \mathcal{P}(A) \longrightarrow \mathcal{P}(\varphi(A))$$

for each $A \in |\underline{E}|$. Because φ preserves epimorphisms, it will preserve existential quantification in the sense that for each $f: A \longrightarrow B$ in $\underline{E}$, the following diagram

$$\text{(2.3)} \qquad \begin{array}{ccc} \mathcal{P}(A) & \xrightarrow{\exists_f} & \mathcal{P}(B) \\ \varphi'_A \downarrow & & \downarrow \varphi'_B \\ \mathcal{P}(\varphi A) & \xrightarrow[\exists_{\varphi(f)}]{} & \mathcal{P}(\varphi B) \end{array}$$

commutes.

We shall say that "φ <u>preserves universal quantification</u>" if the diagram (2.3) with $\exists$ replaced by $\forall$ commutes. Finally, if φ preserves Ω and existential and universal quantification, it is natural to call φ <u>first-order logic-preserving</u>.

The canonical isomorphism we have because φ preserves products

$$\tilde{\varphi} : \varphi(A) \times \varphi(B) \longrightarrow \varphi(A \times B)$$

will be denoted $\tilde{\varphi}_{A,B}$ or $\tilde{\varphi}$, as indicated.

A functor $\varphi: \mathcal{S} \longrightarrow \mathcal{S}$ (where $\mathcal{S}$ is the category of sets) preserves first order logic if it preserves finite inverse limits and if $\varphi(2) = 2$. For, φ preserves the notion of complement in the subobject lattices (which are here boolean algebras) because it preserves 2. It preserves existential quantification because it preserves surjective maps (which have left inverses in $\mathcal{S}$).

Finally, universal quantification can in this case be expressed in terms of complement and existential quantification in the familiar way:
$\forall_f(X') = \neg \exists_f(\neg X')$.

Thus the functor "ultrapower" described in (0.1) preserves first order logic.

We now come to the key structure. Namely, for each $A,B \in |\underline{E}|$ we have a canonical map, generalizing (0.0),

$$(2.4) \qquad \hat{\varphi}_{A,B}: \varphi(A \pitchfork B) \longrightarrow \varphi(A) \pitchfork \varphi(B).$$

This we get by exponential adjointness from the composite

$$\varphi(A \pitchfork B) \times \varphi(A) \xrightarrow{\cong} \varphi((A \pitchfork B) \times A) \xrightarrow{\varphi(ev)} \varphi(B)$$

where the first map is the canonical isomorphism coming from the assumption that φ preserves products. In the language of closed categories, $\hat{\varphi}$ makes into a closed functor. In [7], we proved the following:

THEOREM 2.1. Let $\varphi : \underline{E} \longrightarrow \underline{E}_o$ be a functor between elementary toposes which preserves finite inverse limits and Ω. Then the following two statements are equivalent:

(i) φ preserves universal quantification

(ii) for each A,B, the map $\hat{\varphi}_{A,B}$ in (2.4) is monic.

The theorem will allow one to think of $\varphi(A \pitchfork B)$ as a subset of $\varphi(A) \pitchfork \varphi(B)$ (in case φ preserves universal quantification), namely the "set of <u>internal</u> maps from φA to φB". In particular $\varphi(A \pitchfork \Omega) \subseteq \varphi A \pitchfork \Omega$ can be thought of as the set of <u>internal</u> subsets of φA ($\varphi A \pitchfork \Omega$ being the <u>full</u> power set of φA). (Compare with (0.0) of the Introduction.) The fact that $\hat{\varphi}$ is monic will not be used in an essential way except in the proof of Proposition 3.3. We do not know whether the assumption can be avoided.

We now come to the construction, and consider a fixed 1st order logic preserving functor $\varphi : \underline{E} \longrightarrow \underline{E}_o$.

DEFINITION 2.2. Call a map $f: \varphi(A) \longrightarrow \varphi(B)$ <u>internal</u> if its name factors through $\hat{\varphi}_{A,B}$:

$$(2.5) \qquad \ulcorner f \urcorner = 1 \xrightarrow{f_o} \varphi(A \pitchfork B) \xrightarrow{\hat{\varphi}} \varphi A \pitchfork \varphi B$$

for some f_o. Call f a <u>standard</u> map if it is of the form $\varphi(g)$ for some $g: A \longrightarrow B$ in $\underline{E}$.

A subobject $A_o \rightarrowtail \varphi A$ of φA is an <u>internal subobject</u> provided its characteristic map $\varphi A \longrightarrow \Omega = \varphi\Omega$ is an internal map. Note that the map a itself is not internal as a <u>map</u> since for the notion of internal map to make sense, also the domain must be of form $\varphi(X)$ for some $X \in |\underline{E}|$.

A map which is standard is also internal; if $f = \varphi(g)$, then $\varphi(\ulcorner g \urcorner)$ will do as f_o in (2.5).

The class of internal maps are closed under composition. This is a standard exercise in closed category theory (see e.g. [2], Theorem 6.6, p. 449). Also, from closed category theory, one derives a canonical 1-1 correspondence between internal maps

$$(2.6) \qquad \frac{\varphi(A) \longrightarrow \varphi(B \pitchfork C)}{\varphi(A \times B) \longrightarrow \varphi(C)} ,$$

both sets of internal maps being indexed by the set of maps

$$1 \longrightarrow \varphi(A \pitchfork (B \pitchfork C)) \cong \varphi((A \times B) \pitchfork C).$$

Since φ preserves pull-backs and internal maps compose, it is easy to see that pulling internal subobjects back along internal maps yields internal subobjects. For existential or universal quantification along internal maps, the same is true, but non-trivial. We shall only use and prove a weaker statement in this direction.

PROPOSITION 2.3. Existential quantification along a standard morphism preserves the notion of internal subobjects. Explicitely, if $f\colon X \to Y$ is a map in $\underline{E}$, then the following diagram commutes:

$$(2.7)\qquad \begin{array}{ccc} \varphi(X)\pitchfork\Omega & \xrightarrow{\exists(\varphi(f))} & \varphi(Y)\pitchfork\Omega \\ \hat{\varphi}\uparrow & & \uparrow\hat{\varphi} \\ \varphi(X\pitchfork\Omega) & \xrightarrow[\varphi(\exists(f))]{} & \varphi(Y\pitchfork\Omega) \end{array}$$

We need a Lemma.

LEMMA 2.4. The functor φ preserves the $\in$-relation in the sense that, inside $(\varphi(X)\pitchfork\Omega)\times\varphi(X)$

$$\varphi(\in_X) = (\varphi(X\pitchfork\Omega)\times\varphi(X)) \cap \in_{\varphi(X)},$$

$\varphi(X\pitchfork\Omega)\times\varphi(X)$ being viewed as a subobject of $(\varphi(X)\pitchfork\Omega)\times\varphi(X)$ by means of $\hat{\varphi}\times 1$.

Proof. This is an easy consequence of the fact that φ preserves pull-backs.

Proof of Proposition 2.3. To prove (2.7) commutative we pass to exponential adjoints; the desired equality is then the total equaltity in the string

$$\begin{array}{ll} \hat{\varphi}\times 1 \cdot \exists(\varphi(f))\times 1 \cdot \mathrm{ev} & \\ = \hat{\varphi}\times 1 \cdot \mathrm{ch}(\exists_{1\times\varphi(f)}(\in_{\varphi(X)})) & \text{by definition of } \exists(\varphi(f)) \\ = \mathrm{ch}(\exists_{1\times\varphi(f)}(\hat{\varphi}\times 1)^{-1}(\in_{\varphi(X)})) & \text{by pull-back naturality of ch} \\ = \mathrm{ch}(\exists_{1\times\varphi(f)}\ (\varphi(\in_X))) & \text{by Lemma 2.4} \\ = \tilde{\varphi}\cdot \mathrm{ch}(\exists_{\varphi(1\times f)}(\varphi(\in_X))) & \text{by pull-back naturality of ch} \\ = \tilde{\varphi}\cdot \mathrm{ch}(\varphi(\exists_{1\times f}(\in_X))) & \text{since } \varphi \text{ preserves existential quantification} \\ = \tilde{\varphi}\cdot \varphi(\mathrm{ch}(\exists_{1\times f}(\in_X))) & \text{since } \varphi \text{ preserves pull-backs} \\ = \hat{\varphi}\cdot \varphi(\exists(f)\times 1 \cdot \mathrm{ev}) & \text{by definition of } \exists(f) \\ = \varphi(\exists(f))\times 1 \cdot \hat{\varphi}\times 1 \cdot \mathrm{ev}\ . & \end{array}$$

A relation from X to Y is a subobject R of $X \times Y$. If S is another relation from Y to Z then the composite relation from X to Z is defined as the $\exists_{proj_{13}}$ (image along projection to first and third factor) of that subobject of $X \times Y \times Z$ which one gets by intersection $proj_{12}^{-1}(R)$ with $proj_{23}^{-1}(S)$. In the set case, this is usual composition of relations. An internal relation from $\varphi(X)$ to $\varphi(Y)$ is a relation $R \rightarrowtail \varphi(X) \times \varphi(Y)$ from $\varphi(X)$ to $\varphi(Y)$ such that

$$R \rightarrowtail \varphi(X) \times \varphi(Y) \cong \varphi(X \times Y)$$

is an internal subobject of $\varphi(X \times Y)$.

Since composing relations involve pull-backs, intersection, and existential quantification along projections, which are of course standard maps, it is easy to conclude from Proposition 2.3 and the fact that φ preserves pull-backs that

PROPOSITION 2.5. The class of internal relations is closed under relational composition.

Those relations we actually are interested in, are certain "partial maps" (the pseudomaps of the definition below). Relational composition of such does not require application of any existential quantification. That existential quantification is preserved may therefore be a redundant hypothesis on φ.

We are now ready to describe the factorization mentioned in the introduction.

We define the category $\underline{E}^*$ as follows:

The objects are triples (A_o, a, A) where $A_o \in |\underline{E}_o|$, $A \in |\underline{E}|$, and $a \in Hom_{\underline{E}_o}(A_o, \varphi(A))$ is a monomorphisms such that

$$A_o \overset{a}{\rightarrowtail} \varphi(A)$$

is an internal subobject of $\varphi(A)$. (One should see A_o as the main aspect of such an object; $\varphi(A)$ plays only the role of "atlas".)

The morphisms between (A_o,a,A) and (B_o,b,B) in $\underline{E}^*$ are the morphisms in $\underline{E}_o$: $f: A_o \longrightarrow B_o$ with the property that "the graph of a", $\langle a, f \cdot b\rangle \cdot \tilde{\varphi}_{A,B}$ is an internal subobject of $\varphi(A \times B)$. The morphisms of $\underline{E}^*$ are called <u>pseudo-morphisms</u>, and its objects <u>pseudo-objects</u>, in analogy with [14].

It follows from Proposition 2.5 that the composite of two pseudo-morphisms is a pseudomorphism. Also, identity maps of pseudo-objects are pseudo-maps.

If $f \in \mathrm{Hom}_{\underline{E}}(A,B)$ then $\langle 1_{\varphi(A)}, \varphi(f)\rangle \cdot \tilde{\varphi}_{A,B} = \varphi(\langle 1_A, f\rangle)$ is a standard subobject.

Hence we have a well-defined functor

$$\bar{\varphi}: \underline{E} \longrightarrow \underline{E}^*$$
$$\bar{\varphi}(A) = (\varphi(A), 1_{\varphi(A)}, A)$$
$$\bar{\varphi}(f) = \varphi(f).$$

Also we have a canonical functor

$$\psi: \underline{E}^* \longrightarrow \underline{E}_o$$
$$\psi(A_o, a, A) = A_o$$
$$\psi(f) = f.$$

Clearly we have that

$$\varphi \text{ equals } \bar{\varphi} \text{ followed by } \psi.$$

The properties of this factorization will be investigated in the next section. We shall need the following Lemma

LEMMA 2.6. Suppose that we have a commutative square

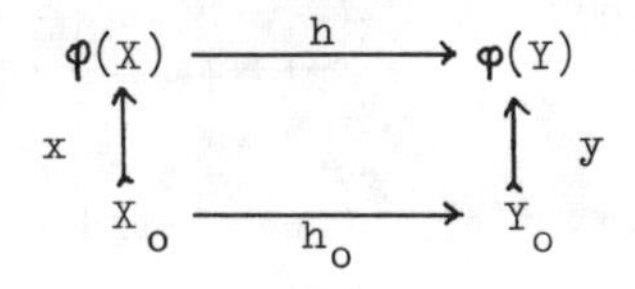

where x and y are internal subobjects and where h is an internal map. Then h_o is a pseudomorphism (between the objects $X_1 \in |\underline{E}^*|$ and $Y_1 \in |\underline{E}^*|$ displayed as left and right hand column in the diagram).

The proof is straightforward and omitted.

3. Properties of the factorization

The category of objects of form $\varphi(X)$ and with internal maps $\varphi(X) \longrightarrow \varphi(Y)$ as morphisms can be proved to be cartesian closed. But it is not a topos; it is not even finitely left complete because in general it will have too few equalizers. The introduction of pseudoobjects and pseudomaps remedies this.

PROPOSITION 3.1. The category $\underline{E}^*$ has finite inverse limits.

Proof. By general category theory, it suffices to produce a terminal object, equalizers and binary products. All these things are immediate to construct using Proposition 2.3. - in fact, in such a way that $\psi : \underline{E}^* \longrightarrow \underline{E}_o$ preserves and reflects them.

Also, ψ preserves and reflects isomorphisms and monomorphisms. Any subobject of a pseudo-object $b: B_o \rightarrowtail \varphi B$ can be represented by a monomorphism in $\underline{E}^*$ of the form

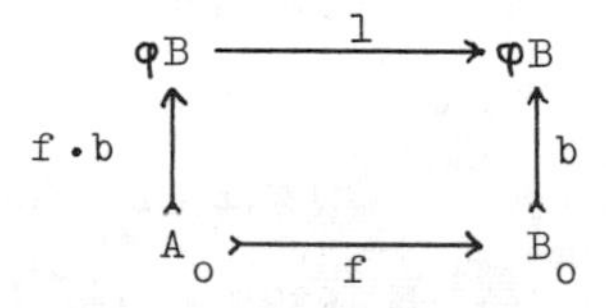

This is used together with Lemma 2.6 in deriving

PROPOSITION 3.2. The category $\underline{E}^*$ has a subobject classifier Ω^*, namely the pseudo-object

$$\Omega \xrightarrow[\gamma]{\cong} \varphi(\Omega),$$

and identifying Ω^* with Ω, true^* is identified with true: $1 \longrightarrow \Omega$.

The more difficult thing about the category $\underline{E}^*$ is that it has higher order structure, (that is, exponentials). As we know by Corollary 1.2, it suffices to show that it has intrinsic power "set" formation, that is, if

$$B_1 = B_o \rightarrowtail^{b} \varphi(B)$$

is an object of $\underline{E}^*$, there is another object in $\underline{E}^*$, $B_1 \overset{*}{\pitchfork} \Omega^*$ with the correct universal property: that subobjects of a product object $A_1 \times B_1$ in $\underline{E}^*$ are in a natural 1-1 correspondence with maps in $\underline{E}^*$

$$A_1 \longrightarrow B_1 \overset{*}{\pitchfork} \Omega .$$

We define $B_1 \overset{*}{\pitchfork} \Omega$ to be the left hand column in the pull-back diagram

$$(3.1) \qquad \begin{array}{ccc} \varphi(B \pitchfork \Omega) & \xrightarrow{\hat{\varphi}} & \varphi B \pitchfork \Omega \\ \uparrow & & \uparrow \exists_b \\ D & \longrightarrow & B_o \pitchfork \Omega \end{array}$$

It is true but not at all obvious that D becomes an internal subobject of $\varphi(B \pitchfork \Omega)$. (The proof of this fact will be sketched below). The idea in proving the universal property consists in the observation (Proposition 3.3 below) that although it is not <u>in general</u> true that pseudo-maps $h_o: X_1 \rightarrow Y_1$ are restrictions of internal maps, (that is, come about by the procedure given in Lemma 2.6) this is true if Y is of the form $B \pitchfork \Omega$. Our proof of this special kind of injectivity property of $\varphi(B \pitchfork \Omega)$ hinges on φ preserving universal quantification, that is, by Theorem 2.1, on the monicness of the canonical map

$$\hat{\varphi}_{B,\Omega}: \varphi(B \pitchfork \Omega) \longrightarrow \varphi(B) \pitchfork \Omega .$$

For, $\hat{\varphi}_{B,\Omega}$ being monic, the construction of a map into $\varphi(B \pitchfork \Omega)$ is equivalent to the construction of a map into $\varphi B \pitchfork \Omega$ satisfying certain conditions; by the universal properties of $\pitchfork$ and Ω, of course one has good chances of constructing maps ending in $\varphi B \pitchfork \Omega$. We have in fact

PROPOSITION 3.3. Every pseudomap $h: A_1 \rightarrow B_1 \overset{*}{\pitchfork} \Omega^*$ (bold arrows in the diagram below) can be extended to an internal map h' (dotted)

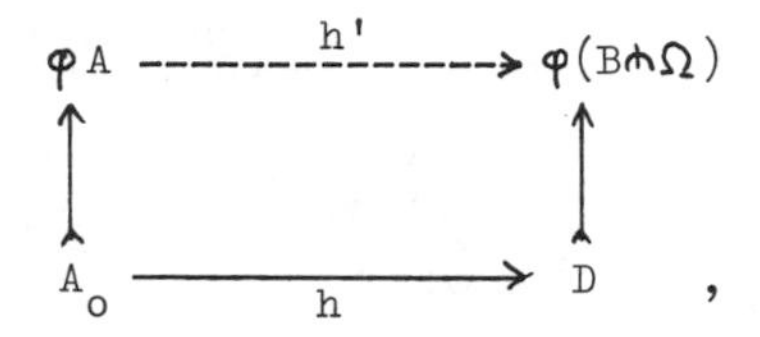

(and conversely, the bottom map in such a square is necessarily a pseudo-map).

The universal property, to be proved for $B_1 \overset{*}{\pitchfork} \Omega^*$ now essentially goes via the chain of relations

$$h: A_1 \longrightarrow B_1 \overset{*}{\pitchfork} \Omega^*$$

extend to h': $\quad \varphi A \longrightarrow \varphi(B \pitchfork \Omega)$

pass by (2.6) to: $\quad \varphi(A \times B) \longrightarrow \varphi(\Omega) = \Omega$

restrict along $A_o \times B_o \rightarrow \varphi(A \times B)$ to: $\quad A_1 \times B_1 \longrightarrow \Omega^*$.

The passage the other way uses extendability of pseudomaps into Ω^* to internal maps; this extendability also follows from Proposition 3.3.

Of course a good deal of checking is required to actually prove that the relations give rise to a 1-1 correspondence. We omit them.

We are now going to sketch the proof that the "power set pseudo object" $\beta: D \rightarrowtail \varphi(B \pitchfork \Omega)$ constructed above is actually internal. In general, proving $X_o \rightarrowtail \varphi X$ an internal subobject means displaying its characteristic function $\varphi X \rightarrow \varphi \Omega = \Omega$ as an internal map, which is again achieved by displaying a map $1 \rightarrow \varphi(X \pitchfork \Omega)$, a "witness of internalness of the subobject". Now, in the data for construction of D, we have $b: B_o \rightarrowtail \varphi B$, an internal subobject. If the witness of its internalness is $\overline{b}: 1 \to \varphi(B \pitchfork \Omega)$, then we can consider the composite map

$$(3.2) \qquad 1 \xrightarrow{\overline{b}} \varphi(B \pitchfork \Omega) \xrightarrow{\varphi(seg)} \varphi((B \pitchfork \Omega) \pitchfork \Omega).$$

where "the segment map" seg: $B \pitchfork \Omega \longrightarrow (B \pitchfork \Omega) \pitchfork \Omega$ is the exponential adjoint of the characteristic map of the inclusion-order-relation on $B \pitchfork \Omega$ (this order-relation can be viewed as a subobject of $(B \pitchfork \Omega) \times (B \pitchfork \Omega)$). Set-theoretically, seg assigns to a subset B' of B a certain <u>family</u> of subsets of B, namely the family of all B'' with $B'' \subseteqq B'$. Now, the map (3.2) is a witness of internalness of <u>some</u> subobject of $\varphi(B \pitchfork \Omega)$, and it turns out to be precisely a witness of internalness of the subobject D. To see this requires essentially two "segment" theoretic Lemmas, the one expressing $\varphi(seg)\cdot$ in terms of the segment mapping for $\varphi B \pitchfork \Omega$, the other one being the equation $\ulcorner ch(b) \urcorner \cdot seg = \ulcorner ch(\exists_b) \urcorner$. The former is obvious to formulate and prove, the latter is obvious for the category of sets, a little harder for a general topos.

We omit details; they may partly be found in the preliminary versions of the paper; partly the construction is now subsumed under the more general and slightly smoother formulation of Volger [17]. His proof also requires the functor to preserve universal as well as existential quantification.

Combining these remarks with Corollary 1.2 yields

PROPOSITION 3.4. $\underline{E}^*$ has exponentiation.

Recall that we called a functor which preserves finite inverse limits, existential and universal quantification, and finite coproducts a <u>1st order logic preserving functor</u>. We called it <u>higher order logic preserving</u> if it furthermore preserves exponentiation.

THEOREM 3.5. Let $\varphi : \underline{E} \longrightarrow \underline{E}_o$ be a 1st order logic preserving functor between toposes. Then there is a factorization

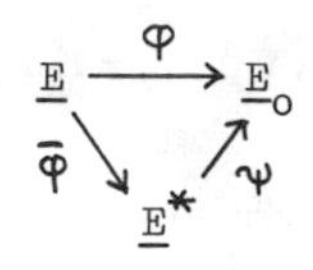

with $\underline{E}^*$ a topos, $\bar{\varphi}$ and ψ both 1st order logic preserving and with the properties

$\bar{\varphi}$ preserves higher order logic

ψ preserves elements (i.e., the map (3.3) below is bijective).

Proof. Most of the work has been done. We take $\underline{E}^*$, $\bar{\varphi}$, ψ as described in section 2. It is a topos, by the Propositions 3.2, 3.2, 3.4 and Corollary 1.2. By construction, $\bar{\varphi}$ and ψ preserve Ω, and it is easy to see that they preserve finite inverse limits. The fact that $\bar{\varphi}$ preserves epics follows because φ preserves epics and ψ reflects isos (using also that epi-mono factorizations exist in $\underline{E}^*$ because it is a topos). To see that ψ preserves epics, we note that if h is a pseudomorphism from (A_o,a,A) to (B_o,b,B), then its graph Γ is an internal subobject of $\varphi(A \times B)$. If we apply $\exists_{\varphi(proj_2)}$ to it we get an internal subobject of $\varphi(B)$, according to Proposition 2.3, which is actually the image of $\psi(h)\colon A_o \longrightarrow B_o$. Thus epi-mono factorization of $\psi(h)$ can be lifted back to a factorization in $\underline{E}^*$. Now we use that ψ preserves isomorphisms.

The fact that $\bar{\varphi}$ preserves exponentiation of the form $B \pitchfork \Omega$ is immediate from the constructing diagram (3.1), with

$$\exists_b - \exists_{id_{\varphi(B)}} = id_{\varphi B \pitchfork \Omega} .$$

From this and the construction (Corollary 1.2) of general exponential objects out of "power-set" objects, it easily follows that $\bar{\varphi}$ preserves all exponential objects.

Next, ψ preserves points. Let $A_1 = (A_o, a, A) \in |\underline{E}^*|$. Then the map given by

$$(3.3) \qquad \hom_{\underline{E}^*}(1, A_1) \longrightarrow \hom_{\underline{E}_o}(1, A_o)$$

$(A_o = \psi(A_1))$ is injective since ψ is faithful. It is surjective since every $1 \longrightarrow A_o$ is a pseudo-map; for, its graph is a map

$$\varphi(1) = 1 \longrightarrow \varphi(1 \times A),$$

and every map out of $\varphi(1)$ is internal; this follows from

$$\varphi(1 \pitchfork A) \cong \varphi(A) \cong \varphi(1) \pitchfork \varphi(A),$$

the composite isomorphism being $\hat{\varphi}_{1,A}$. Now we use the general fact that a monic internal map is also an internal subobject. This fact follows because if $\Gamma_{X,Y}\colon X \pitchfork Y \longrightarrow (X \times Y) \pitchfork \Omega$ denotes "graph formation",

$$\varphi(\Gamma_{X,Y}) \cdot \hat{\varphi} = \hat{\varphi} \cdot \Gamma_{\varphi X, \varphi Y} \,.$$

That $\overline{\varphi}$ and ψ preserve coproducts is easy.

Finally we must prove that $\overline{\varphi}$ and ψ preserve universal quantification. By Theorem 2.1, $\hat{\varphi}$ is monic. Therefore, the constructing diagram (3.1) for power set objects in $\underline{E}^*$ has a monic map as its bottom arrow; but this arrow can easily be seen to be

$$\hat{\psi} : \psi(B_1 \overset{*}{\pitchfork} \Omega^*) \longrightarrow \psi B_1 \pitchfork \Omega \,.$$

From this, one deduces that all instances of $\hat{\psi}$ are monic, and then again by Theorem 2.1 we get that ψ preserve universal quantification. Also, $\hat{\overline{\varphi}}$ has all its instances mono - (even iso-) morphisms, so again by Theorem 2.1, $\overline{\varphi}$ preserves universal quantification. The theorem is proved.

REFERENCES

1. M. Barr, Exact Categories, in Barr, Grillet, and van Osdol: Exact Categories and Categories of Sheaves, Springer Lecture Notes, Vol. 236 (1971).

2. S. Eilenberg and G.M. Kelly, Closed Categories, Proc. Conf. Categorical Algebra (La Jolla 1965), Springer Verlag 1966.

3. J. Gray, The Meeting of Midwest Category Seminar in Zürich August 24-30, 1970, Reports of the Midwest Category Seminar V, Springer Lecture Notes Vol. 195 (1971).

4. G.M. Kelly, Adjunction for Enriched Categories, Reports of the Midwest Category Seminar III, Springer Lecture Notes Vol. 106 (1969).

5. A. Kock, Introduction to Functorial Semantics, mimeographed notes, The Bertrand Russell Memorial Logic Conference; Uldum, Denmark, August 1971.

6. A. Kock and C.J. Mikkelsen, Non-standard extensions in the theory of toposes, Aarhus Universitet Preprint Series 1971/72 No. 25.

7. A. Kock and C.J. Mikkelsen, Strongly faithful functors between toposes, in preparation.

8. A. Kock and G.C. Wraith, Elementary Toposes, Aarhus Universitet Lecture Notes Series No. 30 (1971).

9. F.W. Lawvere, An elementary theory of the category of sets, mimeographed, University of Chicago 1963.

10. F.W. Lawvere, Reports of the Conference of Mathematical Logic, Leicester 1965.

11. F.W. Lawvere, Equality in Hyperdoctrines and Comprehension Scheme as an Adjoint Functor, Proceedings of Symposia in Pure Mathematics, Vol. 17, AMS (1970).

12. F.W. Lawvere, Quantifiers and Sheaves, Actes, Congress International Math. 1970 (Nice), Tome 1, p. 329-334.

13. F.W. Lawvere and M. Tierney, to appear.

14. M. Machover and J. Hirschfeld, Lectures on Non-standard Analysis, Springer Lecture Notes Vol. 94 (1969).

15. C.J. Mikkelsen, Colimits in toposes, in preparation.

16. A. Robinson, Non-standard Analysis, North Holland 1966.

17. H. Volger, Logical categories, semantical categories, and topoi, to appear.

A NONSTANDARD REPRESENTATION OF BOREL MEASURES AND σ-FINITE MEASURES

Peter A. Loeb

Yale University and University of Illinois, Urbana, Illinois

In this paper, we extend the results in [3] and [4] by considering some special * finite partitions of measurable spaces. We will show that an extension of the real line contains a linearly ordered * finite collection of compact sets on which every finite Borel measure is essentially concentrated. We will also show that a σ-finite measure can be transformed into a counting measure by using partitions.

In general, we use Robinson's notation [7] with the exception that the monad of a number a will be denoted by $m(a)$. If we have an enlargement of a structure that contains the set R of real numbers, then $*R$ denotes the set of nonstandard real numbers and $*N$, the set of nonstandard natural numbers. A set S is called * finite if there is an internal bijection from an initial segment of $*N$ onto S; a * finite set has all of the "formal" properties of a finite set. Given b and $c \in *R$ we write $b \simeq c$ if $b - c$ is in $m(0)$; when $|b|$ is bounded above by a standard real number, we write ${}^0b$ for the unique standard real number in $m(b)$. For brevity, we write $\gamma \in m^+(0)$ if $\gamma \in m(0)$ and $\gamma > 0$.

Throughout this paper we will work with an infinite set X and an infinite σ-algebra $\mathfrak{M}$ of subsets of X. We will assume

---

This work was supported by N.S.F. Grant NSF GP 14785 and a grant from the University of Illinois Center for Advanced Study.

that we have a fixed enlargement of a structure containing $X, \mathfrak{M}$ and the extended real numbers.

Let $\mathcal{P}$ be the collection of all finite, $\mathfrak{M}$-measurable partitions of X. That is, $P_\alpha \in \mathcal{P}$ if $P_\alpha = \{B_1, B_2, \dots, B_n\}$, $X = \bigcup_{i=1}^{n} B_i$, and for $1 \leq i \leq j \leq n$, $B_i \in \mathfrak{M}$, $B_i \neq \emptyset$, and $B_i \cap B_j = \emptyset$ when $i \neq j$. We write $P_\beta \geq P_\alpha$ for P_α and P_β in $\mathcal{P}$ if for each set $B \in P_\alpha$ we have $B = \bigcup_{\substack{C \in P_\beta \\ C \subset B}} C$, that is, if P_β is a refinement of P_α. The relation $\geq$ is concurrent. Therefore, there are partitions $P \in {}^*\mathcal{P}$ with $P \geq P_\alpha$ for each $P_\alpha \in \mathcal{P}$. We shall write $P \in {}^*\mathcal{P}_0$ if $P \geq P_\alpha$ for each $P_\alpha \in \mathcal{P}$. The results in [3] and [4] deal with the properties of arbitrary partitions $P \in {}^*\mathcal{P}_0$; we here consider some special partitions $P \in {}^*\mathcal{P}_0$.

§1 BOREL MEASURES

Assume that X is a topological space and $\mathfrak{M}$ is the collection of Borel sets in X. Let $\mathcal{B}$ denote the collection of finite, inner-regular Borel measures on $(X, \mathfrak{M})$, and let ${}^*\mathcal{B}_F$ be a $*$ finite internal subset of ${}^*\mathcal{B}$ such that for each $\mu \in \mathcal{B}$, ${}^*\mu \in {}^*\mathcal{B}_F$. That such a $*$ finite set exists follows from the concurrency of the relation $\subset$ in the collection of all finite subsets of $\mathcal{B}$. We may let μ_s be the sum of the measures in ${}^*\mathcal{B}_F$. Clearly $\mu_s \in {}^*\mathcal{B}$. Now by the inner-regularity of μ_s, we have the following result.

1. THEOREM. Choose any $P_0 = \{A_i : 1 \leq i \leq \lambda\} \in {}^*\mathcal{P}_0$ and $\gamma \in m^+(0)$. There is a partition $P \geq P_0$ in ${}^*\mathcal{P}$ such

that to each $A_i \in P_0$, $1 \leq i \leq \lambda$, there corresponds a compact set $K_i \in P$ with $K_i \subset A_i$ and $\mu_s(K_i) > \mu_s(A_i) - \frac{\gamma}{2^i}$. Thus for each standard measure $\mu \in \mathcal{B}$,

$$*\mu(\bigcup_{i=1}^{\lambda} K_i) > \mu(X) - \gamma .$$

<u>Proof</u>. Use the inner-regularity of μ_s, and let P be the common refinement of P_0 and the partition $\{K_1,\ldots,K_\lambda, *X - \bigcup_{i=1}^{\lambda} K_i\}$.

2. COROLLARY. (Standard). Choose any $\mu \in \mathcal{B}$, $\varepsilon > 0$ in R, $n \in N$, and f a <u>real-valued</u> Borel measurable function on X. Then

(i) There is a set $S_n \subset X$ with $\mu(S_n) < \frac{\varepsilon}{2^{n+1}}$ such that to each $x \in X - S_n$ there corresponds an open set O_x with $|f(x) - f(y)| < \frac{1}{n}$ when $y \in O_x \cap (X - S_n)$.

(ii) [Lusin's Theorem] There is a compact set C with $\mu(X - C) < \varepsilon$ such that f is continuous on C.

<u>Proof</u>. Since $\lim_{m \to \infty} \mu(\{x \in X: f(x) > m\}) = 0$, we may assume that f is bounded. Let $P \in *\mathcal{P}_0$ be the partition in Theorem 1. Then for $1 \leq i \leq \lambda$, we have $\sup_{x \in K_i} *f(x) - \inf_{x \in K_i} *f(x) \simeq 0$. (See [4], Theorem 1.3.) Statement (i) is now true when interpreted for $*X$ and therefore X; simply let $K = \bigcup_{i=1}^{\lambda} K_i$, $S_n = *X - K$ and $O_x = *X - (K - K_j)$ where $1 \leq j \leq \lambda$ and $x \in K_j$. Statement (ii) follows from (i) by taking a compact set $C \subset X - \bigcup_{n=1}^{\infty} S_n$ with $\mu(C) > \mu(X - \bigcup_{n=1}^{\infty} S_n) - \frac{\varepsilon}{2}$.

We next consider the cases $X = R$ and $X = (0,1)$ where $(0,1) = \{x \in R: 0 < x < 1\}$. We show that there is a linearly ordered $*$ finite collection of compact sets which is essentially the support of each $\mu \in \mathcal{B}$.

3. THEOREM. Assume $X = R$ (or, respectively, $X = (0,1)$), and choose $\gamma \in m^+(0)$. There is a partition $\tilde{P} \in *\mathcal{P}_0$, an integer $\eta \in *N - N$, a $*$ finite sequence of integers $m_0 < m_1 < \ldots < m_\omega$ in $*R$, and a $*$ finite sequence of compact sets $C_i \subset *X$, $1 \leq i \leq \omega$, such that:

(i) for $1 \leq i \leq \omega$, $C_i \in \tilde{P}$ and $C_i \subset (\frac{m_i-1}{\eta}, \frac{m_i}{\eta})$, and

(ii) for each $\mu \in \mathcal{B}$, $\sum_{i=1}^{\infty} *\mu(C_i) > \mu(X) - \gamma$.

Moreover for the case $X = (0,1)$, we may take $m_0 = 0$, $m_\omega = \eta$ and we have $\frac{m_i}{\eta} - \frac{m_{i-1}}{\eta} \simeq 0$ for $1 \leq i \leq \omega$.

Note. For the closure $[0,1]$ of $(0,1)$, we adjoin $C_0 = \{0\}$ and $C_{\omega+1} = \{1\}$ to $\tilde{P}$.

Proof. Given $\gamma \in m^+(0)$ and a partition $P_0 = \{A_i: 1 \leq i \leq \lambda\} \in *\mathcal{P}_0$, let the partition P and the sequence $\{K_i: 1 \leq i \leq \lambda\}$ be the same as in Theorem 1. Let $K = \bigcup_{i=1}^{\lambda} K_i$, and choose $\eta \in *N - N$ so that for any $i, j \in *N$ with $1 \leq i < j \leq \lambda$, the distance from K_i to K_j is greater than $\frac{1}{\eta}$. Take $m_0 = 0$ if $X = (0,1)$, and let m_0 be the first integer in $*R$ such that $\frac{m_0}{\eta} < y$ for all $y \in K$ if $X = R$. Having chosen m_{i-1} for $i \in *N$, we stop and let $\omega = i - 1$ if $(\frac{m_{i-1}}{\eta}, +\infty) \cap K = \emptyset$. Otherwise, we let m_i be the first integer bigger than m_{i-1} such that $\frac{m_i}{\eta} \notin K$ and $(\frac{m_{i-1}}{\eta}, \frac{m_i}{\eta}) \cap K \neq \emptyset$. If $X = (0,1)$, we may replace m_ω with η. It follows from the choice of η, that for any i, j, k with $1 \leq i \leq \omega$ and $1 \leq j \leq k \leq \lambda$, the conditions $K_j \cap (\frac{m_{i-1}}{\eta}, \frac{m_i}{\eta}) \neq \emptyset$ and $K_k \cap (\frac{m_{i-1}}{\eta}, \frac{m_i}{\eta}) \neq \emptyset$ force the equality $j = k$. We now take $C_i = K \cap (\frac{m_{i-1}}{\eta}, \frac{m_i}{\eta})$ for $1 \leq i \leq \omega$, and we let $\tilde{P}$ be the common refinement of P and the partition $\{C_1, \ldots, C_\omega, {}^*X - K\}$. The rest is clear.

§2 EVEN PARTITIONS FOR σ-FINITE MEASURE SPACES

In this section, we consider a fixed non-atomic, σ-finite measure μ on the arbitrary measurable space $(X, \mathfrak{M})$, and we show that μ can be treated as essentially a counting measure. (Also see the work of Bernstein and Wattenberg [1] and Hensen [2].) If $I = \{i \in {}^*N: 1 \leq i \leq \lambda\}$ is the index set for a partition $P = \{A_1, A_2, \ldots, A_\lambda\} \in {}^*\mathcal{P}$, then for each $C \in {}^*\mathfrak{M}$ we let

$$I(C) = \{i \in I: A_i \subset C \text{ and } {}^*\mu(A_i) > 0\},$$

and we let $|I(C)|$ denote that integer in $^*N$ which corresponds to the number of elements in $I(C)$. If $B \in \mathfrak{M}$, we may write $I(B)$ instead of $I({}^*B)$.

4. LEMMA (Standard). Given a finite set of positive real numbers $\{a_1, a_2, \ldots, a_n\}$ and any $\varepsilon > 0$ in R, there are positive natural numbers $m_1, m_2, \ldots, m_n$ such that for any i, j in N with $1 \leq i \leq n$ and $1 \leq j \leq n$, we have

$$1 - \varepsilon < \frac{a_i}{m_i} \bigg/ \frac{a_j}{m_j} < 1 + \varepsilon.$$

Proof. We may assume that $0 < \varepsilon < 1$. Let $c = (1 + \varepsilon)^{\frac{1}{n}}$ and let $k(0,1) = k(1,1) = 1$. Having chosen integers $k(1,1),\ldots,$ $k(i-1, i-1)$ and $k(0,1),\ldots,k(i-2,i-1)$ for $1 < i \leq n$, let $k(i,i)$ and $k(i-1,i)$ be positive integers such that

$$\frac{1}{c} < \frac{a_{i-1}}{k(i-1,i-1)\ k(i-1,i)} \bigg/ \frac{a_i}{k(i,i)} < c.$$

For $1 \leq i \leq n - 1$, let m_i be given by the product

$$m_i = k(i,i) \cdot \prod_{j=i}^{n-1} k(j,j+1),$$

and let $m_n = k(n,n)$. By cancelling, we obtain for $1 < i \leq n$, $\frac{1}{c} < \frac{a_{i-1}}{m_{i-1}} \bigg/ \frac{a_i}{m_i} < c$. Thus for $1 \leq i < k \leq n$, $\frac{1}{1+\varepsilon} = \frac{1}{c^n} < \frac{a_i}{m_i} \bigg/ \frac{a_k}{m_k} =$ $\prod_{j=0}^{k-i-1} (\frac{a_{j+i}}{m_{j+i}} \bigg/ \frac{a_{j+i+1}}{m_{j+i+1}}) < c^n = 1 + \varepsilon$, and of course $\frac{1}{1+\varepsilon} <$ $\frac{a_k}{m_k} \bigg/ \frac{a_i}{m_i} < 1 + \varepsilon$. Since $1 - \varepsilon = (1 - \varepsilon)(\frac{1+\varepsilon}{1+\varepsilon}) = \frac{1}{1+\varepsilon}(1 - \varepsilon^2) <$ $\frac{1}{1+\varepsilon}$, we are done.

5. THEOREM. Let μ be a not necessarily finite, non-atomic measure on $(X,\mathcal{M})$, and let Y be an element of $*\mathcal{M}$ with $*\mu(Y) \in \{r \in *R: r > 0\}$. Choose $\gamma \in m^+(0)$ and $P_0 \in *\mathcal{P}_0$, and assume that Y is exactly the union of sets from P_0. Then there is a partition $P = \{A_1,A_2,\ldots,A_\lambda\} \geq P_0$ in $*\mathcal{P}$ with index set $I = \{i \in *N: 1 \leq i \leq \lambda\}$ such that for each $i \in I(Y)$,

$$1) \qquad \left| {}^*\mu(A_i) - \frac{{}^*\mu(Y)}{|I(Y)|} \right| < \frac{\gamma}{|I(Y)|}$$

and for each $B \in \mathfrak{M}$,

$$2) \qquad \left| {}^*\mu({}^*B \cap Y) - {}^*\mu(Y) \frac{|I({}^*B \cap Y)|}{|I(Y)|} \right| < \gamma.$$

Proof. Let J be the index set for P_0 and recall that $J(Y) = \{j \in J: C_j \in P_0, C_j \subset Y, {}^*\mu(C_j) > 0\}$. By Lemma 4, there are nonstandard, positive integers $\{m_j \in {}^*N: j \in J(Y)\}$ such that for every $i, j \in J(Y)$ we have

$$1 - \frac{\gamma}{{}^*\mu(Y)} < \frac{{}^*\mu(C_i)}{m_i} \Big/ \frac{{}^*\mu(C_j)}{m_j} < 1 + \frac{\gamma}{{}^*\mu(Y)}$$

By Lyapunov's Theorem (see [5]), there is for any $B \in \mathfrak{M}$ and $\rho \in (0,1)$ a set $D \in \mathfrak{M}$ with $D \subset B$ and $\mu(D) = \rho\mu(B)$. We can find, therefore, a partition $P \geq P_0$ with index set I so that for each $j \in J(Y)$ and $i \in I(C_j)$,

$${}^*\mu(A_i) = \frac{{}^*\mu(C_j)}{m_j} \quad .$$

It follows that for any $i, k \in I(Y)$, we have ${}^*\mu(A_k) - \frac{\gamma}{{}^*\mu(Y)} {}^*\mu(A_k) < {}^*\mu(A_i) < {}^*\mu(A_k) + \frac{\gamma}{{}^*\mu(Y)} {}^*\mu(A_k)$. Fixing i and summing k over $I(Y)$, we get

$$3) \qquad {}^*\mu(Y) - \gamma < |I(Y)|\,{}^*\mu(A_i) < {}^*\mu(Y) + \gamma.$$

Dividing (3) by $|I(Y)|$ we obtain Equation (1). Now choosing any $B \in \mathfrak{M}$ and summing i over $I({}^*B \cap Y)$, we have

$$(*\mu(Y) - \gamma)\frac{|I(*B \cap Y)|}{|I(Y)|} < *\mu(*B \cap Y) < (*\mu(Y) + \gamma)\frac{|I(*B \cap Y)|}{|I(Y)|},$$

and Equation 2 follows.

6. COROLLARY. Assume that $\mu(X) < \infty$ and $Y = *X$. Then for each $B \in \mathfrak{M}$, $\left|\mu(B) - \mu(X)\frac{|I(B)|}{|I(X)|}\cdot\right| < \gamma$, and if $\mu(X) = 1$, $\mu(B) \simeq \frac{|I(B)|}{|I(X)|}$.

7. COROLLARY. Assume that $\mu(X) = +\infty$ and that $X = \bigcup_{n=1}^{\infty} X_n$ with $\mu(X_n) < \infty$ and $X_n \subset X_{n+1}$ for $n = 1,2,\ldots$. Let Y be a nonstandard element of $*\{X_n\}$, i.e., $Y = X_\eta$ for $\eta \in *N - N$. Then for every $B \in \mathfrak{M}$ with $\mu(B) < \infty$, we have

$$\mu(B) \simeq *\mu(X_\eta)\frac{|I(*B \cap X_\eta)|}{|I(X_\eta)|},$$

and for every $B \in \mathfrak{M}$ with $\mu(B) = +\infty$,

$*\mu(X_\eta)\frac{|I(*B \cap X_\eta)|}{|I(X_\eta)|}$ is an infinite element of $*R$.

References

[1] Bernstein, A.R., and Wattenberg, F., Nonstandard Measure Theory, Applications of Model Theory to Algebra, Analysis, and Probability, Edited by W. A. J. Luxemburg, pp. 171-185, Holt, Rinehart and Winston, 1969.

[2] Henson, C. W., On the Nonstandard Representation of Measures, Trans. Amer. Math. Soc. 172, Oct. 1972, pp. 437-446.

[3] Loeb, P. A., A Nonstandard Representation of Measurable Spaces and L_∞, Bull. Amer. Math. Soc. 77, No. 4, July 1971, pp. 540-544.

[4] ———————, A Nonstandard Representation of Measurable Spaces L_∞, and L_∞^*, Contributions to Non-Standard Analysis, Edited by W. A. J. Luxemburg and A. Robinson, North-Holland, 1972, pp. 65-80.

[5] Robertson, A. P., and Kingman, J. F. C., On a Theorem of Lyapunov, The Journal of the London Mathematical Society, Vol. 43, 1968, pp. 347-351.

[6] Robinson, A., On Generalized Limits and Linear Functions, Pacific J. Math, 14, 1964, pp. 269-283.

[7] Robinson, A., Non-Standard Analysis, North-Holland, 1966.

A NOTE ON CONTINUITY FOR ROBINSON'S PREDISTRIBUTIONS

Peter A. Loeb

Yale University and University of Illinois, Urbana, Illinois

Let R denote the real line and ${}^*R$ a fixed extension of R. [See 1.] Given a and b in ${}^*R$, we write $a \simeq b$ when $a - b$ is in the monad of 0. Let α be a predistribution in the sense of A. Robinson [1, pp. 133-145]. Assume that α is standard at each standard point $x_0 \in R$; for brevity we say that α is point-standard. This means that for each $x_0 \in R$ there is an internal function $g \in \alpha$ with $g(x) \simeq g(x_0)$ for all $x \simeq x_0$. Robinson has shown [1, Theorem 5.3.14] that there is a unique standard function $f: R \to R$ such that if $x_0 \in R$ and $g \in \alpha$ with $g(x) \simeq g(x_0)$ for all $x \simeq x_0$, then $g(x_0) \simeq f(x_0)$. We call f the standard part of α. Given the additional assumption that f is continuous, Robinson has shown [1, pp. 141-143] that ${}^*f \in \alpha$ and α represents a distribution in the sense of the theory of L. Schwartz [2]. It is not necessary, however, to make this additional assumption.

Theorem. The standard part of a point-standard predistribution is continuous.

This work was supported by N.S.F. Grant NSF GP 14785 and a grant from the University of Illinois Center for Advanced Study.

Proof. Let f be the standard part of a point-standard predistribution α, and fix $x_0 \in R$ and $g_0 \in \alpha$ so that $g_0(x) \simeq f(x_0)$ for all $x \simeq x_0$. Given a standard $\varepsilon > 0$, let A_ε be the internal set of positive numbers δ in ${}^*R$ such that $|g_0(x) - f(x_0)| < \varepsilon$ when $|x - x_0| < \delta$. If $\delta \simeq 0$ and $\delta > 0$, then $\delta \in A_\varepsilon$, but the monad of 0 is external. Thus, there is a standard number $\delta_0 \in A_\varepsilon$. Now let x_1 be any standard point with $|x_1 - x_0| < \delta_0$, and choose $g_1 \in \alpha$ so that $g_1(x) \simeq f(x_1)$ for all $x \simeq x_1$. As before, there is a standard number $\delta_1 > 0$ such that the interval $(x_1 - \delta_1, x_1 + \delta_1)$ is contained in $(x_0 - \delta_0, x_0 + \delta_0)$ and $|g_1(x) - f(x_1)| < \varepsilon$ when $|x - x_1| < \delta_1$ in ${}^*R$. Let h be a standard, non-negative C^∞ function with compact support contained in $(x_1 - \delta_1, x_1 + \delta_1)$ such that $\int_{-\infty}^{\infty} h(x)dx = 1$. Since $g_0 \in \alpha$ and $g_1 \in \alpha$, $\int_{-\infty}^{\infty} {}^*h(x)g_0(x)dx \simeq \int_{-\infty}^{\infty} {}^*h(x)g_1(x)dx$. Moreover, $|\int_{-\infty}^{\infty} {}^*h(x)g_0(x)dx - f(x_0)| < \varepsilon$ and $|\int_{-\infty}^{\infty} {}^*h(x)g_1(x)dx - f(x_1)| < \varepsilon$. Therefore $|f(x_0) - f(x_1)| < 3\varepsilon$ when $|x_1 - x_0| < \delta_0$ in R, and we are done.

Note. The assumption of continuity may also be eliminated from Theorem 5.3.17 of [1].

References

[1] A. Robinson, Non-standard Analysis. North-Holland, Amsterdam, 1966.

[2] L. Schwartz, Theorie des Distributions, I, II. Act. Sci. Ind., 1091, 1122, Hermann et Cie. Paris (1951).

FIELD EMBEDDINGS OF GENERALIZED METRIC SPACES*

Louis Narens

School of Social Sciences, University of California, Irvine

In [5] (pp. 133-143) W. Sierpiński proves that each countable metric space is homeomorphic to some subset of the rational numbers. His method of proof makes essential use of field and completeness properties of the real number system. Since the real numbers are the only Dedekind complete ordered field, Sierpiński's proof cannot be generalized to other spaces. In this paper, an appropriate generalization to the above theorem is given for spaces which are defined by a "metric" into an arbitrary ordered abelian group. The method of proof is by model theory.

NOTATION AND CONVENTIONS. If A is a set, $|A|$ will denote the cardinality of A. Ω will always be used to denote a regular cardinal. Regular means that if $\{A_i\}_{i\in I}$ is a family of pairwise disjoint sets such that $|I| < \Omega$ and for each i in I, $|A_i| < \Omega$, then $|\bigcup_{i\in I} A_i| < \Omega$. Ω will also be considered as an ordinal number. In this context, Ω is identified with the smallest ordinal number α such that $|\alpha| = \Omega$. Ω_0 will denote the first infinite cardinal. If $(A,<)$ is an ordered set and $B \subset A$, B is said to be left cofinal with A if and only if for each a in A there is a b in B such that $b \leq a$. There will always be a set, B, that is left cofinal with A and has minimum cardinality. In this case, $|B|$ will be a regular cardinal, and we say that $(A,<)$ is of type $|B|$. L will denote the first order language with equality that has predicates $+ (x,y,z)$, $\cdot\ (x,y,z)$, $< (x,y)$, and an individual constant symbol $\underset{\sim}{0}$. Let K be the set of true

---

*This paper represents a portion of the author's Ph.D. dissertation, which was prepared under the supervision of Professors A. Robinson and A. E. Hurd. The author wishes to thank Miss Judith Ng for her many helpful suggestions in the preparation of this paper.

sentences of the rational numbers where $+$ is interpreted as addition, $\cdot$ multiplication, $<$ the ordering relation of the rationals, and $\underset{\sim}{0}$ the additive identity. Let Q be a model of K. The symbol "Q" will also be used to denote the universe of discourse of the model Q. $|Q|$ will denote the cardinality of Q (that is, the cardinality of the universe of discourse of the model Q). In the expression "$A \subset Q$" it is understood that "Q" stands for the universe of discourse of the model Q. Let $A \subset Q$. Then L(A) is the language L together with new individual constant symbols, $\underset{\sim}{a}$, for each a in A. It is understood that if Q' is an elementary extension of Q, then in the interpretation in Q' of a sentence of L(A), each constant symbol $\underset{\sim}{a}$ - unless explicitly otherwise stated - is interpreted by the object a. Let $\varphi(x_1,\ldots,x_n)$ be a formula of L(A). Then it is understood that the only variables free in $\varphi(x_1,\ldots,x_n)$ are among the variables $x_1,\ldots,x_n$. "$Q \models \varphi(a_1,\ldots,a_n)$" means that φ is satisfied in the model Q by the elements $a_1,\ldots,a_n$. "$Q \equiv Q'$ in the language L(A)" means that the models Q and Q' are elementarily equivalent in the language L(A).

DEFINITION 1. An ordered abelian group is an abelian group G together with a total ordering, $<$, on G such that if $x,y,z,w \in G$ and $x < y$ and $z < w$ then $x + z < y + w$. The set of elements of G that are greater than the additive identity, 0, is called the set of positive elements of G and is denoted by G^+. If $g \in G$ then $|g|$, the absolute value of g, is defined to be g if $g \in G^+$, and $-g$ if $g \notin G^+$. If $(G^+,<)$ is of type Ω then G is said to be of type Ω.

DEFINITION 2. Let X be a nonempty set and G an ordered abelian group. A G-metric on X is a function ρ from $X \times X$ into $G^+ \cup \{0\}$ which satisfies the following three rules:

(1) $\rho(x,y) = 0$ if and only if $x = y$,

(2) $\rho(x,y) = \rho(y,x)$,

(3) $\rho(x,y) \leq \rho(x,z) + \rho(y,z)$.

Note that each ordered abelian group, G, becomes a G-metric space with metric $\rho(x,y) = |x - y|$.

DEFINITION 3. Let ρ be a G-metric on X. For each x in X and each $g \in G^+$ let $S(x,g) = \{y \in X | \rho(x,y) < g\}$. Then $B = \{S(x,g) | x \in X$ and $g \in G^+\}$ forms a base for a topology. The resulting topological space is denoted by (X,ρ) or sometimes (X,ρ,G). If G is of type Ω, then (X,ρ) is said to be of *type* Ω.

G-metric spaces of type $\Omega = 1$ are discrete spaces. G-metric spaces of type $\Omega = \Omega_0$ are metrizable, and are therefore metric spaces. (See [3] pp. 127-129.) A wide variety of uniform spaces are "metrizable" by G-metrics. A characterization of these spaces is given in the following theorem of L. W. Cohen and C. Goffman in [2].

THEOREM. *A uniform space* X *is* G-*metrizable by an ordered abelian group of type* $\Omega \geq \Omega_0$ *if and only if* X *has a neighborhood system* $\eta = \{U_i(x) | i < \Omega$ *and* $x \in X\}$ *with the following properties*:

(1) $\bigcap_{i<\Omega} U_i(x) = \{x\}$,

(2) *if* $i < j < \Omega$ *then* $U_i(x) \supset U_j(x)$,

(3) *if* $i < \Omega$ *then there is a* $j < \Omega$ *such that* $i \leq j$ *and if* $U_j(x) \cap U_j(y) \neq \emptyset$ *then* $U_j(y) \subset U_i(x)$,

(4) *if* Ω' *is a regular cardinal* $< \Omega$ *then* $\bigcap_{i<\Omega'} U_i(x)$ *is open*.

DEFINITION 4. Let (X,ρ) be a G-metric space of type Ω and $\{x_i\}_{i<\Omega}$ an Ω sequence of points of X. Then we say that $\lim_\Omega x_i = x$ if and only if for each $g \in G^+$ there is a $j < \Omega$ such that for all i, $j < i < \Omega$, $\rho(x_i,x) < g$.

LEMMA 1. *Let* (X,ρ) *be a* G-*metric space of type* Ω *and* (Y,ψ) *be a* H-*metric space of type* Ω. *Then the following two statements are equivalent*:

(1) f *is a homeomorphism from* X *onto* Y.

(2) f *is a one-to-one function from* X *onto* Y *such that for each* Ω-*sequence of points* $\{x_i\}_{i<\Omega}$ *of* X, $\lim_\Omega x_i = x$ *if and only if* $\lim_\Omega f(x_i) = f(x)$.

Proof. Suppose (1). Then f is a one-to-one function from X onto Y. Let $\{x_i\}_{i<\Omega}$ be an Ω-sequence of members of X. Suppose $\lim_\Omega x_i = x$. Let $h \in H^+$. Since f is a homeomorphism, $U = \{z|\psi(f(x),f(z)) < h\}$ is an open subset of X. Let g be a member of G^+ such that $\{z|\rho(x,z) < g\} \subset U$. Since $\lim_\Omega x_i = x$, let j be such that for all i, $j < i < \Omega$, $\rho(x,x_i) < g$. Then $\psi(f(x),f(x_i)) < h$ for all i, $j < i < \Omega$. Thus $\lim_\Omega f(x_i) = f(x)$. One can similarly show that $\lim_\Omega f(x_i) = f(x)$ implies that $\lim_\Omega x_i = x$. Thus (1) implies (2).

Suppose (2). Let A be a subset of X and z an accumulation point of A. Then it is easy to show that there is an Ω-sequence, $\{z_i\}_{i<\Omega}$, of points of A such that $\lim_\Omega z_i = z$. Then for each $i<\Omega$, $f(z_i) \in f(A)$. By (2), $\lim_\Omega f(z_i) = f(z)$. It then follows that $f(z)$ is an accumulation point of $f(A)$. Similarly it can be shown that if $y \in X$ is such that $f(y)$ is an accumulation point of $f(A)$ then y is an accumulation point of A. Thus A is closed if and only if $f(A)$ is closed. Therefore f is a homeomorphism from X onto Y. Thus (2) implies (1).

LEMMA 2. _Let_ (X,ρ,G) _be a_ G-_metric space of type_ Ω _and_ H _a subgroup of_ G _such that_ H^+ _is left cofinal with_ G^+. _Suppose that the range of the function_ ρ _is contained in_ H. _Then the identity map of_ X _onto_ X _is a homeomorphism of_ (X,ρ,G) _onto_ (X,ρ,H).

Proof. Since H^+ is left cofinal with G^+, $\lim_\Omega x_i = x$ in (X,ρ,G) if and only if $\lim_\Omega x_i = x$ in (X,ρ,H). By Lemma 1, the identity map is a homeomorphism from (X,P,G) onto (X,P,H).

DEFINITION 5. Let A and B be subsets of G^+. A is said to be _infinitesimal with respect to_ B if and only if for each natural number n, each $a \in A$, and each $b \in B$, $na < b$.

LEMMA 3. _Suppose_ G _is an ordered group of type_ $\Omega < \aleph_0$, $B \subset G^+$, _and_ $|B| < \Omega$. _Then there is a_ $g \in G^+$ _such that_ $\{g\}$ _is infinitesimal with respect to_ B.

Proof. Let $D = \{x \in G^+|x < b$ for each $b \in B\}$. Since G is of type Ω and $|B| < \Omega$, $D \neq \emptyset$. Let $d_0 \in D$. Since D is left cofinal with G^+ and G is of type $\Omega < \aleph_0$, let e be a member of D and such that $e < d_0$. Since

$2(d_0 - e) + 2e = 2d_0$, it follows that $2(d_0 - e) \leq d_0$ or $2e \leq d_0$. Let $d_1 = \min\{(d_0 - e), e\}$. Then $2d_1 \leq d_0$. By repeating this argument for d_1, there is a d_2 such that $2d_2 \leq d_1$. Thus there is a sequence, $d_0, d_1, \ldots, d_n, \ldots$ such that for each $n < \aleph_0$, $2^n d_n \leq d_0$. Let $E = \{d_n | n < \aleph_0\}$. Since $|E| = \aleph_0 < \Omega$, let $g \in G^+$ such that $g < e$ for all e in E. Then for each positive natural number n, $g < d_n$. Thus for each positive natural number n, $ng < nd_n < 2^n d_n \leq d_0$. Since $d_0 < b$ for each $b \in B$, we conclude that $ng < b$ for each $b \in B$. Therefore g is infinitesimal with respect to B.

LEMMA 4. _Suppose_ G _is an ordered group of type_ $\Omega > \aleph_0$. _Then there is an_ Ω-_sequence of members of_ G^+, $\{g_i\}_{i<\Omega}$, _that is left cofinal with_ G^+ _and such that for_ $i < j < \Omega$, $\{g_j\}$ _is infinitesimal with respect to_ $\{g_i\}$.

Proof. Let B be left cofinal with G^+ and $|B| = \Omega$. Let $\{a_i\}_{i<\Omega}$ be a well-ordering of B. Let $g_0 = a_0$. Assume that $0 < j < \Omega$ and for all $i < j$, g_i has been defined. Let $G_j = \{g_i | i < j\}$. By Lemma 3, let g_j be the first member in the well-ordering $\{a_i\}_{i<\Omega}$ of B that is infinitesimal with respect to G_j. Then $\{g_i | i < \Omega\}$ is left cofinal with B and is therefore left cofinal with G^+.

LEMMA 5. _Suppose_ G _is an ordered abelian group of type_ $\Omega > \aleph_0$, $A \subset G^+$, _and_ A _is not left cofinal with_ G^+. _Then there is a_ $g \in G^+$ _such that_ $\{g\}$ _is infinitesimal with respect to_ A.

Proof. Since A is not left cofinal with G^+, there is an element, $b \in G^+$, such that $b < a$ for each $a \in A$. By Lemma 3, let $g \in G^+$ and be such that $\{g\}$ is infinitesimal with respect to $\{b\}$. Then it follows that $\{g\}$ is infinitesimal with respect to A.

LEMMA 6. _Suppose_ G _is an ordered abelian group of type_ $\Omega > \aleph_0$, $A \subset G^+$, _and_ A _is not left cofinal with_ G^+. _Let_ $B = \{g \in G^+ | g$ _is not infinitesimal with respect to_ $A\}$. _Then_ B _is not left cofinal with_ G^+.

Proof. By Lemma 5, let $g \in G^+$ be such that $\{g\}$ is infinitesimal with respect to A. Then $\{g\}$ is infinitesimal with respect to B (and thus B is not left cofinal with respect to G^+). For if not, then for some $b \in B$ and some natural number n, $ng \geq b$. But since b is not infinitesimal with respect to A, for some natural number m, $mb \geq a$ for some a in A. Hence $(m \cdot n)g \geq a$ for some a in A. Thus $\{g\}$ is not infinitesimal with respect to A. A contradiction.

DEFINITION 6. Let (X,ρ,G) and (Y,ψ,H) be metric spaces of type Ω. Let f be a one-to-one function from X into Y. Then f is said to be a <u>homeomorphism from</u> (X,ρ) <u>onto a subset of</u> Y if and only if for each Ω-sequence of points of X, $\{x_i\}_{i<\Omega}$, $\lim_\Omega x_i = x$ if and only if $\lim_\Omega f(x_i) = f(x)$.

THEOREM 1. <u>Let</u> (X,ρ) <u>be a</u> G-<u>metric space of type</u> $\Omega > \Omega_0$ <u>and</u> $|X| \leq \Omega$. <u>Then there is an elementary extension</u>, Q, <u>of the rational numbers such that</u> Q <u>is of type</u> Ω, $|Q| = \Omega$, <u>and</u> (X,ρ) <u>is homeomorphic to a subset of</u> Q.

Proof. Part 1: each point in (X,ρ) is an isolated point. Let Q_0 be the rational numbers and $A_0 = \{a_0\}$ where $a_0 \in Q_0^+$. Suppose Q_j and A_j have been defined for all j, $0 \leq j < i < \Omega$ and have the following properties:

(1) if $k < j$ then Q_j is an elementary extension of Q_k;

(2) $A_j \subset Q_j^+$;

(3) if $k < j$ then there is an element a of A_j such that $\{a\}$ is infinitesimal with respect to Q_k^+;

(4) $|A_j| < \Omega$;

(5) if $a,b \in A_j$ and $a < b$ then $\{a\}$ is infinitesimal with respect to $\{b\}$.

If i is a limit ordinal, let $Q_i = \cup_{j<i} Q_j$ and $A_i = \cup_{j<i} A_j$. If i is not a limit ordinal, let Q_i be an elementary extension of Q_{i-1} and a model of the set of all sentences of the form

$$0 < \underset{\sim}{a} \text{ and } \underset{\sim}{a} < \underset{\sim}{b},$$

where $\underset{\sim}{a}$ is a new constant symbol and b is an element of Q_{i-1}^+. Then in Q_i, if a is the interpretation of $\underset{\sim}{a}$, $\{a\}$ is infinitesimal with respect to Q_{i-1}^+. In this case, let $A_i = A_{i-1} \cup \{a\}$.

In this way, Q_i and A_i are defined for each $i < \Omega$ and have properties (1) to (5). Let $Q' = \cup_{i<\Omega} Q_i$ and $A = \cup_{i<\Omega} A_i$. By property (3), A is left cofinal with $(Q')^+$. By property (4) $|A| = \Omega$. Since no subset of A of smaller cardinality is left cofinal with $(Q')^+$, Q' is of type Ω. By the Downward Löwenheim-Skolem Theorem, let Q be an elementary submodel of Q' such that $A \subset Q$ and $|Q| = |A| = \Omega$. Since $A \subset Q^+$, Q is of type Ω. Let $B = \{1/a | a \in A$ and $a < 1/2\}$. Then $|B| = \Omega$. Let a and a' be members of A such that $a < a'$. From property (5) it follows that $\{a\}$ is infinitesimal with respect to $\{a'\}$. Therefore $1/2\ a' > a$. Thus $a' - a > 1/2\ a' > a \cdot a'$. This gives $\frac{a' - a}{a \cdot a'} > 1$, from which it follows that $1/a - 1/a' > 1$. Therefore B is a set of isolated points. Let f be a one-to-one function from X into B. Then f is a homeomorphism from (X, ρ) onto a subset of Q.

It should be noted that the construction of Q shows the existence of a Q-metric space of cardinality Ω and of type Ω.

Part 2: (X, ρ) contains a point that is not an isolated point. Thus $|X| = \Omega$. Let $\{w_i\}_{i<\Omega}$ be a well-ordering of X. Since G is of type Ω, let $\{g_i\}_{i<\Omega}$ be an Ω-sequence of members of G^+ that is left cofinal with G^+ and such that if $i < j < \Omega$ then $g_j < g_i$. We will construct an elementary extension, Q', of the rational numbers by transfinite induction.

Step 0. Let $A_0 = \emptyset$, $H_0 = \emptyset$, $X_0 = \emptyset$, Q_0 be the standard model of the rational numbers, and K_0 be the set of true sentences of the model Q_0 in the language $L = L(X_0)$.

Step i, $0 < i < \Omega$. Suppose A_j, H_j, X_j, Q_j, and K_j have been defined for all $j < i$ and have the following inductive properties:

(1) $A_j \subset G^+$, $H_j \subset A_j$, and $X_j \subset X$;

(2) for all $k < j$, $A_k \subset A_j$, $H_k \subset H_j$, $X_k \subset X_j$, $K_k \subset K_j$, and Q_k is an elementary submodel of Q_j;

(3) $G^+ - A_j \neq \emptyset$ and is infinitesimal with respect to A_j;

(4) $|H_j| < \Omega$, and for each $k < j$ there is an $h \in H_j$ such that for each $a \in A_k$, $h < a$;

(5) K_j is the set of true sentences of the model Q_j in the language $L(X_j)$;

(6) for all $k < j$, $w_k \in X_{k+1}$;

(7) if $x, y \in X_j$ and $x \neq y$, then $\rho(x,y) \in A_j$;

(8) if j is a successor ordinal and $x \in X - X_j$ then for some $y \in X_j$, $\rho(x,y) \notin A_j$;

(9) for all $k < j$ and all $x,y \in X_j$: $\rho(x,y) \in A_j - A_k$ if and only if for each $r \in Q_k^+$, $Q_j \models 0 < |\underset{\sim}{x} - \underset{\sim}{y}| < r$.

Case I: i is not a limit ordinal. Let w be the member of $\{w_i\}_{i<\Omega}$ with the least index and such that $w \notin X_{i-1}$. (Since, by hypothesis, X has a non-isolated point, it follows from properties (3) and (7) that w exists.) Then, by properties (6) and (2), either $w = w_{i-1}$ or $w_{i-1} \in X_{i-1}$. By property (3), there is an $h' \in G^+ - A_{i-1}$ such that $\{h'\}$ is infinitesimal with respect to A_{i-1}. Suppose $u,v \in X_{i-1}$, $p(u,w) < h'$, and $p(v,w) < h'$. Then $p(u,v) \leq p(u,w) + p(v,w) \leq 2h'$. Since $u,v \in X_{i-1}$ and $2h' \notin A_{i-1}$, by property (7), we conclude that $u = v$. Therefore we may define h'' as follows: if there is a $v \in X_{i-1}$ such that $p(w,v) < h'$, let $h'' = p(w,v)$; otherwise let $h'' = h'$. By property (3), let g be the member of $\{g_i\}_{i<\Omega}$ with the least index and such that $g \in G^+ - A_{i-1}$. Let $h = \min(h'',g)$. Then $\{h\}$ is infinitesimal with respect to A_{i-1}. Let $A_i = \{x \in G^+ | \{x\}$ is not infinitesimal with respect to $\{h\}\}$. By Lemma 6, $G^+ - A_i \neq \emptyset$. By a simple argument it follows that $G^+ - A_i$ is infinitesimal with respect to A_i and $A_{i-1} \subset A_i$. Let $H_i = H_{i-1} \cup \{h\}$. Let $F = \{B | B \subset X, X_{i-1} \subset B$, $w \in B$, and if $x,y \in B$ and $x \neq y$ then $\rho(x,y) \in A_i\}$. If $B_0 \subset B_1 \subset \ldots \subset B_p \subset \ldots$ is a chain in F then $\bigcup_p B_p$ is in F. Hence by Zorn's lemma, F has a maximal member. Let X_i be such a maximal member. Since $w \in X_i - X_{i-1}$, $X_i \neq X_{i-1}$. Define for $x,y \in X_i$, $x \sim y$ if and only if $\rho(x,y) \notin A_{i-1}$. Then "$\sim$" is an equivalence relation. Note that if $x,y \in X_{i-1}$ and $x \sim y$, then $x = y$. Let D be the following set of sentences of $L(Q_{i-1} \cup X_i)$:

$$D = \{\underset{\sim}{0} < |\underset{\sim}{x} - \underset{\sim}{y}| < \underset{\sim}{r} | r \in Q_{i-1}^+, x \in X_i, y \in X_i, x \neq y, \text{ and } x \sim y\}.$$

Then $K = K_{i-1} \cup D$ is consistent. (This can easily be proven by showing that each finite subset of K has an interpretation in Q_{i-1}.) Let Q_i be an elementary extension of Q_{i-1} that is also a model of K. Let K_i be the set of true sentences in the language $L(X_i)$ of Q_i. Then, by construction, properties (1) to (7) hold for A_i, H_i, X_i, Q_i, and K_i.

(8) is shown by contradiction: from the assumption that for some $x \in X - X_i$ and all $y \in X_i$, $\rho(x,y) \in A_i$, it can easily be shown that X_i is not a maximal member of F. (9) is shown as follows: assume $k < i$ and $x,y \in X_i$. (i) If $\rho(x,y) \in A_i - A_k$ then for each $r \in Q_k^+$, $Q_i \models 0 < |\underset{\sim}{x} - \underset{\sim}{y}| < r$, since Q_i is a model of the set of sentences D. (ii) Now suppose that $\rho(x,y) \in A_k$. By (8) we can find $x_1,y_1 \in X_k$ (where possibly $x = x_1$ or $y = y_1$) such that $\rho(x,x_1) \notin A_k$ and $\rho(y,y_1) \notin A_k$. Since $\rho(x,y) \in A_k$, it can easily be shown by using property (3), that $x_1 \neq y_1$. Thus let t be such that $Q_k \models 0 < t = |\underset{\sim}{x}_1 - \underset{\sim}{y}_1|$. Then by (i), $Q_i \models |\underset{\sim}{x} - \underset{\sim}{x}_1| < t/4$ and $Q_i \models |\underset{\sim}{y} - \underset{\sim}{y}_1| < t/4$. By a double application of the triangle inequality,

$$Q_i \models |\underset{\sim}{x}_1 - \underset{\sim}{y}_1| \leq |\underset{\sim}{x}_1 - \underset{\sim}{x}| + |\underset{\sim}{x} - \underset{\sim}{y}| + |\underset{\sim}{y} - \underset{\sim}{y}_1| .$$

Thus.

$$Q_i \models |\underset{\sim}{x} - \underset{\sim}{y}| \geq |\underset{\sim}{x}_1 - \underset{\sim}{y}_1| - |\underset{\sim}{x}_1 - \underset{\sim}{x}| - |\underset{\sim}{y} - \underset{\sim}{y}_1| .$$

Therefore $Q_i \models |\underset{\sim}{x} - \underset{\sim}{y}| \geq t - t/4 - t/4 = t/2$. Since $t/2 \in Q_k^+$, we are done.

Case II: i is a limit ordinal. Let $A_i = \bigcup_{j<i} A_j$, $H_i = \bigcup_{j<i} H_j$, $X_i = \bigcup_{j<i} X_j$, $Q_i = \bigcup_{j<i} Q_j$, and $K_i = \bigcup_{j<i} K_j$. Properties (1), (2), (5), (6), (7), (8), and (9) can easily be seen to hold by inspection. Since $i < \Omega$, and for $j < i$, $|H_j| < \Omega$, the regularity of Ω implies that $|H_i| < \Omega$. Let $j < i$. Then there is an $h \in H_{j+1}$ such that for each $a \in A_j$, $h < a$. Thus there is an $h \in H_i$ such that for all $a \in A_j$, $h < a$. Thus (4) has been verified for H_i. Since $A_i = \bigcup_{j<i} A_j$, (4) implies that H_i is left cofinal with A_i. Since $|H_i| < \Omega$, by Lemma 3, let $g \in G^+$ and such that $\{g\}$ is infinitesimal with respect to H_i. Then $\{g\}$ is infinitesimal with respect to A_i. Thus $G^+ - A_i \neq \emptyset$. It is easily seen that $G^+ - A_i$ is infinitesimal with respect to A_i. Thus property (3) holds for A_i.

Let $Q' = \bigcup_{i<\Omega} Q_i$. From the method of defining the A_i, it immediately follows that $\bigcup_{i<\Omega} A_i = G^+$. By property (6), $\bigcup_{i<\Omega} X_i = X$. Let f be the function from X into Q' such that $f(x)$ is the interpretation in the model Q' of the symbol "$\underset{\sim}{x}$" of the language $L(X)$. It will be shown that Q' is of type Ω and f is a homeomorphism from (X,ρ) onto a subset of Q'. Since Q' is a union of an elementary chain of models, Q' is an elementary extension of Q_j for each $j < \Omega$ and it follows from properties (7) and (9) that f is a one-to-one function. Let $k < \Omega$. By property (4), let h' be such that for each $a \in A_k$, $h' < a$. Since, by hypothesis, there is a point in (X,ρ) that is not an isolated point, let y and y_1 be such that $y \neq y_1$ and $\rho(y,y_1) < h'$. Since $X = \bigcup_{j<\Omega} X_j$, by (2) let j be such that y and y_1 are in X_j. Then $\rho(y,y_1) \in A_j - A_k$. By property (9), $\{|f(y) - f(y_1)|\}$ is infinitesimal with respect to Q_k^+. Since $|f(y) - f(y_1)| \in Q_j^+$, we have shown the following: for each $k < \Omega$, there is a j, $k < j < \Omega$, such that for some $u \in Q_j^+$, $\{u\}$ is infinitesimal with respect to Q_k^+. This fact, combined with the regularity of Ω, implies that $Q' = \bigcup_{j<\Omega} Q_j$ is of type Ω. To show that f is a homeomorphism, let $\{x_i\}_{i<\Omega}$ be an Ω-sequence of members of X. By Definition 6, it must be shown that $\lim_\Omega x_i = x$ if and only if $\lim_\Omega f(x_i) = \lim_\Omega f(x)$. Suppose $\lim_\Omega x_i = x$. Let $t \in (Q')^+$. Since $Q' = \bigcup_{i<\Omega} Q_i$, let α be such that $\alpha < \Omega$ and $t \in Q_\alpha$. By property (4) let $h \in H_{\alpha+1}$ such that $h < a$ for each $a \in A_\alpha$. Choose β so that for all i, $\beta < i < \Omega$, $\rho(x_i,x) < h$. Let i be such that $\beta < i < \Omega$. Then $\rho(x_i,x) < a$ for each $a \in A_\alpha$. We may assume that $0 < \rho(x_i,x)$. Since $G^+ = \bigcup_{j<\Omega} A_j$, let γ be such that $\rho(x_i,x) \in A_\gamma$. Then $\gamma \neq \alpha$. It follows from property (2) that $\alpha < \gamma$. By property (9), for each $r \in Q_\alpha^+$, $|f(x_i) = f(x)| < r$. In particular $|f(x_i) - f(x)| < t$. Therefore $\lim_\Omega f(x_i) = \lim_\Omega f(x)$. Now suppose that $\lim_\Omega f(x_i) = f(x)$. Let $g \in G^+$. Since $G^+ = \bigcup_{i<\Omega} A_i$, let $\alpha < \Omega$ and such that $g \in A_\alpha$. Let $h \in G^+ - A_\alpha$. By property (3), $\{h\}$ is infinitesimal with respect to A_α. Since, by hypothesis, there is a point in (X,ρ) that is not an isolated point, let z and z_1 be in X and such that $z \neq z_1$ and $\rho(z,z_1) < h$. Let $\beta < \Omega$ and such that $z \in X_\beta$ and $z_1 \in X_\beta$. Then, by property (7), $\rho(z,z_1) \in A_\beta$. By (9), $0 < |f(z) - f(z_1)| < r$ for each $r \in Q_\alpha^+$. Since $\lim_\Omega f(x_i) = \lim_\Omega f(x)$, let $\gamma < \Omega$ and such that for all i, $\gamma < i < \Omega$, $|f(x_i) - f(x)| < |f(z) - f(z_1)|$. Let i be such that $\gamma < i < \Omega$. If $f(x_i) = f(x)$ then $x_i = x$ and $\rho(x,x_i) = 0 < g$. So suppose that $f(x_i) \neq f(x)$. By property (9), $\rho(x,x_i) \in A_j - A_\alpha$ for some $j < \Omega$. By property (3) we conclude that $\{\rho(x,x_i)\}$ is infinitesimal with respect to A_α. In particular, $\rho(x,x_i) < g$. Therefore $\lim_\Omega x_i = x$. Thus we have shown that f is a homeomorphism from (X,ρ) onto a subset of Q'.

The construction of the model Q' does not guarantee that the cardinality of Q' is Ω. We can, however, find an elementary submodel, Q, of Q' which has cardinality Ω and is such that f is still a homeomorphism from X onto a subset of Q. This model, Q, is of course an elementary extension of the rationals. Let $V = \{f(x) | x \in X\}$ and $T = \{|u - v| \,|\, u,v \in V$ and $u \neq v\}$. We will first show that T is left cofinal with $(Q')^+$. Let $t \in (Q')^+$. Then for some $\alpha < \Omega$, $t \in Q_\alpha$. Let $h \in G^+ - A_\alpha$. Then by property (3) $h < g$ for each $g \in A_\alpha$. By hypothesis, let z be a point in (X,ρ) such that z is not isolated. Then let $z_1 \in X$ and such that $z \neq z_1$ and $\rho(z,z_1) < h$. Let $\beta < \Omega$ and such that $z \in X_\beta$ and $z_1 \in X_\beta$. Since $\rho(z,z_1) < h < g$ for each $g \in A_\alpha$, it follows that $\rho(z,z_1) \notin A_\alpha$. By property (7), $\rho(z,z_1) \in A_\beta$. By property (2), $\beta < \alpha$. Hence by property (9), $|f(z) - f(z_1)| < r$ for each $r \in Q_\alpha^+$. In particular, $|f(z) - f(z_1)| < t$. Since $|f(z) - f(z_1)| \in T$, we have shown that T is left cofinal with Q'. Since $|X| = \Omega$ and V is the range of f and f is one-to-one, $|V| = \Omega$. By the Downward Löwenheim-Skolem Theorem, let Q be an elementary submodel of Q' that contains V and has cardinality Ω. Since Q contains V, Q contains T. Hence Q^+ is left cofinal with $(Q')^+$. By Lemma 2, f is a homeomorphism from X onto a subset of Q.

In the case of $\Omega = \Omega_0$ each G-metric space of type Ω_0 is metrizable, and by the theorems of [5] mentioned at the beginning of this paper, each countable metric space is homeomorphic to a subset of the rational numbers. Theorem 1 assures us that each G-metric space of type $\Omega > \Omega_0$ and cardinality $\leq \Omega$ is homeomorphic to some subset of a nonstandard model of the rational numbers that has cardinality Ω. But from it we cannot conclude that there is a <u>single</u> nonstandard model of the rational numbers, Q, such that $|Q| = \Omega$ and each G-metric space of type Ω and cardinality $\leq \Omega$ is homeomorphic with a subset of Q. To produce such a model, the notion of saturated model is needed.

DEFINITION 7. Let Q be a nonstandard model of the rational numbers. Let S be a set of formulas of $L(Q)$ such that each formula in S has only the variable "v" free. S is said to be <u>simultaneously satisfiable in</u> Q if and only if there is a $q \in Q$ such that for each $\varphi(v) \in S$, $\varphi(q)$ is true in Q. S is said to be <u>finitely satisfiable in</u> Q if and only if each finite subset of S is simultaneously satisfiable in Q. Q is said to be Ω-<u>saturated</u> if and only if for each $A \subset Q$ such that $|A| < \Omega$, each set of formulas of $L(A)$ with only the variable "v" free that is finitely satisfiable in Q is simultaneously satisfiable in Q.

Let Q' be an Ω-saturated model of the rational numbers and $|Q'| = \Omega$. The generalized continuum hypothesis implies the existence of Q'. (See [1] pp. 219-221.) Q' is a <u>universal</u> model in the sense that if Q is any nonstandard model of the rational numbers such that $|Q| \leq \Omega$, then Q is isomorphically embeddable in Q'. (See [1] pp. 224-227.) However, this embedding, f, may not preserve topological properties: Q may not be homeomorphic with $f(Q)$, $f(Q)$ considered as a subset of Q'. By Lemma 2, we see that in order for the embedding f of Q into Q' to preserve topological properties, it is necessary and sufficient that $(f(Q))^+$ be left cofinal with $(Q)^+$.

THEOREM 2. <u>Let</u> Q <u>and</u> Q' <u>be two elementarily equivalent models of the rational numbers such that</u> $|Q| = |Q'| = \Omega > \Omega_0$, Q <u>is of type</u> Ω, <u>and</u> Q' <u>is</u> Ω-<u>saturated. Then there is an isomorphic embedding</u>, f, <u>of</u> Q <u>into</u> Q' <u>such that</u> $(f(Q))^+$ <u>is left cofinal with</u> $(Q)^+$.

Proof. We will first show that Q' is of type Ω. Suppose that Q' were of type κ, $\kappa < \Omega$. Since Q' is an ordered field, $\kappa \geq \Omega_0$. Let T be a subset of $(Q')^+$ and such that $|T| = \kappa$ and T is left cofinal with $(Q')^+$. Let S be the set of formulas of $L(T)$ of the form

$$\underset{\sim}{0} < v \wedge v < \underset{\sim}{t}$$

where the symbol "$\wedge$" denotes the "and" of $L(T)$ and $t \in T$. Since $\kappa \geq \Omega_0$, S is finitely satisfiable in Q'. Since Q' is Ω-saturated, S is simultaneously satisfiable in Q'. Therefore, let $q \in Q'$ and such that $\varphi(q)$ is true in Q' for each formula $\varphi(v)$ in S. Then $q \in (Q')^+$ and $q < t$ for each $t \in T$. Thus T is not cofinal with $(Q')^+$. A contradiction.

Since Q' is of type Ω, let $\{e_i\}_{i<\Omega}$ be an Ω-sequence of members of $(Q')^+$

that is left cofinal with $(Q')^+$ and such that for $i < j < \Omega$, $e_i > e_j$. Let $Q^+ = \{a_i\}_{i<\Omega}$. We will define the function f by transfinite induction.

Step 0. Let $d_0 = a_0$. Let T_0 be the set of formulas of L, $\varphi(v)$, such that $Q \models \varphi(d_0)$. We will show that T_0 is finitely satisfiable in Q'. Let $\varphi_1(v),\ldots,\varphi_n(v)$ be finitely many members of T_0. Then $Q \models \varphi_1(d_0) \wedge \ldots \wedge \varphi_n(d_0)$ and hence $Q \models \exists v(\varphi_1(v) \wedge \ldots \wedge \varphi_n(v))$. Since $Q \equiv Q'$, $Q' \models \exists v(\varphi_1(v) \wedge \ldots \wedge \varphi_n(v))$. Thus there is a $b \in Q'$ such that $\varphi_1(b),\ldots,\varphi_n(b)$ are true in Q'. Since T_0 is finitely satisfiable in Q' and Q' is Ω-saturated, T_0 is simultaneously satisfiable in Q'. Therefore let c_0 be an element of Q' such that $Q' \models \varphi(c_0)$ for each $\varphi(v)$ in T_0. Let $f_0 = \{(d_0,c_0)\}$. Let $A_0 = \{d_0\}$. Then $Q \equiv Q'$ in the language $L(A_0)$, where it is understood that d_0 is the interpretation of $\underset{\sim}{d}_0$ in the model Q and $c_0 = f(d_0)$ is the interpretation of $\underset{\sim}{d}_0$ in the model Q'.

Case β, $0 < \beta < \Omega$. Suppose that A_γ and f_γ have been defined for all ordinals $\gamma < \beta$. By inductive hypothesis, the following five properties are assumed to hold for all $\gamma < \beta$:

(1) $A_\gamma \subset Q^+$, $|A_\gamma| < \Omega$, and for all $\delta \leq \gamma$, $A_\delta \subset A_\gamma$;

(2) $Q \equiv Q'$ in the language $L(A_\gamma)$;

(3) f_γ is a function from A_γ into Q' such that if $\varphi(v)$ is a formula of $L(A_\gamma)$ and $a \in A_\gamma$ then, $Q \models \varphi(a)$ if and only if $Q' \models \varphi(f(a))$;

(4) if $\delta \leq \gamma$ then $f_\delta \subset f_\gamma$;

(5) if γ is an odd ordinal then for some $a \in A_\gamma$, $f_\gamma(a) < e_\gamma$.

Subcase A: β is a limit ordinal. Let $A_\beta = \bigcup_{\gamma<\beta} A_\gamma$ and $f_\beta = \bigcup_{\gamma<\beta} f_\gamma$. Then properties (1) to (5) hold for β.

Subcase B: β is an even successor ordinal. Let d_β be the first element of $\{a_i\}_{i<\Omega}$ such that $d_\beta \in Q^+ - A_{\beta-1}$. Let $T_{\beta-1}$ be the set of formulas of $L(A_{\beta-1})$, $\varphi(v)$, with only the variable "v" free, such that $Q \models \varphi(d_\beta)$. Then as in Case 0, $T_{\beta-1}$ is finitely satisfiable in Q'. Since (by property (1)) $|A_{\beta-1}| < \Omega$, and Q' is Ω-saturated, let c_β be such that for each $\varphi(v) \in T_{\beta-1}$, $Q' \models \varphi(c_\beta)$. Let $A_\beta = A_{\beta-1} \cup \{d_\beta\}$ and $f_\beta = f_{\beta-1} \cup \{(d_\beta,c_\beta)\}$. Then properties (1) to (5) hold for β. It should be noted that in property (2), the constant symbol $\underset{\sim}{d}_\beta$ of the language

$L(A_\beta)$ is interpreted by d_β in the model Q and by $c_\beta = f_\beta(d_\beta)$ in the model Q'.

Subcase C: β is an odd ordinal. By the Downward Löwenheim-Skolem Theorem, let Q_β be an elementary submodel of Q such that $A_{\beta-1} \subset Q_\beta$ and $|Q_\beta| = \max\{\aleph_0, |A_{\beta-1}|\}$. By Lemma 3, let $d_\beta \in Q^+$ such that $\{d_\beta\}$ is infinitesimal with respect to Q_β^+. Then $d_\beta < r$ for each $r \in Q_\beta^+$. Let $T_{\beta-1}$ be the set of formulas of $L(A_{\beta-1})$ with only the variable "v" free and such that for each $\varphi(v) \in T_{\beta-1}$, $Q \models \varphi(d_\beta)$. Let $\varphi(v)$ be a member of $T_{\beta-1}$. We will show that

$$Q_\beta \models \forall x \exists y (0 < x \to 0 < y \wedge y < x \wedge \varphi(y)).$$

For suppose not. Then,

$$Q_\beta \models \exists x \forall y (0 < x \wedge (0 < y \wedge y < x \to \neg\varphi(y)).$$

Hence for some d in Q_β,

$$Q_\beta \models \forall y (0 < d \wedge (0 < y \wedge y < d \to \neg\varphi(y)).$$

By elementary equivalence,

$$Q \models \forall y (0 < d \wedge (0 < y \wedge y < d \to \neg\varphi(y)).$$

But, by definition of d_β, $0 < d_\beta$ and $d_\beta < d$. Hence,

$$Q \models \neg\varphi(d_\beta).$$

But, by definition of $T_{\beta-1}$, $Q \models \varphi(d_\beta)$. A contradiction. Since

$$Q_\beta \models \forall x \exists y (0 < x \to 0 < y \wedge y < x \wedge \varphi(y)),$$

and $Q_\beta \equiv Q$ and $Q \equiv Q'$,

$$Q' \models \forall x \exists y (0 < x \to 0 < y \wedge y < x \wedge \varphi(y)).$$

Hence

$$Q' \models \exists y (0 < e_\beta \to 0 < y \wedge y < e_\beta \wedge \varphi(y).$$

Thus,

$$Q' \models \exists y(0 < y \wedge y < e_\beta \wedge \varphi(y).$$

Let $T'_{\beta-1} = T_{\beta-1} \cup \{\underset{\sim}{0} < v \wedge v < \underset{\sim}{e}_\beta\}$ be a set of formulas of $L(A_{\beta-1} \cup \{\underset{\sim}{e}_\beta\})$. The constant symbol "$\underset{\sim}{e}_\beta$" is to be interpreted as e_β in the model Q'. Let $\varphi_1(v),\ldots,\varphi_n(v)$ be finitely many formulas of $T'_{\beta-1}$. Without loss of generality, assume that $n > 1$ and $\varphi_1(v) = \underset{\sim}{0} < v \wedge v < \underset{\sim}{e}_\beta$. Let $\varphi_0(v) = \varphi_2(v) \wedge \ldots \wedge \varphi_n(v)$. Then $\varphi_0(v) \in T_{\beta-1}$. Since

$$Q' \models \exists y(\underset{\sim}{0} < y \wedge y < \underset{\sim}{e}_\beta \wedge \varphi_0(y)),$$

$\{\varphi_1(v),\ldots,\varphi_n(v)\}$ is simultaneously satisfiable in Q'. Therefore $T'_{\beta-1}$ is finitely satisfiable. Since Q' is Ω-saturated and $|A_{\beta-1} \cup \{e_\beta\}| < \Omega$, let c_β be an element in Q' such that for each $\varphi(v) \in T'_{\beta-1}$, $Q' \models \varphi(c_\beta)$. Let $A_\beta = A_{\beta-1} \cup \{d_\beta\}$ and $f_\beta = f_{\beta-1} \cup \{(d_\beta,c_\beta)\}$. By inspection, properties (1) to (4) hold for β. Since $f_\beta(d_\beta) = c_\beta < e_\beta$, property (5) also holds for β.

Let $f^+ = \bigcup_{\alpha<\Omega} f_\alpha$. By the definitions of f_β and A_β, f^+ is a function. The method of selecting d_β at even ordinals β guarantees that domain $f^+ = Q^+$. If a, b are in the domain of f^+ and $a \neq b$ then it follows from property (3) that $f^+(a) \neq f^+(b)$. Thus f^+ is one-to-one. Let $f^- = \{(x,y)|(-x,-y) \in f^+)$ and $f = f^+ \cup f^- \cup \{(0,0)\}$. Then from properties (2) and (3) it is easy to verify that f is an isomorphism from Q into Q'. From property (5) we conclude that $f^+(Q)$ - and therefore $(f(Q))^+$ - is left cofinal with $(Q')^+$.

THEOREM 3. <u>Assume the generalized continuum hypothesis. Then for each</u> $\Omega \geq \Omega_0$ <u>there is an elementary extension</u> Q, <u>of cardinality</u> Ω, <u>of the rational numbers such that each</u> C-<u>metric space of type</u> Ω <u>and cardinality</u> $\leq\Omega$ <u>is homeomorphic to some subset of</u> Q.

Proof. The case of $\Omega = \Omega_0$ follows from the above mentioned theorems of [5]. The case of $\Omega > \Omega_0$ follows from Theorems 1 and 2 and Lemma 2.

References

[1] J. L. Bell and A. B. Slomson, Models and ultraproducts: an introduction, Amsterdam, 1969.

[2] L. W. Cohen and C. Goffman, On the metrization of uniform space, Proceedings of the American Mathematical Society 1 (1950), pp. 750-753.

[3] J. L. Kelley, General topology, New York, 1955.

[4] A. Robinson, Introduction to model theory and to the metamathematics of algebra, Amsterdam, 1965.

[5] W. Sierpiński, General topology, University of Toronto Press, 1956.

HOMEOMORPHISM TYPES OF GENERALIZED METRIC SPACES

Louis Narens

School of Social Sciences, University of California, Irvine

In the first issue of the Journal *Fundamenta Mathematicae*, W. Sierpiński [7] proves that if X and Y are denumerable subsets of Euclidean space which have no isolated points, then X and Y are homeomorphic. Combining this result and section 9 of Sierpiński's General Topology [6], one can immediately demonstrate the well-known theorem that every two denumerable metric spaces without isolated points are homeomorphic. Sierpiński's method of proof makes essential use of field and completeness properties of the real number system. Since the real numbers are the only Dedekind complete ordered field, Sierpiński's proof cannot be generalized to other spaces. In this paper, an appropriate generalization to the above theorem will be given for spaces whose topology can be defined by a "metric" into an ordered Abelian group. By [2], a wide variety of uniform spaces are examples of such *generalized metric spaces*. The method of proof uses techniques of model theory including saturated models.

The notation, conventions, and definitions of *Field embeddings of generalized metric spaces* (in this volume) will hold. In addition, the following notation and conventions will be observed. L will denote -- unless otherwise specified -- the first-order language that has predicates $+(x,y,z)$, $\cdot(x,y,z)$, $<(x,y)$, and the individual constant symbol $\underset{\sim}{0}$. L(Q) will denote the first order language L with a new predicate symbol $\underset{\sim}{Q}(x)$. Ω_0 will denote the cardinality of the natural numbers and Ω will denote a regular cardinal that is larger than Ω_0. Q_0 will denote the standard model of the rational numbers. $Q = Q(\Omega)$ will mean that Q is a Ω-saturated model that is elementary equivalent to Ω_0 in the language L.

$R = R(\Omega)$ will mean that R is a Ω-saturated model that is elementary equivalent in the language $L(Q)$ to the real numbers where $\underset{\sim}{Q}$ is interpreted in the standard real numbers as the set of standard rational numbers, Q_o. If $R = R(\Omega)$ and $Q = Q(\Omega)$ then it is easy to show that Q is isomorphic to the interpretation of $\underset{\sim}{Q}$ in R. That is, we may assume that if $Q = Q(\Omega)$ and $R = R(\Omega)$ that Q is the interpretation of $\underset{\sim}{Q}$ in R. We express this by saying, "$R = R(\Omega)$, $Q = Q(\Omega)$, and $Q \subset R$." If $R = R(\Omega)$ and $Q = Q(\Omega)$ then $(a,b)_Q$ will denote $\{x \in Q \mid a < x < b\}$. $(a,b)_R$, $[a,b)_Q$, etc., are defined in the obvious way. If $(A,<)$ is an ordered set and B and D are nonempty subsets of A, then B is said to be left-cofinal with A if and only if for each $x \in A$ there is a $y \in B$ such that $y \leq x$, and D is said to be right-cofinal with A if and only if for each $x \in A$ there is a $y \in D$ such that $x \leq y$. If A is a set then $|A|$ will denote the cardinality of A.

DEFINITION 1. Let Y be a set. (A,B) is said to be a <u>partition of</u> Y if and only if $A \neq \emptyset$, $B \neq \emptyset$, $Y = A \cup B$, and $A \cap B = \emptyset$.

DEFINITION 2. Let $Q = Q(\Omega)$ and $A \subset S \subset Q$. A is said to be a <u>piece of</u> S if and only if the following two conditions hold:

(<u>i</u>) for each partition (E,F) of A,

$$\inf_{\substack{e \in E \\ f \in F}} |e-f| = 0;$$

(<u>ii</u>) if $A \subset B \subset S$ and for each partition (E,F) of B,

$$\inf_{\substack{e \in E \\ f \in F}} |e-f| = 0,$$

then $B = A$.

THEOREM 1. <u>Suppose</u> $Q = Q(\Omega)$ <u>and</u> $S \neq \emptyset$ <u>and</u> $S \subset Q$. <u>Then there is a disjointed family</u> $\mathcal{G}$ <u>of subsets of</u> S <u>such that each member of</u> $\mathcal{G}$ <u>is a piece of</u> S <u>and</u> $\cup\mathcal{G} = S$.

<u>Proof</u>. Let $\mathcal{G}$ be the family of all pieces of S. Since $S \neq \emptyset$, let x be an arbitrary member of S. Let $\mathcal{F}$ be the family of all subsets D of S such that $x \in D$ and for all partitions (E,F) of D,

$$\inf_{\substack{e \in E \\ f \in F}} |e-f| = 0.$$

Let $\mathcal{C}$ be a nonempty chain of members of $\mathcal{F}$. It will be shown that $\cup\mathcal{C}$ is also a member of $\mathcal{F}$. By definition, $\cup\mathcal{C} \subset S$ and $x \in \cup\mathcal{C}$. Let (L,M) be a partition of $\cup\mathcal{C}$. Since $\mathcal{C}$ is a chain, let B be a member of $\mathcal{C}$ such that $L' = B \cap L \neq \emptyset$ and $M' = B \cap M \neq \emptyset$. Then (L',M') is a partition of B. Since $B \in \mathcal{F}$,

$$\inf_{\substack{\ell \in L' \\ m \in M'}} |\ell-m| = 0.$$

Hence

$$\inf_{\substack{\ell \in L \\ m \in M}} |\ell-m| = 0$$

Thus we have shown that $\cup\mathcal{C}$ is a member of $\mathcal{F}$. Therefore by Zorn's lemma, $\mathcal{F}$ has a maximal member A. That is, if $D \in \mathcal{F}$ and $D \supset A$ then $D = A$. Hence, by Definition 2, A is a piece of S. We have thus shown that for each $x \in S$, there is a piece A_x of S such that $x \in A_x$. Note that if $x, y \in S$ then either $A_x \cap A_y = \emptyset$ or $A_x = A_y$. (For if $A_x \cap A_y \neq \emptyset$ and $A_x \neq A_y$ then $B = A_x \cup A_y$ would be a set such that for each partition (E,F) of B,

$$\inf_{\substack{e \in E \\ f \in F}} |e-f| = 0,$$

and such that either A_x or A_y would be a proper subset of B. This would contradict (<u>ii</u>) of Definition 2.) Let $\mathcal{G} = \{A_x | x \in S\}$. Then $\mathcal{G}$ is a disjointed family of pieces of S such that $\cup\mathcal{G} = S$.

DEFINITION 3. Let A be an ordered set and E, F be nonempty subsets of A. We say that $E < F$ if and only if for each $e \in E$ and each $f \in F$, $e < f$.

DEFINITION 4. Let A be a nonempty set. A is said to be <u>Ω-cuttable</u> if and only if for each pair of nonempty subsets E, F of A, if $|E| < \Omega$ and $|F| < \Omega$ and $E < F$, then there is an $x \in A$ such that $E < \{x\} < F$. Note that if $A = \{x\}$ then A is 1-cuttable.

DEFINITION 5. Let A be a nonempty ordered set. $\ell(A)$ is, by definition, the smallest cardinal number of a set that is left-cofinal with A, and $r(A)$ is, by definition, the smallest cardinal number of a set that is right-cofinal with A. In particular, $\ell(A) = 1$ if and only if A has a left endpoint.

DEFINITION 6. A nonempty ordered set A is said to be an η-α _set of cardinality_ Ω if and only if A is Ω-cuttable, $|A| = \Omega$, $\ell(A) = \Omega$, and $\Omega(A) = \Omega$.

Let $Q = Q(\Omega)$. Then it is easy to show that Q is an η-α set of cardinality Ω. It is also easy to show that all η-α sets of cardinality Ω are isomorphic. Note that $(0,1)_Q$ is also an η-α set of cardinality Ω.

THEOREM 2. _Let_ $Q = Q(\Omega)$. _Suppose that_ $S \subset Q$, A _is a piece of_ S, _and_ $|A| > 1$. _Then_ A _is_ Ω-_cuttable_.

Proof. Suppose not. Since A is not Ω-cuttable and $|A| > 1$, let E, F be nonempty subsets of A such that $|E| < \Omega$, $|F| < \Omega$, $E < F$, and for all $x \in A$ it is not the case that $E < \{x\} < F$. Since $Q = Q(\Omega)$, Q is a Q-metric space of type Ω with metric $\rho(x,y) = |x-y|$. Therefore, since Ω is a regular cardinal and $|E| < \Omega$ and $|F| < \Omega$, it follows that $\{a \in Q^+ | a = \rho(e,f)$ for some $e \in E$ and some $f \in F\}$ is not left-cofinal with Q^+. So let $b \in Q^+$ be such that $b < \rho(e,f)$ for each $e \in E$ and each $f \in F$. Then

$$0 < b \leq \inf_{\substack{e \in E \\ f \in F}} |e-f|.$$

Let $E' = \{y \in A | y \leq z$ for some $z \in E\}$ and $F' = \{y \in A | y \geq z$ for some $z \in F\}$. Then (E',F') is a partition of A and $\inf_{\substack{e \in E \\ f \in F}} |e-f| = \inf_{\substack{e \in E \\ f \in F}} |e-f| \geq b > 0$ which contradicts Definition 2.

The following lemma is an immediate consequence of [7].

LEMMA 1. (Sierpiński) _There is a homeomorphism from the half-open unit interval of the (standard) rational numbers,_ [0,1), _onto the open unit interval of the (standard) rational numbers_ (0,1).

LEMMA 2. _Let_ $Q = Q(\Omega)$. _Then_ $[0,1)_Q$, $(0,1]_Q$, _and_ $(0,1)_Q$ _are homeomorphic_.

Proof. Since the function $f(x) = 1-x$ is a homeomorphism from $[0,1)_Q$ onto $(0,1]_Q$, we need only show that $[0,1)_Q$ and $(0,1)_Q$ are homeomorphic. Recall that Q_o is the standard rational numbers. By Lemma 1, let F be a homeomorphism from $[0,1)_{Q_o}$ onto $(0,1)_{Q_o}$. Recall that L is the first order language that has predicates $+(x,y,z)$, $\cdot(x,y,z)$, and $<(x,y)$. Let $\underset{\sim}{F}(x,y)$ be a new predicate symbol. Let L' be the language L together with the new predicate symbol $\underset{\sim}{F}(x,y)$. Let K be the set of true sentences of L' of the model Q_o where $+(x,y,z)$, $\cdot(x,y,z)$, and $<(x,y)$ have their usual interpretations, and where $\underset{\sim}{F}(x,y)$ is interpreted as F: that is,

$$Q_o \models \underset{\sim}{F}(a,b) \text{ if and only if } F(a) = b.$$

Assume the generalized continuum hypothesis. Let Q' be an Ω-saturated model of K such that $|Q'| = \Omega$. Then, with respect to the language L, Q and Q' are isomorphic models. Hence we may assume that $[0,1)_Q = [0,1)_{Q'}$ and $(0,1)_Q = (0,1)_{Q'}$. Let F' be the interpretation of $\underset{\sim}{F}(x,y)$ in the model Q'. It will now be shown that F' is a homeomorphism from $[0,1)_{Q'}$ onto $(0,1)_{Q'}$. Since F is a homeomorphism from $[0,1)_{Q_o}$ onto $(0,1)_{Q_o}$, the following 8 sentences of L' are true in Q_o:

(1) $\forall x(0 \leq x < 1 \rightarrow \exists y\underset{\sim}{F}(x,y))$

(2) $\forall y(0 < y < 1 \rightarrow \exists x\underset{\sim}{F}(x,y))$

(3) $\forall x(\exists y\underset{\sim}{F}(x,y) \rightarrow 0 \leq x < 1)$

(4) $\forall y(\exists x\underset{\sim}{F}(x,y) \rightarrow 0 < y < 1)$

(5) $\forall x\forall y\forall z(\underset{\sim}{F}(x,y) \cap \underset{\sim}{F}(x,z) \rightarrow y = z)$

(6) $\forall x\forall y\forall z(\underset{\sim}{F}(x,z) \cap \underset{\sim}{F}(y,z) \rightarrow x = y)$

(7) $\forall x\forall e\exists d\forall y\forall z\forall w(|x-y| < d \cap \underset{\sim}{F}(x,z) \cap \underset{\sim}{F}(y,w) \rightarrow |z-w| < e)$

(8) $\forall x\forall e\exists d\forall y\forall z\forall w(|x-y| < d \cap \underset{\sim}{F}(z,x) \cap \underset{\sim}{F}(w,y) \rightarrow |z-w| < e)$.

In the model Q_o, sentences (1) through (6) say that $\underset{\sim}{F}$ is a one-to-one function from $[0,1)$ onto $(0,1)$ and sentences (7) and (8) say that $\underset{\sim}{F}$ is bicontinuous. Since sentences (1) through (8) are in the set of sentences K, (1) through (8) are also true in the model Q'. Thus in the model Q' sentences (1) through (6) say that F' is a one-to-one function from $[0,1)_{Q'}$ onto $(0,1)_{Q'}$, and sentences (7) and (8) say that F' is bicontinuous. Hence F' is a homeomorphism from

$[0,1)_{Q'}$ onto $(0,1)_{Q'}$.

LEMMA 3. Let $R = R(\Omega)$, $Q = Q(\Omega)$, and $Q \subset R$. Let κ be a cardinal such that $1 \le \kappa < \Omega$. Suppose that $\{A_i\}_{i<\kappa, i\neq 1}$ is a disjointed collection of subsets of Q such that for each $i < \kappa$, $i \neq 1$, A_i is an η-α set of cardinality Ω. Suppose that $A = \cup_{i<\kappa, i\neq 1}$. Then A, as an ordered subset of Q, is homeomorphic with $(0,1)_Q$.

Proof. Since $1 \le \kappa \le \Omega$, let $\{a_i\}_{i<\kappa}$ be a sequence of irrational points of $(0,1)_R$ such that if $2 \le i < j < \kappa$ then $a_i < a_j < a_o < a_1$. Then it is easy to show that if $i < \kappa$ and $i \neq 1$ then $(a_i,a_{i+1})_Q$ is an η-α set of cardinality Ω. Therefore, let f_o be an order-isomorphism from A_o onto $(a_o,a_1)_Q$ and for $2 \le i < \kappa$, f_i be an order-isomorphism from A_i onto $(a_i,a_{i+1})_Q$. Let $f = \cup_{i<\kappa, i\neq 1} f_i$. We will show that f is a homeomorphism from A into $(0,1)_Q$. Since $\{A_i\}_{i<\kappa, i\neq 1}$ is a disjointed family, f is a one-to-one function. Suppose that $x_i \in A$ for $i < \Omega$ and $x \in A$ and $\lim_{\Omega} x_i = x$. Let j be such that $j \le \kappa$ and $x \in A_j$. Since A_j is an η-α set, A_j does not, as an ordered set, have endpoints. Thus it follows that there is a $p < \Omega$ such that if $p \le i < \Omega$ then $x_i \in A_j$. Since f_j is a homeomorphism and for each $y \in A_j$, $f(y) = f_j(y)$, it follows that $\lim_{\Omega} f(x_i) = f(x)$. Conversely, suppose that $\lim_{\Omega} f(x_i) = f(x)$. Then for some $j < \Omega$, $f(x) \in (a_j,a_{j+1})_Q$ and thus for some $p < \Omega$, if $p \le i < \Omega$ then $f(x_i) \in (a_j,a_{j+1})_Q$. Since f_j is a homeomorphism and for $p \le i \le \Omega$, $x_i = f_j^{-1}(f(x_i))$, it follows that $\lim_{\Omega} x_i = x$. Thus we have shown that f is a homeomorphism from A into $(0,1)_Q$.

Let $B = f(A)$. It will be shown that B (considered as an ordered subset of Q) is an η-α set of cardinality Ω. Since $1 \le \kappa \le \Omega$, $|B| = \Omega$. Since $Q = Q(\Omega)$ and $(a_2,a_3)_Q$ is left-cofinal with B and $(a_o,a_1)_Q$ is right-cofinal with B, $\ell(B) = r(B) = \Omega$. Let E, F be nonempty subsets of B such that $E < F$, $|E| < \Omega$, and $|F| < \Omega$. There are two cases to be considered:

Case 1: For some $j < \kappa$ such that $j \neq 1$, $E' = E \cap (a_j,a_{j+1})_Q \neq \emptyset$ and $F' = F \cap (a_j,a_{j+1})_Q \neq \emptyset$. Since f_j is an order-isomorphism, $G = f_j^{-1}(E')$ and $H = f_j^{-1}(F')$ are nonempty subsets of A_j such that $G < H$, $|G| < \Omega$, and $|H| < \Omega$. Since A_j is an η-α set of cardinality Ω, for some $y \in A_j$, $G < \{y\} < H$. Then $f_j(y) \in B$ and $E < \{f_j(y)\} < F$ thus establishing that B is an η-α set of cardinality Ω.

Case 2: For each $j < \kappa$ such that $j \neq 1$, either $E \cap (a_j, a_{j+1})_Q = \emptyset$ or $F \cap (a_j, a_{j+1})_Q = \emptyset$. If $F \cap (a_i, a_{i+1})_Q \neq \emptyset$ for some $2 \leq i \leq \kappa$, let p be the least i, $2 \leq i \leq \kappa$, such that $F \cap (a_i, a_{i+1})_Q \neq \emptyset$; otherwise let $p = 0$. Since $|F| < \Omega$ and $(a_p, a_{p+1})_Q$ is an η-α set of cardinality Ω, let y be such that $y \in (a_p, a_{p+1})_Q$ and $\{y\} < F$. Since by hypothesis $E \cap (a_p, a_{p+1})_Q = \emptyset$ and $E < F$, $E < \{y\} < F$. Thus B is an η-α set of cardinality Ω.

Since B is an η-α set of cardinality Ω and $Q = Q(\Omega)$, it follows that B is order-isomorphic to the η-α set $(0,1)_Q$. In particular, B is homeomorphic to $(0,1)_Q$. Since A is homeomorphic with B, it follows that A is homeomorphic with $(0,1)_Q$.

THEOREM 3. Let $R = R(\Omega)$, $Q = Q(\Omega)$, and $Q \subset R$. Suppose that $S \subset Q$, A is a piece of S, and $|A| > 1$. Then A is homeomorphic to $(0,1)_Q$.

Proof. Case 1: $\ell(A) \neq 1$ and $r(A) \neq 1$. Consider S as a subset of R. Let $\kappa = \ell(A)$ and $\lambda = r(A)$. Let $I = R - Q$. Then I is dense in R. Thus let $\{a_i\}_{i<\kappa}$ and $\{b_i\}_{i<\lambda}$ be sequences of points of I such that the following conditions hold:

(1) if $i < j < \kappa$ then $a_i > a_j$;

(2) if $x \in A$ then there is an $i < \kappa$ such that $a_i < x$;

(3) if $i < \kappa$ then there is an $x \in A$ such that $x < a_i$;

(4) if $i < j < \lambda$ then $b_i < b_j$;

(5) if $x \in A$ then there is an $i < \lambda$ such that $b_i > x$;

(6) if $i < \lambda$ then there is an $x \in A$ such that $x > b_i$; and

(7) if $i < \lambda$ and $j < \lambda$ then $a_i < b_j$.

Let $\mathcal{G} = \{(a_i, a_{i+1})_Q | i < \kappa\} \cup (a_o, b_o)_Q \cup \{(b_i, b_{i+1})_Q | i < \lambda\}$ and let $\mathcal{F} = \{F | F = B \cap A$ for some $B \in \mathcal{G}\}$. Then, by construction, $\cup\mathcal{F} = A$. Since A is a piece of S, by (ii) of Definition 2 it follows that $F \neq \emptyset$ for each $F \in \mathcal{F}$. Let $F \in \mathcal{F}$. We will show that F is Ω-cuttable. Let $x \in F$. Since $F = A \cap (b,b')_Q$ for some $(b,b')_Q \in \mathcal{G}$, there must be a $y \in (b,b')_Q$ such that $y \neq x$. (If not, then $\inf_{g \in G, h \in H} |g-h| \neq 0$ where $G = \{w | w \leq x$ and $w \in A\}$ and $H = \{w | w > x$ and $w \in A\} =$

$\{w|w > b'$ and $w \in A\}$ which contradicts Definition 2.) Thus $|F| > 1$. Now let M, N be two nonempty subsets of F such that $M < N$ and $|M|, |N| < \Omega$. Since $F \subset A$ and A is a piece of S, by Theorem 2, let $z \in A$ be such that $M < \{z\} < N$. By construction, $z \in F$. We have thus shown that F is Ω-cuttable. We will now show that $\ell(F) = r(F) = \Omega$. As before, let b,b' be such that $(b,b')_Q \in G$ and $F = (b,b')_Q \cap A$. We will show that F is right-cofinal with $(b,b')_Q$. Since $(b,b')_Q$ is an η-α set of cardinality Ω, this implies that $r(F) = \Omega$. (A similar argument will show that $\ell(F) = \Omega$.) Suppose that F is not right-cofinal with $(b,b')_Q$. Then let $d \in (b,b')_Q$ be such that $d \geq x$ for each $x \in F$. Let $G_1 = \{x \in A|x \leq d\}$ and $H_1 = \{x \in A|x > d\} = \{x \in A|x > b'\}$. Then (G_1,H_1) is a partition of A such that $\inf_{g \in G_1, h \in H_1} |g-h| \geq b' - d > 0$. This contradicts Definition 2. Since F is an Ω-cuttable subset of Q such that $\ell(F) = r(F) = \Omega$, F is an η-α set of cardinality Ω. Since $\mathcal{F}$ is a collection of disjointed η-α sets and $|\mathcal{F}| \leq \kappa + \lambda + 1 \leq \Omega$, by Lemma 3, $A = \cup\mathcal{F}$ is homeomorphic with $(0,1)_Q$.

<u>Case</u> 2: $\ell(A) = 1$ and $r(A) \neq 1$. Let a_o be the left endpoint of A. Then by Case 1, $A - \{a_o\}$ is homeomorphic to $(0,1)_Q$. Let f be this homeomorphism. Let g be defined as follows: $g(x) = y$ if and only if ($x \in A - \{a_o\}$ and $f(x) = y$, or $x = a_o$ and $y = 0$). Then it is easy to show that g is a homeomorphism from A to $[0,1)_Q$. Therefore by Lemma 2, A is homeomorphic $(0,1)_Q$.

<u>Case</u> 3: $\ell(A) \neq 1$ and $r(A) = 1$. Similar to Case 2.

<u>Case</u> 4: $\ell(A) = 1$ and $r(A) = 1$. Similar to Case 2.

THEOREM 4. <u>Let</u> $Q = Q(\Omega)$. <u>Suppose that</u> S <u>is a nonempty subset of the</u> <u>Q-metric space</u> Q <u>that has no isolated points.</u> <u>Then</u> S <u>is homeomorphic to</u> $(0,1)_Q$.

<u>Proof</u>. x is an isolated point of S if and only if $\inf_{a \in S - \{x\}} |x-a| \neq 0$ if and only if $\{x\}$ is a piece of S. Since S has no isolated points, each piece of S has at least 2 elements. Let $\mathcal{F} = \{A|A$ is a piece of $S\}$. Let $\mathcal{F} = \{A_i|i < \kappa\}$ be an indexing of the members of $\mathcal{F}$ such that if $i \neq j$ then $A_i \neq A_j$. For each $i < \kappa$ let a_i and b_i be members of Q such that $a_i < b_i$ and such that if $i, j < \kappa$ and $i \neq j$ then $(a_i,b_i)_Q \cap (a_j,b_j)_Q = \emptyset$. Since $(a_i,b_i)_Q$ is homeomorphic to $(0,1)_Q$ for each $i < \kappa$ and A_i is homeomorphic to $(0,1)_Q$ for each $i < \kappa$, let

f_i be a homeomorphism from A_i onto (a_i,b_i) for each $i < \kappa$. Let $f = \bigcup_{i<\kappa} f_i$. Then f is a homeomorphism from S onto $\bigcup_{i<\kappa}(a_i,b_i)$. Since $\bigcup_{i<\kappa}(a_i,b_i)$ is homeomorphic to $(0,1)_Q$ by Lemma 3, S is homeomorphic to $(0,1)_Q$.

THEOREM 5. Let $Q = Q(\Omega)$. Suppose that (X,ρ) is a G-metric space of type Ω and that (X,ρ) has no isolated points and $1 < |X| \leq \Omega$. Then (X,ρ) is homeomorphic to $(0,1)_Q$.

Proof. By Theorem 3 of Field embeddings of generalized metric spaces [4], (X,ρ) is homeomorphic to some subset S of Q. By Theorem 4, S is homeomorphic to $(0,1)_Q$.

References

[1] J. L. Bell and A. B. Slomson, Models and ultraproducts: an introduction, Amsterdam, 1969.

[2] L. W. Cohen and C. Goffman, On the metrization of uniform space, Proceedings of the American Mathematical Society 1 (1950), pp. 750-753.

[3] J. L. Kelley, General topology, New York, 1955.

[4] L. Narens, Field embeddings of generalized metric spaces, in this volume.

[5] A. Robinson, Introduction to model theory and to the metamathematics of algebra, Amsterdam, 1965.

[6] W. Sierpiński, General topology, University of Toronto Press, 1956.

[7] W. Sierpiński, Sur une propriété topologique des ensembles dénombrables denses en soi, Fundamenta Mathematicae, 1 (1920), pp. 11-27.

CONDITIONAL PROBABILITIES AND UNIFORM SETS

ROHIT PARIKH
(Boston University)
and
MILTON PARNES
(SUNY at Buffalo)

INTRODUCTION

It is a classical result of Banach that there exists a translation invariant, finitely additive measure extending Lebesgue measure and defined for all sets of reals. A nonstandard proof of this was given by A. Bernstein and F. Wattenberg [1] in 1967.

An excellent discussion of an application of this result in probability theory can be found in Dubins and Savage [3] pp 8-11. However, in defining conditional probabilities using Banach's result one faces the following problem: the traditional way of defining

$$\underline{p}(A,B) = \text{probability of } A \text{ relative to } B = \frac{\mu(A \cap B)}{\mu(B)}$$

breaks down when $\mu(B) = 0$ or ∞.

Nonetheless, even when $\mu(B) = 0$ or ∞, $\underline{p}(A,B)$ may have a perfectly evident intuitive value eg. the value 1 when $A = B$, and if one takes Renyi's [12] approach to probability theory seriously, one is naturally led to the question of defining $\underline{p}(A,B)$ for all pairs A,B, $B \neq \emptyset$ of sets of reals. An immediate extension of this question is to ask what natural algebraic conditions can be imposed on such a $\underline{p}$, thereby recovering as many intuitive values $\underline{p}(A,B)$ as possible. (Cf §2, §3 for more details of such intuitively evident values).

An obvious possibility to consider here is nonstandard measure theory as in Bernstein and Wattenberg [1]. Unfortunately, while their construction can be made to yield a function $\underline{p}(A,B)$ defined on all pairs A,B of sets of reals, $B \neq \emptyset$, there is no reason to believe that the function $\underline{p}$ obtained this way will be translation invariant, i.e. satisfy the identity $\underline{p}(A + x, B + x) = \underline{p}(A,B)$.

It seems then that the question of existence of conditional probability functions satisfying suitable regularity conditions must be studied directly and not as a byproduct of measure theory. An approach using nonstandard techniques does work and we report the results below in §1. The measure theoretic results can then be obtained simply by fixing the parameter B.

In §1 below we prove two main theorems regarding the existence of conditional probability functions satisfying strong regularity conditions. These functions are not unique, there are 2^c possible ones, where c is the cardinality of the reals.

In §2 we discuss a very natural application of our functions to define the notion of uniformly distributed set. A notion of uniformly distributed (countable) sequence does exist in the literature, and the question of giving a fruitful definition for sets was raised to us by J. Barback. As an exercise we suggest you think of a definition before you look up ours.

In §3 is a second application to the theory of determinate pairs of sets A,B where $\underline{p}(A,B)$ does not depend on the particular $\underline{p}$ chosen. This theory, though it looks promising, is in an embryonic stage and the results we have to report are fragmentary.

In this paper we have confined ourselves to the reals and their subsets. Results for other spaces and other applications will appear in a subsequent paper.

§1. EXISTENCE AND NONUNIQUENESS

This section is devoted to proving the existence of conditional probability functions satisfying suitable regularity conditions. Our main Theorems in this section are Theorems 1.1 and 1.2 below. The techniques used come from nonstandard analysis and the basic method is the use of *-finite samples. Specifically, if A,B are any internal sets and F is *-finite, then m(A,F) is the number of points in $A \cap F$. m(A,F) is clearly in N^* and we let

$$\underline{p}(A,B) = \left(\frac{m(A \cap B, F)}{m(B, F)}\right)^{\circ}$$

provided B is nonempty. Thus the central problem is to find suitable *-finite F such that the resulting function $\underline{p}$ has the required properties.

Theorem 1.1: There exists a function $\underline{p}$, defined for all pairs A,B of reals, $B \neq \emptyset$ such that

(a) $0 \leq \underline{p}(A,B) \leq 1$

(b) $\underline{p}(A,B) = \underline{p}(A \cap B, B)$

(c) $\underline{p}(B, B) = 1$

(d) $\underline{p}(A \cup B, C) = \underline{p}(A,C) + \underline{p}(B,C)$ if $A \cap B = \emptyset$

(e) $\underline{p}(A + x, B + y) = \underline{p}(A,B)$ if $A \subseteq B$ and $A + x \subseteq B + y$

(f) $\underline{p}(A,B) \cdot \lambda(B) = \lambda(A)$ if A,B are measurable, $A \subseteq B$ and $\lambda(B) < \infty$.

(g) $\underline{p}(-A, -B) = \underline{p}(A,B)$

(h) $\underline{p}(A,B) = \lim_{n \to \infty} (\underline{p}(A_n, B_n))$, provided that limit exists, where $A_n = A \cap [-n,n]$
$B_n = B \cap [-n,n]$

(j) $\underline{p}(A,B) = 0$ if A is finite and B is not

(k) $\underline{p}(A,B) = 0$ if A is countable and B is not

(ℓ) $\underline{p}(A,B) \cdot \underline{p}(B,C) = \underline{p}(A,C)$ if $A \subseteq B \subseteq C$.

Remark: Condition e immediately implies that $\underline{p}(A + x, B + x) = \underline{p}(A,B)$ and, moreover, if $A, A + x \subseteq B$, then $\underline{p}(A,B) = \underline{p}(A + x, B) = \underline{p}(A, B - x)$. In (f) above, λ is Lebesgue measure.

Theorem 1.2. Exactly like Theorem 1.1 except that condition ℓ is replaced by

(m) $\underline{p}(A \cdot x, B \cdot x) = \underline{p}(A,B)$ if $x \neq 0$.

We do not know if conditions a-m can be fulfilled simultaneously.

Theorems 1.1 and 1.2 are proved in stages. We first show, Theorem 1.3, that a function $\underline{p}$ satisfying conditions a-j, ℓ can be defined on the interval $[0,1] = I$, with addition mod 1. This is then extended to the entire real line yielding thm. 1.1 except for condition k. An averaging argument now yields thm. 1.2, again without condition k. Finally we show how to obtain condition k without losing the already obtained conditions. An example is given to show that condition k does not follow from the others. Indeed, in the natural course of the construction, a nice countable set will be "larger" than a messy uncountable one. We conclude this section by showing that with either Theorem 1.1 or 1.2, there are 2^c distinct $\underline{p}$ satisfying the given conditions.

Def. 1.1: A sample will always be a finite (*-finite) set. A sample $F \subseteq I$ will be n-invariant if $F + \frac{1}{n} = F$. If F is a sample and p is a natural number, then a new sample F_p may be obtained from F as follows:

$$F_p = \left\{ y \mid y = \Sigma\, n_i x_i \text{ where the } x_i \text{ are distinct, } x_i \in F \text{ and } n_i \leq p \text{ are natural numbers} \right\}$$

If A is a set and F is a sample, then

$$\mu(A,F) = \left(\frac{m(A,F)}{m(F,F)}\right)^{o}$$ where x^o is the standard part of x. If F is finite, then the operation ^{o} is of course not needed.

In lemmas A-D and the main lemma below we are concerned solely with the interval I. Lemmas A-C are standard.

Lemma A: If F is an n-invariant sample and U is an interval, then $\mu(U,F) - \frac{1}{n} \leq \lambda(U) \leq \mu(U,F) + \frac{1}{n}$.

Proof: let $a_o < \dots < a_{k+1}$ be the sequence of points in F such that $\{a_o, \dots, a_{k+1}\} = F \cap [a_o, a_{k+1}]$ and $\{a_1, \dots, a_k\} = U \cap F$. Then $\mu(U,F) = k/\ell$ where ℓ is the number of points in F.

Also $a_k - a_1 \leq \lambda(U) \leq a_{k+1} - a_o$

Consider now the simple case where F is a sequence $\alpha, \alpha + \frac{1}{n}, \dots, \alpha + \frac{n-1}{n}$. In that case $\ell = n$ and $a_k - a_1 = \frac{k-1}{\ell}$, $a_{k+1} - a_o = \frac{k+1}{\ell}$, and the required inequality follows at once.

In the general case F is a finite union of such sequences F_i. For each i we have

$$\mu(U, F_i) - \frac{1}{n} \leq \lambda(U) \leq \mu(U, F) + \frac{1}{n}$$

Hence

$$\lambda(U) - \frac{1}{n} \leq \mu(U, F_i) \leq \lambda(U) + \frac{1}{n} .$$

However, $\mu(U, F)$ is an average of the $\mu(U, F_i)$ and lies in the same interval $[\lambda(U) - \frac{1}{n}, \lambda(U) + \frac{1}{n}]$. This yields the lemma.

QED

Lemma B: Let U be an open set.
$F = \{y_1, \dots, y_m\}$ and $n_1, \dots, n_m, m_1, \dots, m_\ell$ are fixed integers. Then

(a) $$\int_I \chi_U (\sum_{i=1}^m n_i y_i + m_1 x)dx = \begin{cases} \lambda(U) & \text{if } m_1 \neq 0 \\ 1 \text{ or } 0 & \text{if } m_1 = 0 \end{cases} .$$

(b) $$\int_{I^\ell} \chi_U (\sum_{i=1}^m n_i y_i + \sum_{i=1}^\ell m_i x_i)d\vec{x} = \begin{cases} \lambda(U) & \text{if some } m_i \neq 0 \\ 1 \text{ or } 0 & \text{if all } m_i = 0 \end{cases}$$

where χ_U is the characteristic function of U.

Proof: a) is trivial if $m_1 = 0$, since the integrand is independent of x and is a constant 1 or 0. If $m_1 \neq 0$, then note that the integrand as a function of x is the characteristic function of the union of m_1 disjoint translates of U/m_1. Hence the integral is $\lambda(U)$. (The temptation to think that it is $m_1\lambda(U)$ will go away if one notes that the answer is 1 for U = I)

b) is also trivial if all the m_i are 0. Otherwise, say $m_1 \neq 0$. Then

$$\int_{I^\ell} (\chi_U (\sum_{i=1}^m n_i y_i + \sum_{i=1}^\ell m_i x_i) d\vec{x}$$

$$= \int_{I^{\ell-1}} \left\{ \int_I \chi_U(\sum_{i=1}^m n_i y_i + \sum_{i=1}^\ell m_i x_i)dx_1 \right\} d\vec{x}' \quad \text{where } \vec{x}' = (x_2, \dots, x_\ell)$$

By (a) the part in curly brackets is $\lambda(U)$ and independent of $\vec{x}'$

Q.E.D.

<u>Lemma C</u>: Let $U_1, \ldots, U_q$ be a given finite sequence of open subsets of I, $\varepsilon > 0$, $r \in N$. G a given finite subset of I. Then there exists $k > r$ and finite $F \supseteq G$ such that for at least half the integers p between k/2 and k, $\mu(U_i, F_p) < \lambda(U_i) + \varepsilon$.

<u>Proof</u>: Let $\delta = \varepsilon/8$. Each U_i is a countable union of disjoint open intervals. Let $U_i = V_i + D_i$ where V_i, D_i are disjoint and V_i is a finite union of intervals, say k_i intervals, and such that

$$\sum_{i=1}^{q} \lambda(D_i) < \delta. \text{ Let } k_o = \max(k_1, \ldots, k_q).$$

Let n be such that $k_o/n < \varepsilon/2$, $F' = G \cup \left\{\frac{1}{n}, \ldots, \frac{n-1}{n}\right\}$. Then if $F \supseteq F'$, F_p is always n-invariant and $\mu(V_i, F_p) \leq \lambda(V_i) + k/n < \lambda(V_i) + \varepsilon/2$ Let $D = \bigcup D_i$, $\lambda(D) < \delta$. Also if for some p, $\mu(D, F_p) < \varepsilon/2$ then $\mu(D_i, F_p) < \varepsilon/2$ and so $\mu(U_i, F_p) < \lambda(U_i) + \varepsilon$. Let $F = F' \cup \{x_1, x_2\}$ where x_1, x_2 are yet to be chosen.

Write $F' = \{y_1, \ldots, y_m\}$

$\mu(D, F_p)$ depends on x_1, x_2. Write it as $\alpha_p(x_1, x_2)$. Then if $\overline{\overline{F_p}} = L$, then

$$\alpha_p(x_1, x_2) = \mu(D, F_p) = \frac{1}{L} \sum_{y \in F'_p} \sum_{m_i \leq p} \chi_D (y + \sum_{i=1,2} m_i x_i)$$

so

$$a_p = \iint_{I^2} \alpha_p(x_1, x_2) dx_1\, dx_2 \leq \frac{1}{L} \left\{ \overline{\overline{F'_p}} \cdot (1 + \lambda(D)(p^2 + 2p)) \right\}$$

Where the 1 corresponds to the case $m_1 = m_2 = 0$ and $(p^2 + 2p)\lambda(D)$ comes from the other terms with $m_1 + m_2 > 0$. (Apply lemma B with D as the U).

Now we can assume that $L = (p^2 + 2p + 1)\overline{\overline{F'_p}}$, since this will happen for most x_1, x_2. Thus

$$a_p \leq \frac{1 + \lambda(D)(p^2 + 2p + 1)}{p^2 + 2p + 1}.$$

$$\sum_{p=k/2}^{k} a_p = \iint_{I^2} (\sum_{p=k/2}^{k} \alpha_p(x_1, x_2)) dx_1\, dx_2$$

$$\leq \sum_{p=k/2}^{k} \lambda(D) + \sum_{p=k/2}^{k} \frac{1}{p^2 + 2p + 1}$$

$$\leq \frac{k}{2} \lambda(D) + \frac{2}{k}.$$

Thus there exist x_1, x_2 such that $\sum_{p=k/2}^{k} a_p < k \cdot \lambda(D) = k \cdot \delta$

Hence for at least half the p between $k/2$ and k, $a_p < 4\delta = \varepsilon/2$.

Now we have picked x_1, x_2 and F has the required property. Q.E.D.

Lemma D: Let F be a *-finite set such that $I \subseteq F \subseteq I^*$. Let $\overline{\overline{F}} = \ell$. $A_1, \ldots A_m$ is a *-finite enumeration including all standard subsets of I. Let $k/\log(k+1) \geq 8\ell m\, n$ where $n \in N^* - N$. $X \subseteq \{k/2, \ldots, k\}$ contains at least half the natural numbers between k/2 and k. Then there exists $p \in X$ such that for all $i \leq m$, $m(A_i^*, F_{p+1}) / m(A_i^*, F_p) < 1 + \frac{1}{n}$.

Proof: We may as well assume that each A_i is infinite, otherwise $A_i^* = A_i \subseteq F \subseteq F_1$. So suppose that $\overline{\overline{A_i \cap F_1}} > 2$ for all i. Now if $\alpha = \prod_i (m(a_i, F_k)) / \prod_i (m(A_i, F_1))$, then

$$\alpha < \frac{1}{2^m} (\prod_i (m(A_i^*, F_k)) < \frac{(\overline{\overline{F}}_k)^m}{(2^m)} < (1 + k)^{\ell m} \quad \text{since } \overline{\overline{F}}_k < (1 + k)^{\ell}.$$

Suppose there is no $p \in X$ satisfying the condition. Then for all $p \in X$, there is an i such that $A_i \cap F_{p+1} / A_i \cap F_p \geq 1 + \frac{1}{n}$. Hence $\alpha \geq (1 + \frac{1}{n})^{k/4}$. But $\alpha < (1 + k)^{\ell m}$, $(^k/4) \log (1 + \frac{1}{n}) = \log \alpha < \ell \cdot m \cdot \log(k + 1)$. Also $\log(1 + \frac{1}{n}) > \frac{1}{2n}$. Thus $\frac{k}{8n} < \ell \cdot m \cdot \log(k + 1)$, so $k/\log(k + 1) < 8\ell \cdot m \cdot n$, a contradiction. Q.E.D.

Main Lemma: There exists a *-finite set F and $n, p \in N^* - N$ such that

(1) $[0,1] \subseteq F \subseteq [0,1]^*$

(2) For all standard open $A \subseteq [0,1]$ $\mu(A^*, F_p) = \lambda(A)$

(3) For all standard $A \subseteq [0,1]$, $m(A, F_{p+1}) / m(A, F_p) < 1 + \frac{1}{n}$.

Proof: Let $U_1, \ldots, U_q$ be a *-finite enumeration of open subsets of I^* including all standard ones. Let $k_o = \max(k_1, \ldots, k_q)$ be as defined in the proof of Lemma C and n such that $^{k_o}/_n < {}^{\varepsilon}/2$ where ε is infinitesimal. Let $A_1, \ldots, A_m$ be an enumeration including all (standard) subsets of I and let $G = \{y_1, \ldots, y_s\} \supseteq I$. Take r such that $r/\log(r + 1) > 8(s + n + 2) \cdot m \cdot n$. Take the k provided in lemma C. Put $\ell = s + n + 2$. Now apply lemma D to the collection $A_1, \ldots, A_n$, the set F and the $k > r$ provided by lemma C. Take X to be $\{p \mid \mu(U_i, F_p) < \lambda(U_i) + \varepsilon$ for all $i \leq q\}$. Lemma D gives us a $p \in X$ such that for all $i \leq m$, $m(A_i^*, F_{p+1})/m(A_i^*, F_p) < 1 + \frac{1}{n}$. This p,F are the required ones. Q.E.D.

Theorem 1.3. There exists a function p defined on all pairs of subsets A,B of I, $B \neq \emptyset$, satisfying conditions a-g, j, ℓ of Theorem 1.1.

Proof: Take $H = F_p \cup - F_p$ and define $\underline{p}(A,B) = (m(A^* \cap B^*, H) / m(B^*, H))^o$ where F_p is as in main lemma.

Properties a-d, g, ℓ are immediate. Also j is a consequence of e since all one point sets will have the same measure relative to B and d applies. f follows from ℓ, letting C = I provided we show that $\underline{p}(A, I) = \lambda(A)$ for measurable A. Since measurable sets are a boolean algebra and d applies, it is enough to show that $\underline{p}(A, I) \leq \lambda(A)$. But for all $\varepsilon > 0$, there is an A' open such that $A \subseteq A'$ and $\lambda(A') \leq \lambda(A) + \varepsilon$. However, by (2) of the main lemma,

$$\underline{p}(A, I) \leq \underline{p}(A', I) \leq \lambda(A') < \lambda(A) + \varepsilon.$$

Hence certainly $\underline{p}(A, I) \leq \lambda(A)$. This leaves only condition e.

We first show that for all standard A, x, $1 - \frac{1}{n} \leq m(A + x, H)/m(A,H) \leq 1 + \frac{1}{n}$

Note that (suppressing the * on A)

$$\frac{m(A + x, H)}{m(A, H)} = \frac{m(A + x, F_p) + m(A + x, - F_p)}{m(A, F_p) + m(A, - F_p)}$$

$$= \frac{m(A, F_p - x) + m(A, - F_p - x)}{m(A, F_p) + m(A, - F_p)}$$

$$\leq \frac{m(A, F_{p+1}) + m(A, - F_{p+1})}{m(A, F_p) + m(A, - F_p)}$$

$$\leq 1 + \frac{1}{n}$$

Since $F_p - x \subseteq F_{p+1}$ etc. and using (3) of main lemma.

The other inequality ($\geq 1 - \frac{1}{n}$) is similar, use -x instead of x. Thus

$$1 - 2n + \frac{1}{n^2} \leq \frac{m(A + x, H)}{m(A, H)} \cdot \frac{m(B, H)}{m(B + y, H)} \leq 1 + \frac{2}{n} + \frac{1}{n^2}$$

Hence

$$1 - \frac{2}{n} + \frac{1}{n^2} \leq \frac{m(A + x, H)}{m(B + y, H)} \Big/ \frac{m(A, H)}{m(B, H)} \leq 1 + \frac{2}{n} + \frac{1}{n^2}$$

Hence if $A \subseteq B$, $A + x \subseteq B + y$, we get $\underline{p}(A + x, B + y) = \underline{p}(A,B)$. (Two standard numbers whose ratio is within $\frac{2}{n} + \frac{1}{n^2}$ of unity, must be equal.)

Q.E.D.

<u>Theorem 1.4</u>: There is a function $\underline{p}$ satisfying conditions a-j, ℓ of Theorem 1.1 defined on the entire real line.

<u>Proof</u>: Let n be infinite. Since I has a $\underline{p}$ function as required, so does every interval $[-\alpha, \alpha]$ for $\alpha > 0$. We consider the case $\alpha = k \cdot n$ where k is not yet fixed. Let $A_1, \ldots, A_m$ be a *-enumeration including all standard sets of reals. Just as in Lemma D we can show there is a k such that for all i,

$$1 - \frac{1}{n} < \frac{\underline{p}(A_i, [-(k+1)n, (k+1)n]) \cdot 2(kn+n)}{\underline{p}(A_i, [-(k-1)n, (k-1)n]) \cdot 2(kn-n)} < 1 + \frac{1}{n}$$

for all i. I.e. the shift from k-1 to k+1 doesn't make much difference. Now take $\underline{p}'(A,B)$ to be $\underline{p}(A \cap [-kn, kn], B \cap [-kn, kn])^{o}$. This $\underline{p}'$ has the required property. Roughly the reason is we have to convert translation mod 2kn into straight translation. However "most of" $A \cap [-kn, kn]$ lies in $A \cap [-(k-1)n, (k-1)n]$ Hence a translation by a standard number doesn't change $\underline{p}'$ more than an infinitesimal amount. But $\underline{p}'$ is standard and the only standard infinitesimal is zero.

Q.E.D.

Theorem 1.5: There is a translation invariant functional Φ defined on all bounded functions from the reals to reals such that

(1) $\text{glb } f \leq \Phi(f) \leq \text{lub } f$

(2) $\Phi(\alpha f + \beta g) = \alpha \Phi(f) + \beta \Phi(g)$.

Proof: Let F_p be as in the main lemma, $n \in N^* - N$. Let

$$H_k = \{-nk, \dots, 0, \dots, nk\} + F_p = \{x+y \mid x \in N^*, |x| \leq nk, y \in F_p\}$$

Let $$\Phi_k(f) = \left(\frac{\Sigma f(x_i) : x_i \in H_k}{\underline{\underline{H}}_k}\right)^{o}$$

Just as in lemma D we can choose a value of k which makes Φ_k translation invariant. We then take $\Phi = \Phi_k$. 1,2 are trivial.

Q.E.D.

Theorem 1.6: There is a $\underline{p}$ satisfying conditions a-j,m of Theorem 1.2 on the real line.

Proof: Fix $\underline{p}'$ as in Theorem 1.4. Given A, B, define $f_{A,B}(x) = \underline{p}'(Ae^{x}, Be^{x}) = f(x)$ (suppressing A, B). Now take $\underline{p}(A,B) = \Phi(f)$ where Φ is as in Theorem 1.5.

Now $\underline{p}(A,A) = 1$ since $f_{A,A}$ is the constant 1. $0 \leq \underline{p}(A,B) \leq 1$ since $0 \leq \underline{p}'(A,B) \leq 1$ always. $\underline{p}(A,B) = \underline{p}(A \cap B, B)$ etc. $\underline{p}(A+\alpha, B+\beta) = \underline{p}(A,B)$ if $A \subseteq B$, $A + \alpha \subseteq B + \beta$ because

$$\underline{p}(A+\alpha, B+\beta) = \Phi(f_{A+\alpha, B+\beta}(x)) .$$

Now

$$f_{A+\alpha, B+\beta}(x) = \underline{p}'((A+\alpha)e^x, (B+\beta)e^x)$$

$$= \underline{p}'(Ae^x + \alpha e^x, Be^x + \beta e^x)$$

$$= \underline{p}'(Ae^x, Be^x)$$

$$= f_{A,B}(x)$$

Since $Ae^x + \alpha e^x \subseteq Be^x + \beta e^x$, and $Ae^x \subseteq Be^x$.

Thus $\quad f_{A+\alpha, B+\beta} = f_{A,B}$

Hence $\quad \Phi(f_{A+\alpha, B+\beta}) = \Phi(f_{A,B})$

i.e. $\quad \underline{p}(A + \alpha, B + \beta) = \underline{p}(A,B)$.

Now let $A' = A\alpha$, $B' = \beta\alpha$, $\alpha \neq 0$. Can assume $\alpha > 0$ by condition g.

Now

$$f_{A',B'}(x) = \underline{p}'(A' e^x, B' e^x)$$

$$= \underline{p}'(A \alpha e^x, B \alpha e^x)$$

$$= \underline{p}'(A e^{x+\log\alpha}, Be^{x+\log\alpha})$$

$$= f_{A,B}(x + \log \alpha)$$

Hence $\Phi(f_{A',B'}) = \Phi(f_{A,B})$ by the translation invariance of Φ. But that gives $\underline{p}(A', B') = \underline{p}(A,B)$.

Q.E.D.

Theorems 1.4, 1.6 are just Theorems 1.1, 1.2 without condition k. If we show how to obtain this condition without losing any of the others, we will be finished. We first prove a lemma which will be useful also in Theorem 1.7.

Lemma E: Let $\mathcal{J}$ be an ideal of sets of reals, closed under linear transformations and countable unions. Let $\underline{p}_1$ be a function satisfying some of the conditions a-m of Theorems 1.1, 1.2. Then there exists another function $\underline{p}$ satisfying the same conditions as $\underline{p}_1$ and such that for all $A \in \mathcal{J}$, $B \notin \mathcal{J}$, $\underline{p}(A, A \cup B) = 0, \underline{p}(B, A \cup B) = 1$.

Proof: Note that every countable set is in $\mathcal{F}$. Let $\mathcal{F}_1 \subseteq \mathcal{F}^*$ be a *-countable family containing $\mathcal{F}$ and closed under standard linear transformations. (We use the fact that there is a *-finite set containing $\mathcal{F}$ and a *-finite set of linear transformations containing all standard ones. The *-closure of the first under the second is *-countable) Let X be the union of $\mathcal{F}_1$. X is in $\mathcal{F}^*$ and is invariant under standard linear transformations. Now we let

$$\underline{p}(A,B) = \underline{p}_1(A,B) \qquad \text{if } B^* \subseteq X$$

$$= (\underline{p}_1(A-X,\ B-X))^o \quad \text{if } B^* \not\subseteq X.$$

This new $\underline{p}$ has the needed properties.

Proof of Theorems 1.1, 1.2: They follow immediately from Lemma E, taking $\mathcal{F}$ to be the family of all countable sets.

Theorem 1.7: There exist 2^c distinct functions satisfying the conditions of Theorem 1.1 and Theorem 1.2.

Proof: Let X be a set of cardinality $c = 2^{\aleph_0}$ containing algebraically independent elements. (E.g. they would be mutually transcendent. However it is sufficient if none of them is in the subfield generated by the others).

We claim that if L is a linear transformation, then $X \cap L[X]$ has at most two elements. For suppose $a_1, a_2, a_3 \in X \cap L[X]$ and $L(x) = r \cdot x + s$ then there are $b_1, b_2, b_3 \in X$ such that $a_i = r \cdot b_i + s$, $i = 1,2,3$. However, $r = \frac{a_1 - a_2}{b_1 - b_2}$ and $s = a_1 - \left(\frac{a_1 - a_2}{b_1 - b_2}\right) b_1$. Thus a_3 can be expressed in terms of a_1, a_2, b_1, b_2, b_3, a contradiction.

Now $\overline{\overline{c}} = 2 \circ c^2$. Hence for each real number r we can find sets A_r, B_r such that $\overline{\overline{A_r}} = \overline{\overline{B_r}} = c$ and for $r \neq s$, A_r, B_r, A_s, B_s are all disjoint. Now if Z is a subset of the reals, let $C_Z = \bigcup C'_r$ where $C'_r = A_r$ when $r \in Z$ and $C'_r = B_r$ when $r \notin Z$. Let $\underline{p}_Z$ be the measure obtained by applying lemma E to our original $\underline{p}$ of Theorem 1.1 (or 1.2). Now if $Z \neq Z'$ then there is an r which is in one but not the other. Say $r \in Z - Z'$. Then $A_r \subseteq C_Z - C_{Z'}$, $\underline{p}_Z(A_r, A_r \cup B_r) = 0$ and $\underline{p}_{Z'}(B_r, A_r \cup B_r) = 0$. Hence $\underline{p}_Z \neq \underline{p}_{Z'}$. $\{\underline{p}_Z | Z \subseteq \text{reals}\}$ is the required set of 2^c functions. Q.E.D.

Now we indicate why condition k had to be imposed in Theorem 1.1 and wouldn't follow from the others. In the proof of lemma D and the main lemma, the sample used is of the form F_p where F has ℓ elements and $p \geq k/2$ where $k/\log(k+1) \geq 8 \ell m n$. Hence p is much larger than ℓ. Now given a standard irrational number r, consider A = all integral multiples of r. A will have an intersection with F_p

of size at least p. Take B to be a basis for the reals as a vector space over the rationals, $B \subseteq [0,1]$. Then $\overline{\overline{B \cap F_p}}$ can be at most ℓ since F_p is contained in an ℓ-dimensional subspace over the rationals. Thus we will get $\underline{p}(A, A \cup B) = 1$ and $\underline{p}(B, A \cup B) = 0$ at first, even though A is countable and B is not.

§2. UNIFORM AND ALMOST INVARIANT SETS

Consider the following intuitive question. A rational number x lies in the interval [0,1]. What is the probability that $x \in [a,b]$ where $0 \leq a < b \leq 1$? It seems clear that the answer ought to be $\frac{1}{2}$ when $a = 0$, $b = \frac{1}{2}$, and a little more thought convinces one that it ought to be $b - a$ in general. This question has been discussed in the literature, Cf. Dubins and Savage [3 p. 11] and Bernstein and Wattenberg [1]. However, without a general discussion one cannot answer the question whether the same should hold for every countable dense set. In the remainder of §2, $\underline{p}$ is some conditional probability function satisfying the conditions of Theorem 1.1.

<u>Definition 2.1</u>: A set $A \subseteq I$ is $\underline{p}$-uniform if for all a,b, $0 \leq a < b \leq 1$, $\underline{p}([a,b],A) = b - a$. A is <u>uniform</u> if it is uniform for all $\underline{p}$.

<u>Definition 2.2</u>: A set $A \subseteq I$ is <u>almost invariant</u> if A is infinite and there exist positive $\varepsilon_n \to 0$ such that for all n, $A\delta(A + \varepsilon_n)$ = the symmetric difference of A, $A + \varepsilon_n$, is finite.

<u>Theorem 2.1</u>: Every almost invariant set is uniform.

Before we prove theorem 2.1, we shall prove three simple lemmas.

<u>Lemma 2.1</u>: For all A, B, $C \subseteq I$,

(a) $\underline{p}(A \cup C, B \cup C) \geq \underline{p}(A, B)$

(b) $\underline{p}(A - C, B - C) \leq \underline{p}(A, B)$

<u>Proof</u>: (a) We may assume that $A \subseteq B$ since otherwise we can work with $A \cap B$. Let $C' = C - B$. Then we have $A \cup C' \subseteq A \cup C$, $B \cup C' = B \cup C$. Hence

$$\begin{aligned}
\underline{p}(A \cup C, B \cup C) &\geq \underline{p}(A \cup C', B \cup C') \\
&= \underline{p}(A, B \cup C') + \underline{p}(C', B \cup C') \\
&= \underline{p}(A, B) \circ \underline{p}(B, B \cup C') \\
&+ \underline{p}(C', B \cup C') \\
&\geq \underline{p}(A, B) \cdot \underline{p}(B, B \cup C') \\
&+ \underline{p}(A, B) \cdot \underline{p}(C', B \cup C') \\
&= \underline{p}(A, B) \left\{ \underline{p}(B, B \cup C') + \underline{p}(C', B \cup C') \right\} \\
&= \underline{p}(A, B).
\end{aligned}$$

(b) is an immediate consequence of (a) Q.E.D.

Lemma 2.2: If $\underline{p}(C, B) = 0$ then $\underline{p}(A, B \cup C) = \underline{p}(A, B - C) = \underline{p}(A \cup C, B) = \underline{p}(A - C, B) = \underline{p}(A, B)$.

Proof: Completely straightforward using finite additivity and condition ℓ.

Lemma 2.3: Let B_1, B_2 be disjoint. Then

$$\underline{p}(A, B_1 \cup B_2) = \underline{p}(A, B_1) \cdot \underline{p}(B_1, B_1 \cup B_2) + \underline{p}(A, B_2) \cdot \underline{p}(B_2, B_1 \cup B_2)$$

Proof: let $A_1 = A \cap B_1$, $A_2 = A \cap B_2$. Then A is the disjoint union of A_1, A_2. Since $\underline{p}(A, B_1 \cup B_2) = \underline{p}(A_1, B_1 \cup B_2) + \underline{p}(A_2, B_1 \cup B_2)$ the result follows at once by condition ℓ.

Proof of Theorem 2.1: Suppose A is almost invariant. Let $\varepsilon_n \to 0$ such that for all n, $\varepsilon_n > 0$ and $(A + \varepsilon_n)\, \delta\, A$ is finite for all A. Then for any X, $\underline{p}(X, A) = \underline{p}(X + \varepsilon_n, A + \varepsilon_n) = \underline{p}(X + \varepsilon_n, A)$ by lemma 2.2. From this it immediately follows that if k_n = the largest integer $\leq \frac{1}{\varepsilon_n}$, then $\frac{1}{k_n + 1} \leq \underline{p}([\alpha, \alpha + \varepsilon_n], A) \leq \frac{1}{k_n}$.
Now given a, b, ε, $0 \leq a < b \leq 1$, choose $\varepsilon_n < \varepsilon/3$. Find k such that

$a + k\, \varepsilon_n \leq b < a + (k + 1)\, \varepsilon_n$. Then clearly, $\underline{p}([a, a + k\, \varepsilon_n], A) \leq \underline{p}([a, b], A) \leq \underline{p}([a, a + (k + 1)\, \varepsilon_n], A)$. Hence $\frac{1}{k_n + 1} \leq \underline{p}([a, b], A) \leq \frac{k + 1}{k_n}$
Replacing A by I in the above calculation we also get $\frac{k}{k_n + 1} \leq b - a \leq \frac{k + 1}{k_n}$

Hence $\quad |\underline{p}([a, b], A) - (b - a)|$

$$\leq \frac{k_n + k + 1}{k_n(k_n + 1)} \leq \frac{3k_n}{k_n(k_n + 1)} \leq \frac{3}{k_n + 1} < \varepsilon .$$

Since ε was arbitrary, $\underline{p}([a,b],A) = b - a$.

Q.E.D.

We now consider two generalisations of the notions "almost invariant" and "uniform".

Definition 2.3: A set $A \subseteq I$ is weakly invariant if there exist positive $\varepsilon_n \to 0$ such that for all n, $A \delta (A + \varepsilon_n)$ is countable.

Theorem 2.1a: Every uncountable weakly invariant set is uniform.

Proof: Identical to that of Theorem 2.1.

Definition 2.4: A set A of reals is uniform if for all a,b,c,d, $a \leq b < c \leq d$, $\underline{p}([b,c] \cap A, [a,d] \cap A) = \frac{c-b}{d-a}$.

Theorem 2.1b: Every almost invariant infinite set is uniform. Every uncountable weakly invariant set is uniform.

If $A \subseteq$ reals is uniform then $A \cap I$ is a uniform subset of I.
Examples of uniform sets.

(1) The rational numbers. More generally any dense subgroup of the reals.

(2) The set $\left\{ r \in I | r \equiv n\theta \bmod 1 \text{ for some } n \geq 0 \right\}$ where θ is irrational.

The following example shows that not every countable dense subset of I is uniform. Let θ be irrational. X = all even multiples of θ mod 1 Y = all odd multiples of θ mod 1. Then $X \cup (Y \cap [0, 1/2])$ is not uniform. In fact "two thirds" of that set lies in $[0, \frac{1}{2}]$, regardless of the choice of $\underline{p}$.

We close this section by stating two easy theorems without proof.

Theorem 2.2: The disjoint union of finitely many uniform sets is uniform.

Theorem 2.3: The translates A + x of a uniform set A are all uniform.

§3. DETERMINATE PAIRS OF SETS

Let C be a set of conditions on conditional probability functions. E.G. C might be the set of conditions of Theorem 1.1. Then $\mathcal{M}_C$ will be the class of all functions $\underline{p}$ satisfying conditions C. Of particular interest will be the cases $\mathcal{M}_1$, $\mathcal{M}_2$ corresponding to Theorems 1.1, 1.2 respectively.

Definition 3.1: A pair (A,B) (of sets of reals) is C-determinate if for all $\underline{p}, \underline{q} \in \mathcal{M}_C$, $\underline{p}(A,B) = \underline{q}(A,B)$.

Theorem 3.1: If the pair (A,B) is 1-determinate, and $\alpha \neq 0$ then the pair $(A\alpha, B\alpha)$ is also 1-determinate and has the same determined value.

Proof: For $p \in \mathcal{M}_1$, define $\underline{p}'(X,Y) = \underline{p}(X\alpha, Y\alpha)$. Then $\underline{p}' \in \mathcal{M}_1$. Hence for all $\underline{p}, \underline{q} \in \mathcal{M}_1$, we have $\underline{p}(A\alpha, \beta\alpha) = \underline{p}'(A,B) = \underline{q}'(A,B) = \underline{q}(A\alpha, \beta\alpha)$ Q.E.D.

Corollary: The elements of $\mathcal{M}_1$, restricted to 1-determinate pairs of sets, satisfy Theorems 1.1 and 1.2 simultaneously.

The corollary is not of much use since we don't know very much about 1-determinate sets.

Theorem 3.2: If B is a subgroup of the reals under addition and A is a proper subgroup of B, then the pair (A,B) is 1-determinate and $\underline{p}(A,B) \equiv 0$.

Proof: Obvious, there are infinitely many cosets, all with the same probability relative to B.

We conjecture that if (A,B) is determinate for all A, then B is finite. However we do not know how to prove this. Neither do we know about possible values of $\underline{p}(A,B)$ when (A,B) is not a determinate pair. It seems reasonable to suppose that these values form an interval.

Of particular interest are those sets A such that the pair (A,B) is determined whenever B is an interval. We close the paper by pointing out that the usual example A of a non-measurable set, where $(\forall x)(\exists ! y)(y \in A$ and $x - y$ is rational), is determined relative to intervals B, and $\underline{p}(A,B) \equiv 0$.

BIBLIOGRAPHY

[1] A. Bernstein and F. Wattenberg, "Nonstandard Measure Theory" in Applications of Model Theory, Ed. W.A.J. Luxemburg; Holt, Rinehart and Winston, 1969.

[2] R. Bumby and E. Ellentuck, "Finitely Additive Measures and the First Digit Problem" Fundamenta Mathematicae, Vol. 65 (1959) pp. 33-42.

[3] L. Dubins and L. J. Savage, "How to Gamble if You Must", McGraw-Hill, 1965.

[4] N. Dunford and J. Schwartz, "Linear Operators", Vol. 1, Interscience 1967.

[5] P. Loeb, "A Nonstandard Representation of Measurable Spaces and L_{∞}", Bulletin of Amer. Math. Soc., Vol. 77 (1971) pp. 540-544.

[6] R. Parikh, "A Nonstandard Theory of Topological Groups", in Applications of Model Theory, Ed. Luxemburg; Holt, Rinehart and Winston, 1969.

[7] R. Parikh, "Nonuniqueness of Relative Probabilities", Abstract 701-60-12, Notices of the Amer. Math. Soc., Jan. 1973, p. A201.

[8] R. Parikh and M. Parnes, "Relative Probabilities and Uniform Sets", Abstract 72-B129, Notices of the Amer. Math. Soc., Feb. 1972, p. A446.

[9] R. Parikh and M. Parnes, "Conditional Probabilities, Determinate Sets and Indices", Abstract 696-60-2, Notices of the Amer. Math. Soc., Aug. 1972, p. A660.

[10] R. Parikh and M. Parnes, "Conditional Probability Can Be Defined for All Pairs of Sets of Reals", Advances in Mathematics, Vol. 9 (1972) pp. 313-315.

[11] M. Parnes, "On the Measure of Measurable Sets of Integers" to appear in Acta Arithmetica.

[12] A. Renyi, "Foundations of Probability", Holden Day Inc. 1970.

[13] A. Renyi, "On a New Axiomatic Theory of Probability", Acta Mathematica, Budapest, Vol. 6 (1955) pp. 285-335.

[14] H. Royden, "Real Analysis", 2nd Edition, MacMillan.

OMITTING TYPES IN ARITHMETIC AND CONSERVATIVE EXTENSIONS

R. G. Phillips

University of South Carolina

1. Introduction. Let L be a countable first-order language whose non-logical relation symbols include symbols for addition, multiplication, and order and whose constant symbols are the natural numbers. Let W be some consistent extension of Peano's Axioms, with induction, formulated in L. Given a first-order model M for W, L_M denotes the language obtained from L by adding the elements of M to L as constant symbols. An M-formula is any formula of L_M and an n-ary relation on M is called M-definable if it is represented by an M-formula in M with n free variables. If M^* is an extension of M and R is an n-ary relation on M^*, R restricted to M is $\{(x_1,\ldots,x_n) : x_1 \in M, \ldots, x_n \in M, \text{ and } M^* \models R(x_1,\ldots,x_n)\}$.

2. Conservative Extensions. If M models W, a proper elementary extension M^* of M, with respect to L_M, is called a conservative extension of M if each M^*-definable relation restricted to M is M-definable. The purpose of this section is to prove the following theorem:

2.1 Theorem. There exists a conservative extension of each model of W.

This theorem is motivated by the MacDowell-Specker theorem [5] on end extensions of models of W (i.e. proper elementary extensions in which each new element is larger than each element of M) and is in fact a generalization of that theorem. That is, 2.1 has the following corollary:

2.2 Theorem. (MacDowell-Specker) There exists an end extension of each model of W.

We shall use a modification of a proof of 2.2 by C. C. Chang [2] for 2.1. For this purpose we need the following definition:

2.3 Definition. A set of formulas of L_M with the same free variables $x_1, \ldots ,$

x_n, is called a type. A model M realizes a type T if there are elements $a_1, \ldots, a_n$ in M such that $M \models q(a_1,\ldots,a_n)$ for each q in T. M omits T if it does not realize T. M omits a family U of types T if M omits each type T in U.

2.4 Definition. Let D be an ultra-filter on M and let E be a set of functions from M into M which is closed under functions in L_M. Then D-Prod|E denotes the substructure of the ultra-power ($\prod_{i\in M} M/D$) consisting of only those functions in E.

Chang shows that there is an ultra-filter D on M such that $M^* =$ D-Prod|F, where F is the set of M-definable functions from M into M, omits the family of types $t_0 = \{T_a : a \in M\}$ where $T_a = \{x < a \wedge x \neq b : b \in M\}$. From this it is clear that M^* is an end extension of M. Similarly we shall show that a conservative extension of M exists of the form D-Prod|F where D is chosen so that M^* omits the family of 0-types which we now define. In what follows, M models W.

2.5 Definition. Let $q(x_1,\ldots,x_n,y_1,\ldots,y_m)$ be an n+m-ary M-formula and let A be an m-ary relation on M. We set $\Delta_{qA} = \{q(x_1,\ldots,x_n,a_1,\ldots,a_m):(a_1,\ldots,a_m) \in A\} \cup \{\neg q(x_1,\ldots,x_n,a_1,\ldots,a_m): (a_1,\ldots,a_m) \notin A\}$. If A is not M-definable, then we call Δ_{qA} an 0-type.

For example, $t_1 = \{\Delta_{qA} : A \in P\}$ where P is the totality of subsets of M with upper bounds in M but without a largest member and q(x,y) is "$x < y$" is a subset of the set of all 0-types. It is easy to see that an extension of M omits t_0 iff it omits t_1. This and the next two theorems constitute a proof of 2.1 and 2.2.

2.6 Theorem. A proper elementary extension of M is a conservative extension of M iff it omits the set of all 0-types.

2.7 Theorem. There exists a non-principal ultra-filter D on M such that $M^* =$ D-Prod|F omits all 0-types.

Proof. Let p be a ternary formula of L and let X be an unbounded M-definable subset of M. Hence, " $\forall x \exists y \geq x (y \in X)$" is a formula of L_M which holds in M. First, we define two operators D(X,p,a) and R(X,p,a), $a \in M$, so that R(X,p,a) is an M-definable subset of M and D(X,p,a) is an unbounded M-definable subset of X; both operators depending on formulas of L_M whose only parameters are a, the ones involved in the definition of X, and of course the ternary formula p.

To do this, we define a decreasing M-sequence (i.e. a function whose domain is M) of sets X_{-1}, X_0, ... , X_b, ... , as b varies over M so that $X_{-1} = X$ and $X_b = \{x \in X_{b-1}: M \models p(x,a,b)\}$ if this set is unbounded, otherwise $X_b = \{x \in X_{b-1}: M \models \neg p(x,a,b)\}$. We show that each M_b is M-definable and that the M-sequence $\{X_b\}$ is itself M-definable.

Let $\beta(x,y)$ be the Godel Beta function. β has the following property: for each M-definable subset of M of the form $\{x_0, x_1, \ldots, x_b\}$, where $b \in M$, there is in M an a such that $M \models \beta(i,a) = x_i$ for each i in M so that $M \models 0 \leq i \leq b$. Define the function "$\lambda(u) = v$" by the following M-formula:

$$"\exists_x [\beta(u,x)=v \wedge \forall_{i \leq u}[\beta(i,x)=0 \vee \beta(i,x)=1] \wedge [\forall_y \exists_{z \geq y}(z \in X \wedge p(z,a,0)) \leftrightarrow \beta(0,x)=0] \wedge [\forall_{i<u} \forall_{j \leq i} \forall_y \exists_{z \geq y}(z \in X \wedge p^{\beta(j,x)}(z,a,j) \wedge p(z,a,i+1)) \leftrightarrow \beta(i+1,x)=0]]"$$

where p^{β} is the formula "$[\beta(j,x)=0 \rightarrow p(z,a,j) \wedge \beta(j,x)=1 \rightarrow \neg p(z,a,j)]$", i.e. $p^0 = p$ and $p^1 = \neg p$.

Clearly "$\lambda(u)=v$" is an M-definable function whose definition depends only on the parameter a, the parameters in the definition of X, and the ternary formula p. We can show $M \models \forall_u \exists !_v (\lambda(u)=v)$ by induction on u.

Now let the relation "$\Psi(u,v)$" be defined by the M-formula

$$"\forall_{x \leq v}\, p^{\lambda(x)}(u,a,x)".$$

Then $c \in X_b$ iff $M \models \Psi(c,b)$ which shows that $\{X_b\}$ is an M-definable M-sequence of M-definable unbounded subsets of X.

Now we define $R(X,p,a) = \{b: X_b = \{x \in X_{b-1}: M \models p(x,a,b)\}\} = \{b: M \models \lambda(b)=0\}$, and $D(X,p,a) = \{x_0, \ldots, x_b, \ldots\}$ where $x_b = \mu z[z > x_{b-1} \wedge z \in X_b]$. Clearly $R(X,p,a)$ has the required properties stated above. To show that this is true also for $D(X,p,a)$, we note that this latter set is the image of M under the M-function "$\zeta(u)=v$" defined by the formula

$$"\exists_x [\beta(u,x)=v \wedge \beta(0,x) = \mu z[z \in X \wedge \Psi(z,0)] \wedge \forall_{0<i \leq u}[\beta(i,x) = \mu z[z \in X \wedge z > \beta(i-1,x) \wedge \Psi(z,i)]]]"$$

where again β is the Godel beta function.

Now, in a similar manner, we define $\overline{R}(X,p) = \{R(X,p,a) : a \in M\}$ and $\overline{D}(X,p) = \{x_0, \ldots, x_b, \ldots\}$ where $x_b = \mu z[z > x_{b-1} \wedge z \in D(X,p,b)]$. $\overline{R}(X,p)$ is a family of M-definable subsets of M and $\overline{D}(X,p)$ is an unbounded M-definable subset of X.

$\overline{R}$ and $\overline{D}$ have the following property:

2.8 For each a in M there is a set A in $\overline{R}$, namely R(X,p,a), such that for each b in M there is a d in M so that $b \in A \rightarrow \{x \in M : x \geq d\} \cap \overline{D} \subset \{x \in X : M \models p(x,a,b)\}$ and $b \notin A \rightarrow \{x \in X : x \geq d\} \cap \overline{D} \subset \{x \in X : M \models \neg p(x,a,b)\}$. To see this, note that $\{x \in X : x \geq d\} \cap \overline{D} \subset X_b$ for sufficiently large d.

We now let $p_1, p_2, \ldots, p_n, \ldots$ be an enumeration of the ternary formulas of L and we define two sequences thusly:

$D_1 = \overline{D}(M,p_1)$, and given D_n, $D_{n+1} = \overline{D}(D_n,p_{n+1})$. Thus we have $M \supset D_1 \supset D_2 \supset \ldots$ and each D_n is M-definable and unbounded in M.

$H_1 = \overline{R}(M,p_1)$ and $H_{n+1} = \overline{R}(D_n,p_{n+1})$.

We let D be a non-principal ultra-filter generated by the D_n and all sets of the form $\{x \in M : x \geq d\}$, as d varies over M, and we set $M^* = \text{D-Prod}|F$ where F is the set of all M-definable functions on M into M. In [2] it is shown why M^* is a proper elementary extension of M but this is essentially the construction of Skolem in [6].

Let Δ_{qA} be an 0-type where q is n+m-ary and A is m-ary. Let $f = (f_1, \ldots, f_n)$ where each f_i is in F and let g be an M-definable bijection of M onto the M-fold cartesian product of M. Then there is a p_k and an a in M so that

$$M \models \forall_x \forall_y [q(f(x),g(y)) \leftrightarrow p_k(x,a,y)].$$

Let A_0 be the preimage of A under g; then A_0 is not M-definable. Thus A_0 does not belong to H_k. However, there is a set A_k in H_k satisfing 2.8. If b is in $A_k - A_0$, by 2.8 it is easy to see that $M^* \models q(f,g(b))$ and g(b) is not in A. If on the other hand b is in $A_0 - A_k$ then $M^* \models \neg q(f,g(b))$ and g(b) is in A. Therefore M^* omits Δ_{qA}.

The existence of conservative extensions for the standard model of Peano's Axioms was first shown by Alan Cantor [1] in an attempt to generalize the result of George Zahn [7] which we mention in the next section.

3. <u>The Additive Group of a Conservative Extension</u>. Although we did not state specifically that our version of Peano's Axioms was to be those for the natural numbers, the foregoing exposition would indicate that was our intention. However, it should be clear that our results hold also when we consider Peano's Axioms for the ring of Integers. Thus, in this section, our models M of W will always be rings.

3.1 <u>Definition</u>. For each model M of W let M_0 denote the set of M-definable functions f mapping $\{x \in M : x \geq 2\}$ into M such that

$$M \models \forall_x \exists_y \forall_i \left[2 \leq i \leq x \rightarrow Res(y,i)=f(i)\right]$$

where Res(y,i)=z is defined by the formula "$\exists_w \left[y=w \cdot i + z \wedge 0 \leq z < i\right]$".

3.2 <u>Theorem</u>. The additive group of each conservative extension M^* of M is $K \oplus M_0$, where K is the vector-space over the rational numbers Q whose dimension is the cardinality of M^*.

When M is the standard model of the Integers, this is the result of Zahn [7]. In this case, M_0 is all arithmetically definable functions which are initially the residue sequence of an integer. The proof we give of 3.2 is due to MacDowell-Specker [5].

Let M^* be a conservative extension of M and define ρ on M^* as follows: if a belongs to M^*, then $\rho(a) = f$, where $M^* \models [Res(a,i)=f(i)]$ for each $i \geq 2$ in M. Since f is the M^*-definable function Res(a,x) restricted to M, f is M-definable. The fact that $\rho(M^*) = M_0$ now follows easily from the fact that M^* is an elementary extension of M. As in [5], the kernel K of ρ is seen to be the vector-space over Q with dimension the cardinality of M^*. Finally, from $M^*/K = M_0$, we get $M^* = K \oplus M_0$ from the well known theorem (see for example Kaplanski's <u>Infinite Abelian Groups</u>) that states that a divisible subgroup of a group is a direct summand of that group.

As in [7], it is easy to show that M_0 is minimal in the sense that if M^* is any proper elementary extension of M and if ρ is defined on M^* as above, then $M_0 \subset \rho(M^*)$. Since M is a proper subgroup of M_0, we now have another proof that Kemeny's conjecture is false. In [4], Kemeny conjectured the existence of a non-standard model of the Integers whose additive group was $K \oplus Z$, Z the standard model of the Integers.

4. Minimal Extensions. In [3] , Gaifman defines and constructs minimal extensions of models of W. Briefly this goes as follows: if M is a model of W, then a proper elementary extension M^* of M is called a minimal extension of M if there is no proper elementary substructure of M^* which properly extends M. Equivalently, if M^*_a denotes the elementary substructure of M^* generated by M and $a \in M^*$, then M^* is a minimal extension of M if $M^*_a = M^*$ for every a in $M^* - M$.

In Gaifman's construction of minimal extensions, we let U be all binary terms in L. Gaifman then shows that for each t in U and each unbounded M-definable subset X of M there is an unbounded M-definable subset $D_t(X)$ of X, a unary term $\tilde{t}$ of L, and a term t^* of U such that for each a in M there is a d in M so that

4.1 $M \models \forall x [x > d \wedge x \in D_t(X) \rightarrow [t(a,x) = \tilde{t}(a) \vee t^*(a,t(a,x)) = x]]$.

Then, if $U = \{t_1, \ldots, t_n, \ldots\}$ is an enumeration of U, we define a sequence $D_1 \supset D_2 \supset \ldots \supset D_n \supset \ldots$ of unbounded M-definable sets thusly:

$$D_1 = D_{t_1}(M), \ \ldots \ , D_{n+1} = D_{t_{n+1}}(D_n), \ \ldots .$$

We let D be a non-principal ultra-filter generated by the D_n and sets of the form $\{x \in M : x \geq d\}$. We then have the following theorem:

4.2 Theorem (Gaifman). $M^* = D\text{-Prod}|F$ is a minimal extension of M.

Proof. If $b \in M^*$, then $b \in F$ and hence $b(x) = t_k(a,x)$ for some a in M. Using 4.1, we have

$$M^* \models [b = \tilde{t}_k(a) \vee t^*_k(a,b) = i]$$

where i is the identity function from M to M. Since $\tilde{t}_k(a) \in M$, if $b \notin M$, then

$$M^* \models t^*_k(a,b) = i .$$

In other words, $i \in M^*_b$. But since $M^*_i = M^*$, it follows that $M^* = M^*_b$.

The method of constructing the $D_t(X)$ is completely analogous to the construction of the $\overline{D}(X,p)$ of Theorem 2.7 and also because of the analogy between Theorems 2.7 and 4.2 the motivation for the next theorem is evident.

4.3 Theorem. Each model of W has an extension which is both conservative and minimal.

Proof. We simply alternate the construction in 2.7 and 4.2. Thus we set

$$D_1 = \overline{D}(M,p_1),\ D_2 = D_{t_1}(D_1)$$

$$D_{2n+1} = \overline{D}(D_{2n},p_{n+1}),\ D_{2n+2} = D_{t_n}(D_{2n+1})$$

and we set $H_1 = \overline{R}(M,p_1)$, $H_n = \overline{R}(D_{2n},p_{n+1})$. Then, as usual, we let D be a non-principal ultra-filter generated by the D_n and the sets $\{x \in M : x \geq d\}$ and then $M^* =$ D-Prod|F is the required extension.

Other questions concerning the relationship between conservative extensions and minimal extensions are:

4.4 There exist conservative extensions which are not minimal.

This is easily seen by letting M^* be a conservative extension of M and M^{**} be a conservative extension of M^*. Then M^{**} is a conservative extension of M but not a minimal extension of M.

4.5 There exist minimal extensions which are not conservative.

The proof depends on a construction of the form $N^* =$ D-Prod|F where D is generated by a sequence $D_1 \supset D_2 \supset \ldots$ formed by interchanging Gaifman's construction with one that guarantees that N^* will have an element such that the set of standard primes which divide it in N^* is not arithmetical. Here, F will be all arithmetical functions from N into N and then N^* will of course be an elementary extension of N. The details of this construction will be found in [8].

We conclude with two results of John Gregory (unpublished). Gaifman [3] defines an anti-minimal extension M^* of M as a proper elementary extension in which every elementary substructure of M^* which properly extends M is not a minimal extension of M. Gregory's results are:

4.6 Each model M of W has a proper elementary extension which is both conservative and anti-minimal.

4.7 Assuming the consistency of ZF, there is a countable model M of ZFC such that M has no proper elementary end extension which is conservative with respect to the definable relations on the ordinals of M. Here, ZF denotes the Zermelo-Frankel axioms for set theory and ZFC denotes ZF with the axiom of choice.

REFERENCES

1. A. Cantor, Doctoral Dissertation, University of South Carolina, 1972.
2. C.C. Chang, Ultra-products and other Methods of Constructing Models, Sets, Models, and Recursion Theory, Proceedings of the Summer School in Mathematical Logic, Leicester, 1965, pp. 85-121.
3. H. Gaifman. On Local Arithmetical Functions and their Application for Constructing Types of Peano's Arithmetic, Mathematical Logic and Foundations of Set Theory, Proceedings of an International Colloquium, Jerusalem, 1968, pp. 105-121.
4. J. Kemeny, Undecidable Problems of Elementary Number Theory, Mathematische Annalen, 135, pp. 160-169.
5. R. MacDowell and E. Specker, Modelle der Arithmetik, Infinitistic Methods, Proceedings of the Symposium on the Foundations of Mathematics, Warsaw, 1959, (1961) pp. 257-263.
6. T. Skolem, Peano's Axioms and Models of Arithmetic, Mathematical Interpretations of Formal Systems, Amsterdam, 1955.
7. G. Zahn, Doctoral Dissertation, University of South Carolina, 1971.
8. R. Phillips, A Minimal Extension which is not Conservative, in preparation.

The Strength of the Hahn-Banach Theorem

David Pincus (Seattle, Wash.)

I. Introduction

This paper discusses the strength of the Hahn-Banach theorem (HB) considered as an axiom of Zermelo-Fraenkel set theory (ZF) without the axiom of choice (AC). Interest in this question was generated when it was found by Los and Ryll-Nardzewski [10] and independently by Luxemburg [11] that HB follows from the prime ideal theorem for Boolean algebras (PI). AC is independent of PI ([7])[1], hence of HB so the most natural conjecture was that PI and HB are equivalent.

Luxemburg [12] provided the first negative evidence for this conjecture. Using nonstandard methods he exhibited a number of equivalents to HB each a slight weakening of an equivalent to PI[2]. We will see in §II of this paper that PI is in fact independent of HB. This result was announced in [17] and an independence proof of PI from HB was given there for the weak set theory ZFA of Fraenkel and Mostowski. The ZF independence differs from the ZFA independence in the following respect. The ZFA model of [17] can be closely "approximated" by submodels satisfying AC. The ZF model of §II is approximated by submodels satisfying only PI. The proof of HB in the ZF model is therefore accomplished by nonstandard arguements.

---

[1] As is usual in set theory, "A is independent of B" means "If ZF is consistent then so is ZF+A+~B."

[2] A large sampling of the literature on equivalents to PI may be found in [12] and the bibliography of [13].

In order to discuss the other results of this paper we state HB and some related principles. The following terminology relates to a vector space, X, over the real numbers, $\mathbb{R}$. A functional $p:X \to \mathbb{R}$ is <u>sublinear</u> if $p(x+y) \leq p(x)+p(y)$ and $p(\lambda x) = \lambda p(x)$ for $\lambda \geq 0$ in $\mathbb{R}$. f is a <u>partial functional</u> on X if f is a linear functional defined on a subspace of X. <u>$f \leq p$</u> means $f(x) \leq p(x)$ for every x in the common domain of f and p.

<u>HB</u> Let X, p, and f be respectively a real vector space, sublinear functional $X \to \mathbb{R}$, and partial functional on X satisfying $f \leq p$. f extends to a linear functional $g:X \to \mathbb{R}$ such that $g \leq p$.

For terminology of topological vector spaces see [4].

<u>KM</u> (Krein-Milman Theorem) A compact convex subset of a locally convex vector space has an extreme point.

A stronger form of KM is often given involving convex hulls. It is proved from the above form using HB.

<u>SKM</u> In the statement of KM replace "compact" by "convex-compact" (see [1] and [12].).

Let B be a Boolean algebra. A function $\mu: B \to \mathbb{R}$[3] is a <u>measure</u> if $\mu(a) \geq 0$, $\mu(1) = 1$, and $\mu(a \vee b) = \mu(a)+\mu(b)$ whenever $a \wedge b = 0$.

<u>M</u> Every Boolean algebra has a measure.

<u>2VM</u> Every Boolean algebra has a measure taking only the values 0 and 1.

---

[3]The distinction between the algebra, B , and its underlying set, Field B , is made in § II but not elsewhere.

M(ω) P(ω) (P denotes power set, ω denotes the nonnegative integers) has a measure which is 0 on finite sets.

2VM(ω) P(ω) has a measure, 0 on finite sets and taking only the values 0 and 1.

LM Every set of reals is Lesbague measurable.

BP Every set of reals has the Baire property (has first category difference with an open set).

The following diagram exhibits the known interrelationships among the above statements. Equivalents are connected by a 2-way arrow. Irreversable implications are connected by a 1-way arrow. A starred arrow indicates some question of reversability. The arrows are numbered for discussion.

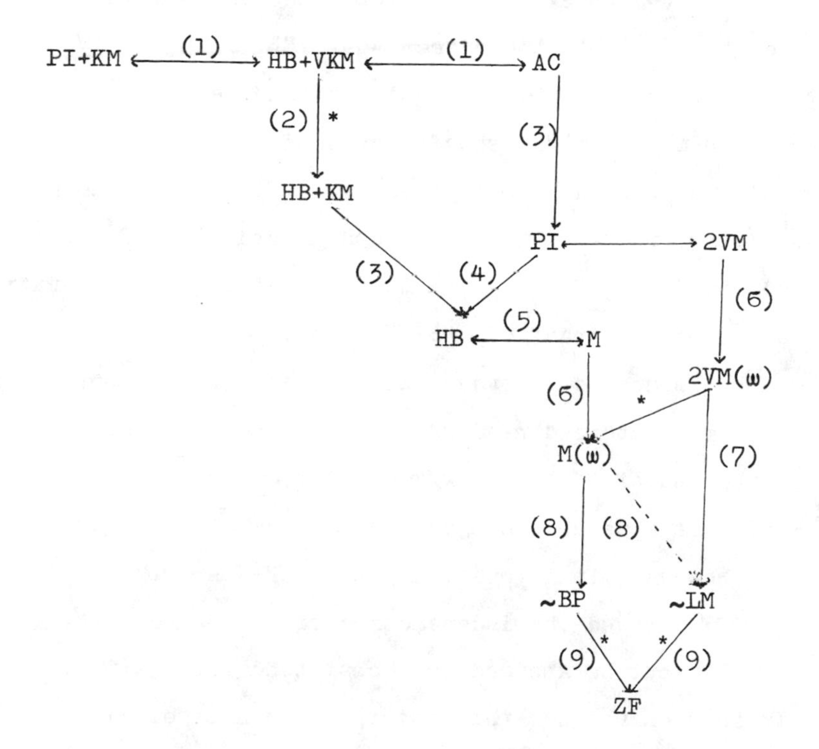

(1) PI+KM $\longleftrightarrow$ AC and HB+VKM $\longleftrightarrow$ AC are theorems of [1]. The first equivalence was also proved by Renz and Luxemburg.

(2) The independence of AC from HB+KM is proved for ZFA in [17]. The ZF independence remains open. We prove ~KM in the model of § II.

(3) PI $\not\rightarrow$ AC is proved in [7]. HB $\not\longrightarrow$ HB+KM is immediate from this, PI $\rightarrow$ HB, and PI+KM $\longrightarrow$ AC.

(4) HB $\not\longrightarrow$ PI is proved in § II.

(5) HB $\longleftrightarrow$ M is one of the main results in [12]. It was independently known to C. Ryll Nardzewski.

(6) M $\rightarrow$ M(ω) and 2VM $\rightarrow$ 2VM(ω) are obtained by looking at the algebra P(ω)/I where I is the ideal of finite subsets of ω. M(ω) $\not\longrightarrow$ M and 2VM(ω) $\not\longrightarrow$ 2VM will be obtained in § III as a consequence of a theorem to the effect that HB is a "truly universal" application of AC. i.e. in some sense HB can't be obtained from any mathematical special case of AC.

(7) 2VM $\rightarrow$ ~LM was proved by Sierpinski [22]. In [21] he gave another interesing proof of ~LM based on an ordering of PP(ω). ~LM $\not\longrightarrow$ 2VM(ω) is due to Solovay and is obtained from considerations similar to those in our proof of ~BP $\not\longrightarrow$ M(ω).

(8) M(ω) $\rightarrow$ ~BP is due to Solovay. It is stated without proof in [20]. Since it may interest readers of this paper we include our own proof at the end of § I. ~BP $\not\rightarrow$ M(ω) is proved in § III.

M(ω) $\rightarrow$ ~LM is a natural conjecture in view of the proven implications, however, it is open. In fact HB+KM $\rightarrow$ ~LM is open.

(9) Solovay [20] proved the independence of ~BP and ~LM from the axioms of ZF but he assumed the consistency of an inaccessable cardinal. It is open whether this assumption can be eliminated. Our proof of the independence of M(ω) in § III does not use an inaccessable cardinal. Thus the consistency of ZF is sufficient to prove the independence of all statements in the diagram except

possibly $\approx$BP and $\sim$LM.

In summation: §II exhibits a model of ZF satisfying HB+$\sim$PI+$\sim$KM. §III gives a theorem on the generality of HB (see the discussion of (6).) and a proof of $\approx BP \not\rightarrow M(\omega)$ (See the discussion of (7),(8), and (9).).

Proof of $M(\omega) \rightarrow \approx BP$ (Stated by Solovay [20].)

A subset of the Cantor space 2^ω is found without the Baire property. It can be transferred to a subset of the unit interval by the map taking binary decimals to their real numbers.

Lemma 1 Let $A \subset 2^\omega$ have meager (first category) symmetric difference with a non-empty open set. There are $a, b \in A$ such that b differs from $1-a$ at finitely many values.

Lemma 2 Let $A \subset 2^\omega$ be a co-meager ($2^\omega - A$ is meager). There are $a, b \in 2^\omega$ such that $a, a \cdot b, a \cdot (1-b) \in A$.

The lemmas quickly lead to a set without the Baire property. Let μ be the measure on 2^ω (identifying subsets of ω with their characteristic functions). Neither $\{a \in 2^\omega : \mu(a) < 1/2\}$ nor $\{a \in 2^\omega : \mu(a) > 1/2\}$ differ from a non-empty open set by a meager set. This is so because the a and b of Lemma 1 would satisfy

$$\mu(a) + \mu(b) = \mu(a) + \mu(1-a) = 1,$$

a contradiction. Thus one of the above 2 sets hasn't the Baire property or both are meager. In the latter case $\{a \in 2^\omega : \mu(a) = \frac{1}{2}\}$ is comeager and for the a and b of Lemma 2:

$$1/2 = \mu(a) = \mu(a \cdot b) + \mu(a \cdot (1-b)) = 1/2 + 1/2 = 1,$$

a contradiction.

It therefore remains only to prove the lemmas. Lemma 2 is considered first. If $f \in 2^n$, $n \in \omega$ we let $U(f)$ denote the basic clopen subset of 2^ω consisting of the extensions of f to ω.

Let $A \subset 2^\omega$ be given a co-meager set. There are open dense sets V_n, $n \in \omega$, such that $A \supset \bigcap_{i\in\omega} V_i$. We define sequences f_n, g_n, $n \in \omega$ such that:

(i) For each $n \in \omega$ there is an $m > n$ in ω such that f_n, $g_n \in 2^m$.

(ii) If $m \geq n$ then f_m extends f_n and g_m extends g_n.

(iii) For each $n \in \omega$ there are $j, k, \ell \geq n$ such that

$$U(f_j) \subset \bigcap_{i\leq j} V_i$$

$$U(f_k \cdot g_k) \subset \bigcap_{i\leq k} V_i$$

$$U(f \cdot (1-g\ )) \subset \bigcap_{i\leq \ell} V_i.$$

Lemma 2 follows immediately from (i), (ii), and (iii). Let a be the limit of the f_n and b be the limit of g_n. These will be well defined members of 2^ω by (i) and (ii). We show that, e.g., $a \cdot b \in A$. For any $n \in \omega$

$$a \cdot b \in U(f_k \cdot g_k) \subset \bigcap_{i\leq k} V_i \subset V_n$$

for the k of (iii). Thus $a \cdot b \in \bigcap_{n\in\omega} V_n \subset A$.

f_n and g_n are defined inductively in 3 cases.

<u>Case 1</u> <u>$n \equiv 0 \bmod 3$</u> f_n is defined to be the least (in some ordering) extension of f_{n-1} such that $n \in \mathrm{Dom}\, f_n$ and $U(f_n) \subset \bigcap_{i\leq n} V_i$. f_n exists since $\bigcap_{i\leq n} V_i$ is dense and open. g_{n-1} is extended to g_n by letting $g_n(j) = 1$, $\mathrm{Dom}\, g_{n-1} < j < \mathrm{Dom}\, f_n$.

<u>Case 2</u> <u>$n \equiv 1 \bmod 3$</u> Let h be the least extension of $f_{n-1} \cdot g_{n-1}$

with $n \in \text{Dom } h$ and $U(h) \subset \bigcap_{i \leq n} V_i$. Extend f_{n-1} and g_{n-1} to f_n and g_n by letting $f_n(j) = 1$ and $g_n(j) = h(j)$ for $\text{Dom } f_n \leq j \leq \text{Dom } h$. Our choice guarantees $f_n \cdot g_n = h$ so $U(f_n \cdot g_n) \subset \bigcap_{i \leq n} V_i$.

Case 3 $n \equiv 2 \bmod 3$ h is the least extension of $f_{n-1} \cdot (1-g_{n-1})$ with $n \in \text{Dom } h$ and $U(h) \subset \bigcap_{i \leq n} V_i$. f_{n-1} and g_{n-1} are extended via $f_n(j) = 1$, $g_n(j) = 1-h(j)$ for $\text{Dom } f_{n-1} < j < \text{Dom } h$. Our choice guarantees $U(f_n \cdot (1-g_n)) = U(h) \subset \bigcap_{i \leq n} V_i$.

(i) and (ii) are immediate. (iii) follows since j, k, and ℓ can be any numbers $\geq n$ with $j \equiv 0$, $k \equiv 1$, and $\ell \equiv 2 \bmod 3$.

Lemma 1 is similar and simpler. The hypotheses guarantee $A \supset U(f_o) \cap \bigcap_{n \in \omega} V_n$ for some finite f_o and open, dense V_n. a and b are found as limits of f_n and g_n respectively where $f_o = g_o$ and $U(f_n)$ and $U(g_n)$ are alternately in $U(f_o) \cap \bigcap_{i \leq n} V_i$.

II. ZF Independence of the Prime Ideal Theorem from the Hahn Banach Theorem

The forthcoming argument was sketched in [17]. There the model was obtained by appeal to a general theorem of [14] concerning embeddings of Fraenkel-Mostowski models in Cohen models. Here we construct the model directly using Cohen's notion of generic set, as discussed in [3], [19], or [20]. This should make our argument self contained and provide an example of the general embedding theorem. The theorems of [14] found application in [15] where metatheorems were given on the transfer of Fraenkel-Mostowski consistency results to ZF set theory. None of these metatheorems are strong enough to handle HB. Thus, at present, there is no way to avoid direct consideration of a Cohen model.

The remainder of section II is subdivided as follows. The model is constructed in IIA. Those properties of the model which will be needed in the sequel are listed in Theorem 2.1. IIB contains set theoretic proofs of ~PI, ~KM, and HB based on Theorem 2.1.

IIA. Construction of the Model

Let L denote Godel's constructible universe.[1] In L let $\mathbb{P}$ denote the set of functions from countable subsets of ω_1[2] to $\{0,1\}$ partially ordered under extension. $D \subset \mathbb{P}$ is dense if any $f \in \mathbb{P}$ has an extension in D. $g \subset (\omega_1)^L$ (but not in L) is generic if it (i.e. its characteristic function) extends some member

---

[1]With some modification we could use any universe with AC here.

[2]With some additional difficulty we could use finite functions from subsets of ω here. We would thus obtain the 4th of Cohen's original models of ZF set theory [2]. His models therefore satisfy: Model 1; GCH but not V = L, Model 2; AC but not GCH; Model 3; PI but not AC, Model 4; HB but not PI.

of every dense $D \subset \mathbb{P}$ such that $D \in L$. It is consistent, [19], that a generic subset of ω_1 exists.

We begin in a universe, $L[g]$, of sets constructible from a generic g. From [19] we note the following facts. L and $L[g]$ have the same ordinal class, O_n, and the same cardinal function $\alpha \to \omega_\alpha$. Both universes satisfy GCH, hence AC, and if $a \in L$ and $b \subset a$ is countable in $L(g)$ then $b \in L$ and is countable there. In particular $\mathbb{P}$ and $\mathbb{R}$ (the real numbers) have the same meaning in L and $L(g)$. In the sequel it will be understood that such terms as ω_α, $\mathbb{P}$, or $\mathbb{R}$ are unambiguous when relativized to either L, $L(g)$, or, for that matter, any universe M with $L \subset M \subset L(g)$.[3]

The following fact about generic sets, [19], is crucial in obtaining universes $M \subset L(g)$ in which AC fails.

(2.1) Let $\Phi(x_1,\ldots,x_{n_0} y)$ be a formula of set theory with exactly $x_1,\ldots,x_n,y$ free. If $\alpha_1,\ldots,\alpha_n \in O_n$ and $\Phi(\alpha_1,\ldots,\alpha_n,g)$ is true in $L(g)$ then there is some $f \in \mathbb{P}$ such that g extends f and $\Phi(\alpha_1,\ldots,\alpha_n,g')$ is true in $L(g)$ where g' is <u>any</u> generic set extending f such that $L(g') = L(g)$.

Let φ denote a fixed, constructible, 1:1, onto function, $\omega_1 \times \omega_1 \times \omega \times 2 \to \omega_1$. For $(i,j,k) \in \omega_1 \times \omega \times 2$ define $g_{ijk} \subset \omega_1$ by its characteristic function: $g_{ijk}(\alpha) = g(\varphi(\alpha,i,j,k))$. g is thus the result of "interlacing" the g_{ijk}. Also define $a_{jk} = \{g_{i,j,k} : i \in \omega\}$, $P_j = \{a_{j,0}, a_{j,1}\}$ and $P = \{\{j,P_n\}\ j \in \omega\}$. The following remarks follow from the genericity of g.

(2.2) The g_{ijk} are distinct.

(2.3) For any $f \in \mathbb{P}$ and indices $(j,k) \in \omega \times 2$ there are

---

[3] A <u>universe</u> is a transitive model of ZF set theory containing all ordinals.

uncountably many $i \in \omega_1$ such that g_{ijk} extends f.

In the next remark we identify $2 = \{0,1\}$ with the group of integers mod 2. The group 2 has a unique nontrivial action on the set 2.

(2.4) Let $\tau \in 2^{\omega}$.[4] Let σ be a constructible permutation of the (i,j,k) such that $\sigma(i,j,k) = (i',j'.k') \rightarrow j' = j \wedge k' = (\tau(j))(k)$. Let g^{σ} be obtained by interlacing the $g_{\sigma(1,j,k)}$ (i.e. $g^{\sigma}(\alpha) = g_{\sigma(i,j,k)}(\beta)$ where $\varphi(\beta,i,j,h) = \alpha$). Then g^{σ} is generic, $L(g^{\sigma}) = L(g)$, and P can be defined from g^{σ} exactly as it is defined from g.

To prove (2.2)-(2.4) define, for $f \in \mathbb{P}$, the function $f_{ijk} \in \mathbb{P}$ via $f_{ijk}(\alpha) = f(\varphi(\alpha,i,j,k))$ for $\varphi(\alpha,i,j,k) \in$ Domain f ($f_{ijk}(\alpha)$ is undefined for $\varphi(\alpha,i,j,k) \notin$ Domain f).

(2.2) follows from the fact that for $(i,j,k) \neq (i',j',k')$)
$\{f \in \mathbb{P}: f_{ijk}(\alpha) \neq f_{i',j',k'}(\alpha)$, some $\alpha \in$ Domain $f_{ijk} \cap$ $\cap$ Domain $f_{i'j'k'}\}$ is in L and dense in $\mathbb{P}$.

(2.3) follows from the fact that for $f \in \mathbb{P}$, $(j,k) \in \omega \times 2$, and $\gamma \in \omega_1$,

$$\{f' \in \mathbb{P}: (\exists i \in \omega_1)[i \geq \gamma \wedge f'_{ijk} \text{ extends } f]\}$$

is in L and dense in $\mathbb{P}$.

In (2.4) we notice that for any constructible permutation σ of the (i,j,k) we have $L(g) = L(g^{\sigma})$ (g and g^{σ} can be defined from each other using the parameter $\sigma \in L$). Also g^{σ} is generic since a constructible dense subset of $\mathbb{P}$, one of whose elements is to be extended by g^{σ}, can be mapped via $\sigma^{-1}(\sigma^{-1}(f)$ is obtained by interlacing f_{ijk} in the $\sigma^{-1}(i,j,k)$th position) to a dense

---

[4] ${}^{A}B$ denotes the set of functions from B to A.

constuctible subset of $\mathbb{P}$, one of whose elements will be extended by g. An examination of the stipulations put on σ in (2.4) will reveal that when defined from g^σ, a^σ_{jk} may either be a_{jk} or $a_{j(1-k)}$. The indexing of the elements of a_{jk} may also be changed but each P_j and P itself will be the same defined from g^σ or g.

(2.5) <u>Definition</u>. The model M is $L(\overline{P})$, the subuniverse of L(g) generated by L and $\overline{P} = \{P\} \cup P \cup \bigcup P \cup \bigcup\bigcup P \cup \ldots$ [5]

M can be explicitly described by setting $M_0 = \overline{P}$, $M_{\alpha+1} = \{A \subset M_\alpha : A$ is first-order definable over M_α from parameters in $M_\alpha\}$, $M_\alpha = \bigcup_{\beta<\alpha} M_\beta$ for limit α, and $M = \bigcup_{\alpha \varepsilon O_n} M_\alpha$. We now state the main theorem of this section.

2.1 <u>Theorem</u>. The following statements hold in M.

1) The axioms of Z F set theory.

2) There is a function P with domain ω such that each P_j (shorthand for P(j)) is a disjoint pair of sets. Other properties of P are given below.

3) There is a relation $n \nabla x$ (read $x \varepsilon \nabla n$) definable from P and satisfying:

   a) Every set is in some ∇n, $n \varepsilon \omega$.

   b) Each ∇n contains all ordinals, P, and the transitive closure of $\bigcup_{j<n} P_j$.[6]

   c) If y is definable from parameters in ∇n then $y \varepsilon \nabla n$.

   d) If $m \leq n$ then $\nabla m \subset \nabla n$ and $\nabla n \cap P_n = \emptyset$.

   e) $\mathbb{R} \subset \nabla 0$.

---

[5] $\overline{X} = \{X\} \cup X \cup \bigcup X \cup \bigcup\bigcup X \cup \ldots$ is called the <u>transitive closure</u> of X.

[6] <u>Theorem 2.1</u> has more clauses than the list in [17]. The additional properties are useful in our proof of HB but are essential only in our proof of ~KM.

4) There is a function $T(Q, B)$ definable from P such that if Q is a choice function for $\{P_j\}_{j<n}$ and B is a Boolean algebra with $B \in \nabla n$ and Field $B \subset \nabla n$ then $T(Q, B)$ is a prime ideal for B.[7]

5) a) There is a 1:1 function $Z \in \nabla_0$ with Domain $Z = \omega$ and $\{Z_i\}_{i\in\omega}$ is a disjoint partitioning of $\cup P_0$.

b) $\cup P_0$ has no countable subset.

The rest of IIA is devoted to a proof of Theorem 2.1. The axioms of ZF are verified in M exactly as they are verified in L. The function P in 2) has already been defined. $\{j, P_j\}$ can be interpreted as an ordered pair since $j \in \omega$ and $P_j \notin \omega$. The properties of P stipulated in 2) are easily seen to be absolute[8] and they hold in L(g).

To verify the remainder of Theorem 2.1 we set up a combined support structure on M (see [14]). We first remark that the function $\alpha \to M_\alpha$ is absolute and that in M the universe is $L(\overline{P})$. The following definition is in M but absolute.

(2.6) Definition. Let $n \in \omega$ and let G be a finite subset of $\bigcup_{j<n} \cup P_j$. Define $d^\alpha(G,n) \subset M_\alpha$ as follows.

---

[7]Our proof applies in universes satisfying the combined support theory of [14] Theorem 4.2. We have there:

"There is a function $t(H, B)$, definable from the support structure, such that if $H \in \mathcal{H}$ well orders the support G and B is Boolean algebra with $B \in \nabla_G$ and Field $B \subset \nabla_G$ then $t(H, B)$ is a prime ideal for B ."

[8]If M,N are universes and $M \subset N$ we say that a property (statement or term) is absolute from M to N if it means the same thing relative to each. We omit reference to M and N if they are clear in context (Here $N = L(g)$.).

$$d^0(G,n) = \omega_1 \cup G \cup \bigcup_{j<n} \cup P_j \cup \{P_k\}_{k\varepsilon\omega} \cup \{k,P_k\}_{k\varepsilon\omega} \cup \{P\}.$$

$$d^{\alpha+1}(G,n) = \{A \varepsilon M^{\alpha+1}: A \text{ satisfies a first order definition over } M_\alpha \text{ with parameters in } d^\alpha(G,n)\}$$

$$d^\alpha(G,n) = \bigcup_{\beta<\alpha} d^\beta(G,n) \quad \text{for limit } \alpha.$$

Now let $d(G,n) = \bigcup_{\alpha\varepsilon O_n} d^\alpha(G,n)$. Note that P is the only parameter of our definition. Finally let

$$\nabla_n = \{x:(\exists G \subset \bigcup_{j<n} \cup P_j)[x \varepsilon d(G,n)]\}.$$

We can quickly verify all the clauses of <u>Theorem 2.1</u> except 3)d), 4) and 5)b). 3)a) follows from an inductive argument that every element of M_α is in some $d^\alpha(G,n)$. Similarly $O_n \subset d(G,n)$ all G,n. Since P, each P_j, each $\{j,P_j\}$, and each element of $\bigcup_{j<n} P_j$ and G are defined to be in $d^0(G,n)$ we have 3)b). 3)c) holds since a definition relative to M can be further relationized to some M_α by the reflection principle. This immediately shows that each $d(G,n)$ is closed under definability. ∇n is closed under definability since parameters from $d(G_1,n),\ldots,d(G_k,n)$ are all in $d(\bigcup_{i=1}^{k} G_i,n)$. By induction it is clear that $m \leq n \rightarrow d^\alpha((G,m) \subset d^\alpha(G,n)$ and the first part of 3)d) follows. We have already mentioned that $L \subset M \subset L(g)$ so $\mathbb{R}$ and $\mathbb{P}$ are absolute. It follows that their elements are definable from ordinal parameters alone thus $\mathbb{R} \subset \nabla n$, proving 3)e), and also $\mathbb{P} \subset \nabla_0$. To obtain 5)a) let $f_1,f_2,\ldots$ be a constructible sequence of elements of $\mathbb{P}$ such that no two of them have a common extension, e.g. $f_1 = \langle 0,1\rangle$, $f_2 = \langle 0,0,1\rangle$, $f_3 = \langle 0,0,0,1\rangle$ etc. Define:

$$Z_i = \{g \varepsilon \cup P_0 : g \text{ extends } f_i\} \quad Z_0 = \cup P_0 - \bigcup_{i=1}^{\infty} Z_i.$$

Each Z_i is nonempty by (2.3) and $Z \in \nabla_0$ by 3)c).

3)d) and 5)b) are consequences of the following 2 lemmas.

<u>2.2 Lemma</u>. Let $\Phi(w,x_1,\ldots,x_m,y_1,\ldots,y_n)$ be a formula of set theory with ordinal parameters. Suppose $\Phi(P,g_1,\ldots,g_m,a_1,\ldots,a_n)$ holds in M where the a_r are in $\cup\cup$Range P, each g_q is in some (unique) a_r, and the g_q and a_r are distinct.

Then there are $f_1,\ldots,f_n \in \mathbb{P}$ such that g_q extends f_q and for <u>any</u> distinct $g_1',\ldots,g_n',a_1',\ldots,a_n'$ with $a_r' \in P_j \leftrightarrow a_r \in P_j$ and $g_q' \in a_r' \leftrightarrow g_q \in a_r$ we have that $\Phi(P,g_1',\ldots,g_m',a_1',\ldots,a_n')$ holds in M.

<u>Proof</u>. In L(g) we can express $g_q = g_{i_q j_q k_q}$ and $a_r = a_{j_r^* k_r^*}$. Let Ψ be the statement in L(g) which says "$\Phi(P,g_{i_1 j_1 k_1},\ldots,a_{j_1^* k_1^*},\ldots,)$ holds in $L(\overline{P})$." By (2.1) there is an $f \in \mathbb{P}$ such that if g' extends f, g' is generic, L(g') = L(g), and P is defined from g' exactly as it is defined from g ; "$\Phi(P,g'_{i_1 j_1 k_1'},\ldots,a'_{j_1^* k_1^*},\ldots)$ holds in $L(\overline{P})$." is true in L(g).

We let $f_q = f_{i_q j_q k_q}$ $q = 1,\ldots,m$. Since g extends f it is clear that $g_q = g_{i_q j_q k_q}$ extends $f_q = f_{i_q j_q k_q}$. If $\{g_q'\}_{q=1}^n$ and $\{a_r'\}_{r=1}^n$ are as in the conclusion of the lemma it remains to find a generic g' such that L(g') = L(g), P is defined from g' exactly as from g, $a'_{j_r^* k_r^*} = a_r'$, and $g'_{i_q j_q k_q} = g_q'$, $q = 1,\ldots,m$. This will be done by (2.4).

τ is defined by noting that a_r' is $a_{j_r^{**} k_r^{**}}$ where $j_r^{**} = j_r^*$ and k_r^{**} is either k_r^* or $1-k_r^*$. Set

$$\tau(j_r^*) = \begin{cases} 0 & \text{if } k_r^{**} = k_r^* \\ 1 & \text{if } k_r^{**} = 1-k_r^* \end{cases}$$

$\tau(j_r^*)$ is well defined since if $j_r^* = j_s^*$, $r \neq s$ we can only have $\{a_{j_r^*k_r^*}, a_{j_r^*k_s^*}\} = P_{j_r^*}$. Extend the above definition to ω by letting $\tau(j)$ be arbitrary for $j \neq j_r^*$, any r.

σ is first defined by

$$\sigma(i_q,j_q,k_q) = (i_q',j_q,(\tau(j_q)(k_q)).$$

To extend σ first enumerate the countably many other triples (i,j,k) with $f_{ijk} \neq \emptyset$. Inductively set

$$\sigma(i,j,k) = (i',j,(\tau(j))(k))$$

where $(i',j,\tau(j)(k))$ is one of the uncountably many, by (2.3), triples which has not yet been put in the range of σ. Thus far σ is a countable subset of L, hence it's in L. For each (j,k) there are ω_1 triples (i,j,k) on which σ hasn't been defined and ω_1 triples $(i',j,\tau(j)(k))$ not in the range of σ. We extend σ by a constructible choice of 1:1 onto maps from the (i,j,k) to the $(i',j,\tau(j)(k))$ as yet unused.

It is now immediate to check that g^σ has all the properties we required of g'. Since $g'_{i_qj_qk_q} = g_q'$, $a'_{j_r^*k_r^*} = a_r'$, we have $\Phi(P,g_1',\ldots,g_m',a_1',\ldots,a_n')$ true in M.

<u>2.3 Lemma</u>. There is a function $t(G,Q,\alpha)$ definable from P such that if Q is a choice function on $\{P_j\}_{j<n}$ and $G \subset \bigcup_{j<n} \bigcup P_j$ is finite, then $\{(\alpha,t(G,Q,\alpha)):\alpha \in O_n\}$ is 1:1 and onto $d(G,n)$.

Proof. We argue in M.

Given G and Q we define a well ordering of $d^O(G,n)$. The elements of ω_1 are first, in their natural order, followed by the elements of G, in their order as elements of 2^{ω_1}. Next are the elements of each P_j, $j < n$, with $Q(P_j)$ listed before the other element of P_j, next the P_j themselves, in order of increasing j, the $\{j, P_j\}$ in order of increasing j, and finally P itself. The well ordering is inductively extended to $d^{\alpha}(G,n)$ by "least definition" as in Godel's well ordering of L. Each $d^{\alpha}(G,n)$ is a set so every segment of the ordering on all of $d(G,n)$ extending the orderings on $d^{\alpha}(G,n)$ is a set. We have thus defined a type O_n ordering of $d(G,n)$. Let $t(G,Q,\alpha)$ be the $\alpha^{\underline{th}}$ element of this ordering.

<u>Proof of 3)d</u>) We have remarked $m \leq n \rightarrow \nabla_m \subset \nabla_n$. Assume $a \in \nabla_m \cap P_n$. Let Q be a choice function for $\{P_j\}_{j<m}$. Since $a \in \nabla_m$ there are finite $G \subset \bigcup_{j<m} \cup P_j$ and $\alpha \in O_n$ such that $a = t(G,Q,\alpha)$. t is defined from P and Q is definable from the elements of $\bigcup_{j<m} P_j$ (by designating those which are in the range). Thus "$a = t(G,Q,\alpha)$" is a statement $\Phi(P,g_1,\ldots,g_k,a_1,\ldots,a_{2(m-1)},a)$ where $G = \{g_1,\ldots,g_k\}$ and $\bigcup_{j<m} P_j = \{a_1,\ldots,a_{2(m-1)}\}$. No element of G is in a so by <u>Lemma 2.2</u> $\Phi(P,g_1,\ldots,g_k,a_1,\ldots,a_{2(m-1)},a')$ is true where $\{a,a'\} = P_n$ (<u>Lemma 2.2</u> says we can replace the g_q's by g_q''s extending f_q and can permute the a_r's consistently with this. In particular we can leave the g_q's alone and $a_1,\ldots,a_{2m}$ alone). This gives us $a' = t(G,Q,\alpha)$ and $a = t(G,Q,\alpha)$ contradicting our assertion that t is a function. Therefore $a \notin \nabla_m$ hence $\nabla_m \cap P_n = \Phi$.

<u>Proof of 5)b</u>) Assume $\cup P_0$ has a countable subset. Again for some n, some choice function Q for $\{P_j\}_{j<n}$, some finite $G \subset \bigcup_{j<n} \cup P_j$, and some $\alpha \in O_n$ we have that $t(G,Q,\alpha)$ is a 1:1 function $\omega \rightarrow \cup P_0$. Since G is finite there must be a $g \in \text{Range } t(G,Q,\alpha) - G$. Thus for some $m \in \omega$ "$t(G,Q,\alpha)(m) = g$" is true in M. This sentence can be thought of as $\Phi(P,g_1,\ldots,g_k,g,a_1,\ldots,a_2\ )$ where $G = \{g_1,\ldots,g_k\}$, $m = \text{Max}(n-1,1)$ and $\{a_1,\ldots,a_m\ \} = P_0 \cup \bigcup_{j<n} P_j$.

Renumber if necessary so that $g \in a_m$. Again by Lemma 2.2 there is some $f \in \mathbb{P}$ such that g extends f and if; g' extends f, $g' \notin G$, $g' \in a_1$, then $\Phi(P, g_1, \ldots, g_k, g', a_1, \ldots, a_2)$. Some $g' \neq g$ exists by (2.3). Thus $t(G,Q,\alpha)(m) = g'$ is also true and $t(G,Q,\alpha)$ is not a function, a contradiction.

It remains to prove 4). We adapt the main arguement of [7].

2.4 Lemma. Let $B \in d(G,n)$ be a Boolean algebra where Field $\subset \nabla n$. Let I be an ideal of B, maximal among those in $d(G,n)$ (Such an I exists since $d(G,n)$ is well orderable). I is then a prime ideal of B.

Proof. Fix an indexing $a_0, \ldots, a_k$ of the elements of $\bigcup_{j<n} P_j$. If I is not prime there is an $x \in$ Field B such that $x \notin I$ and $\sim x \notin I$. $x \in \nabla n$ so $x \in d(G',n)$ for some G'. By considering $G' \cup G$ we can assume, without loss of generality, that $G' \supset G$. The elements of $(G'-G) \cap a_r$ can be indexed $g_1^r, \ldots, g_{s(r)}^r$. Since the a_r are disjoint, the g_t^r are distinct.

For some choice function Q on $\{P_j\}_{j<n}$, $\alpha, \beta \in O_n$ we have $I = t(G,Q,\alpha)$, $x = t(G',Q,\beta)$. Q and β will be fixed in the following discussion thus for any G^* we will denote $t(G^*,Q,\beta)$ by $x(G^*)$.

Consider the statement:

$$\text{"}x(G') \in \text{Field } B \wedge x(G') \notin t(G,Q,\alpha) \wedge \sim x(G') \notin t(G,Q,\alpha).\text{"}$$

We obtain from Lemma 2.2:

(2.4.1) There exist $f_t^r \in \mathbb{P}$, $r = 1, \ldots, k$, $t = 1, \ldots, s(r)$ such that for any $G^* = \{g_t^r\}_{t-1,\ldots,s(r)}^{r-1\ldots k} \cup G$ where g_t^r extends f_t^r we have $x(G^*) \in$ Field B, $x(G^*) \notin I$, and $\sim x(G^*) \notin I$.

Without loss of generality each f_t^r of (2.4.1) can be replaced by an extension. In particular we may assume that distinct f_t^r's take different values on some element of their common domain.

Let $H \subset \cup\cup$ Range P. If $F \subset \mathbb{P}$ we let $H(F)$ denote the set of extensions of elements of F in H. If f_t^r, $r = 1,\ldots,k$, $t = 1,\ldots,s(r)$ are as in (2.4.1), F contains exactly one extension of each f_t^r, H contains exactly one extension of each element of F, and $g \supset f \supset f_t^r \rightarrow g \in a_r$ for $g \in H$, $f \in F$ then $H(F)$ satisfies the hypotheses on G^* in (2.4.1). Thus under these circumstances $x(H(F)) \in B$, $x(H(F)) \notin \mathbb{I}$, and $\sim x(H(F)) \notin \mathbb{I}$.

(2.4.2) Let f_t^r, $1 \leq r \leq k$, $1 \leq t \leq s(r)$ be as in (2.4.1). There are finite sets $K_t^r \subset \mathbb{P}$ such that:

(a) K_t^r has at least 2 elements.

(b) $f \in K_t^r \rightarrow f$ extends f_t^r. $f \neq f'$, $f,f' \in K_t^r \rightarrow \exists \alpha \in \omega_1$ $[f(\alpha) \neq (\alpha)]$.

(c) Let χK denote the set of choice sets from the K_t^r. Let $H \subset \cup\cup$ Range P have exactly one extension of each $f \in \bigcup_{r,t} K_t^r$ and for $g \in H$, $g \supset f \in K_t^r \rightarrow g \in a_r$. Then

$$\bigwedge_{F \in \chi(K)} x(H(F)) \in \mathbb{I}, \qquad \bigwedge_{F \in \chi(K)} \sim x(H(F)) \in \mathbb{I}.$$

<u>Proof of (2.4.2)</u>. It suffices to find a finite $H \subset \cup\cup$ Range P such that:

(i) Each f_t^r is extended by at least 2 elements of H.

(ii) Each element of H extends a (unique) $f_t^r \in H$. If $g \in H$ and g extends f_t^r then $g \in a_r$.

(iii) Let $\mathbb{M}$ denote the set of $G^* \subset H$ where G^* contains exactly one extension of each f_t^r. Then:

$$\bigwedge_{G^* \in \mathbb{M}} x(G^*) \in \mathbb{I}, \qquad \bigwedge_{G^* \in \mathbb{M}} \sim x(G^*) \in \mathbb{I}$$

The existence of the K^r_t follows from the existence of H by Lemma 2.2. To obtain H we first find a finite H_1 satisfying ii) and the first equation of iii). Let $\mathbb{J}$ be the ideal generated by $\mathbb{I}$ and

$$R = \{\sim x(G^*) : G^* = \{g^r_t\}_{t=1,\ldots,s(r)}^{r=1,\ldots,k} \cup G,\ g^r_t \in a_r, g^r_t \text{ extends } f^r_t\}.$$

$\mathbb{J} \in d(G,n)$ and $\mathbb{J} \supset \mathbb{I}$ so $\mathbb{J}$ must contain 1. Thus for some $G^*_1,\ldots,G^*_m \in R$ and some $y \in \mathbb{I}$ we have:

$$1 \leq \sim x(G^*_1) \vee \ldots \vee \sim x(G^*_m) \vee y \leq 1$$
$$0 = x(G^*_1) \wedge \ldots \wedge x(G^*_m) \wedge y$$
$$x(G^*_1) \wedge \ldots \wedge x(G^*_m) \leq y$$
$$x(G^*_1) \wedge \ldots \wedge x(G^*_m) \in \mathbb{I}$$

Let $H_1 = G^*_1 \cup \ldots \cup G^*_m$. ii) is clearly satisfied since $G^*_1 \ldots G^*_m \in R$. The first equation of iii) is satisfied since the larger intersection includes $x(G^*_1) \wedge \ldots \wedge x(G^*_m)$.

A similar arguement shows the existence of an H_2 satisfying ii) and the second equation of iii). Let $H^* = H_1 \cup H_2$ and let H be an extension of H^* satisfying i). H^* clearly satisfies ii) and it satisfies iii) since the intersections there will contain intersections over H_1 and H_2. This concludes the proof of (2.4.2).

We now define finite ${}_iS^r_t \subset \mathbb{P}$, $i \in \omega$, $S^r_t = \{f^r_t\}$. We inductively assume each ${}_{i-1}S^r_t$ is defined. Let $\chi_i S$ denote the family of choice sets from the ${}_iS^r_t$. If $F \in \chi_{(i-1)}S$ its elements inductively satisfy the hypotheses of (2.4.1). By (2.4.2) we obtain sets ${}_F K^r_t$ with properties listed in (2.4.2) for each F_1 set ${}_iS^r_t = \bigcup_{F \in \chi_{(i-1)}S} {}_F K^r_t$. Our definition of ${}_iS^t_r$ apparently used the axiom of choice. We avoid this however by letting ${}_F K^r_t$ be the least set satisfying (2.4.2) in the natural well ordering of L.

Let $S_t^r = \bigcup_{i\in\omega} {}_iS_t^r$. Partially order S_t^r by extension of functions. We immediately check that S_t^r is a finitistic tree in L with at least binary branching.[9] ${}_iS_t^r$ is the ith level of S_t^r. The S_t^r are disjoint. If $K_t^r \subset S_t^r$ we let $\chi(K_t^r)$ be the set of choice sets from the K_t^r. We now cite:

<u>Theorem</u> (Halpern-Lauchli [6]). There is a level q such that for any disjoint partitioning $\chi({}_qS_t^r) = I \cup II$ there is an $m < q$ and an $F \in \chi({}_mS_t^r)$ such that:

> Let K_t^r be the set of immediate successors of the element of $F \cap {}_mS_t^r$. For each $h \in K_t^r$ there exists $h' \in {}_qS_t^r$ extending h such that if
>
> $K'^r_t = \{h' : h \in K_t^r\}$ then either $\chi(K'^r_t) \subset I$ or $\chi(K'^r_t) \subset II$.

Let q be as in the Halpern-Lauchli theorem. Let $H \subset \bigcup\bigcup \text{Range } P$ be a finite set consisting of exactly one extension of each element of each ${}_qS_t^r$, this extension being in a_r. If $F \in \chi({}_qS_t^r)$ we let $H(F)$ be the set of extensions of elements of F in H (as before).

In **B** we can write, using the distributive associative and commutative laws of **B** :

$$(2.4.3) \qquad 1 = \bigwedge_{F\in\chi({}_qS_t^r)} (x(H(F)) \vee \sim x(H(F)))$$

$$= \bigvee_{\substack{I\cup II=\chi({}_qS_t^r)\\ I\cap II=\emptyset}} \Big(\bigwedge_{F\in I} x(H(F)) \wedge \bigwedge_{F\in II} \sim x(H(F))\Big)$$

---

[9] The set of predecessors of each element is finite and totally ordered. The set of immediate successors of each element has at least two elements and is finite. The <u>i^{th} level</u> denotes the set of elements with i predecessors.

For each disjoint partitioning $\chi({}_qS^r_t) = I \cup II$ we have by the theorem a $m < q$, an $F_m \in \chi({}_mS^r_t)$, and an $K'^r_t \subset {}_qS^r_t$ such that $\chi(K'^r_t) \subset I$ or $\chi(K'^r_t) \subset II$ and each immediate successor of an element of F_m is extended by an h' in some K'^r_t. It follows that the K'^r_t satisfy the conclusions of (2.4.2) with respect to F_m. Thus

$$\bigwedge_{F\in\chi(K'^r_t)} x(H(F)) \in \mathbb{I}, \qquad \bigwedge_{F\in\chi(K'^r_t)} \sim x(H(F)) \in \mathbb{I} .$$

$$\bigwedge_{F\in I} x(H(F)) \wedge \bigwedge_{F\in II} \sim x(H(F))$$

contains one of these products so, since $\mathbf{I}$ is an ideal,

$$\bigwedge_{F\in I} x(H(F)) \wedge \bigwedge_{F\in II} \sim x(H(F)) \in \mathbf{I} .$$

Using (2.4.3) this gives the conclusion $1 \in \mathbf{I}$, contradicting our assumption that $\mathbf{I}$ is not prime.

__2.5 Lemma__. $d(G_1,n) \cap d(G_2,n) = d(G_1 \cap G_2,n)$.

__Proof__. Clearly $d(G_1,n) \cap d(G_2,n) \supset d(G_1 \cap G_2,n)$. For the converse inclusion fix a choice function Q for $\{P_j\}_{j<n}$ and index the elements a of $\bigcup_{j<n} P_j$ according to the Q-induced ordering of $d(\Phi,n)$.

If $x \in d(G_1,n) \cap d(G_2,n)$ for appropriate $\beta,\gamma \in O_n$ we have $t(G_1,Q,\beta) = x = t(G_2,Q,\gamma)$. Write $G_1 = \{g^r_t\}^{r=1,\dots,k}_{t=1,\dots,s(r)} \cup (G_1 \cap G_2)$ where $g^r_t \in a_r$ and $s(r)$ is the number of elements of $(G_1 - G_2) \cap a_r$. From __Lemma 2.2__ there are $f^r_t \in \mathbb{P}$ such that g^r_t extends f^r_t and for __any__ $G' = \{g'^r_t\}^{r=1,\dots,k}_{t=1,\dots,s(r)}$ with g'^r_t extends f^r_t and $g'^r_t \in a_r$ we have

$$t(G' \cup (G_1 \cap G_2), Q, \beta) = t(G_2\ Q, \gamma) = x.$$

Thus $x = \cup\{t(G' \cup (G_1 \cap G_2), Q, \beta) : G' = \{g'^r_t\}^{r=1,\ldots,k}_{t=1,\ldots,s(r)}, g'^r_t \supset f^r_t, g'^r_t \varepsilon a_r\}$.

This definition of x has parameters in $d(G_1 \cap G_2, n)$ so $x \varepsilon d(G_1 \cap G_2, n)$.

Proof of 4). If $B \varepsilon \nabla n$ is a Boolean algebra with Field $B \subset \nabla n$ and if Q is a choice function for $\{P_j\}_{j<n}$ we define $t(Q,\beta)$ as follows. Let G be the $G \subset \bigcup \bigcup_{j<n} P_j$ of minimal size with $B \varepsilon d(G,n)$. G is uniquely defined by Lemma 2.5. Let $t(Q,\beta)$ be the least ideal of B in $d(G,n)$ (least in the well ordering induced by Q) which is maximal among the ideals of B in $d(G,n)$. $t(Q,\beta)$ is prime by Lemma 2.4.

The proof of Theorem 2.1 is now complete. We conclude IIA with a note on the embedding of the Fraenkel model, [17], in M. Let $F_0 = \cup$ Range P, $F_{\alpha+1} = \{A \subset F_\alpha : (\exists n \varepsilon \omega)[A \varepsilon d(\phi, n)]\}$, and $F_\alpha = \bigcup_{\beta<\alpha} F_\beta$ for limit α. It is easily verified that $\bigcup_{\alpha \varepsilon On} F_\alpha$, with the restricted ε relation, is isomorphic to the Fraenkel model. This isomorphism is, of course, defined in $L(g)$, not in M.

IIB. Functional Analysis in the Model

We reason entirely in M and assume the results of Theorem 2.1 as set theoretic axioms.

PI is False

Proof. [17] PI implies the ordering theorem, [10], which implies the axiom of choice for pairs. If $\{P_j\}_{j \varepsilon \omega}$ had a choice function, g, it would be in some ∇_n. $g(P_{n+1})$ would be a member of P_{n+1} defined from the parameters g, P, and $n+1$ in ∇_n, a contradiction of 3)c) and d).

KM is False

Proof. Define the subuniverse N as follows. $N_0 = \{\phi\}$, $N_{\alpha+1} = \{A \subset N_\alpha : A \in \nabla_1\}$, $N_\alpha = \bigcup_{\beta<\alpha} N_\beta$ for limit α, and $N = \bigcup_{\alpha\in O_n} N_\alpha$. $O_n \subset \nabla_1$ and ∇_1 is closed under definability so N is easily seen to model ZF set theory. Boolean algebras in ∇_1 with fields $\subset \nabla_1$ have prime ideals in ∇_1 so PI holds in N. On the other hand, from 3)b) and 5), $\{Z_i\}_{i\in\omega}$ is in N and has no choice set because the Z_i are disjoint and $\bigcup_{i\in\omega} Z_i$ has no countable subset. This leads to a counterexample for KM in N since PI + KM → AC. We will see that this N-counterexample remains a counterexample in the universe.

The exact nature of the counterexample is shown in [1]. Consider the Banach spaces:

$$X = \Sigma_{M_i} \ell_1(Z_i), \qquad Y = \Sigma_{1_i} c_o(Z_i).$$ [10]

X is the dual space of Y and the existence of an extreme point in the unit ball of X is equivalent to the existence of a

---

[10] If X_i, $i \in I$ are Banach spaces let $\prod_{i\in I} X_i$ denote their product. If $x \in \prod_{i\in I} X_i$, x_i denotes the ith coordinate of x.

$\Sigma_{M_i} X_i = \{x \in \prod_{i\in I} X_i : \{\|x_i\|_i : i\in I\}$ is bounded$\}$. $\|x\| = \operatorname{Sup}_{i\in I} \|x_i\|_i$.

$\Sigma_{1_i} X_i = \{x \in \prod_{i\in I} X_i : \operatorname{Sup}_{\substack{y\subset I \\ y \text{ finite}}} \sum_{i\in y} \|x_i\|_i$ is finite$\}$. $\|x\|$ is this sup.

If x is a set:

$\ell_1(x) = \{\phi : x \xrightarrow{\phi} \mathbb{R}\ \operatorname{Sup}_{\substack{y\subset x \\ y \text{ finite}}} \sum_{x\in y} |\phi(x)|$ is finite$\}$. $\|\phi\|$ is this sup.

$c_o(x) = \{\phi : x \xrightarrow{\phi} \mathbb{R}\ (\forall\varepsilon > 0)\{x\in X : |\phi(x)| > \varepsilon\}$ is finite. $\|\phi\| = \operatorname{Max}_{x\in X} |\phi(x)|$.

choice function on $\{Z_i\}_{i\epsilon\omega}$.

If N^* is a subuniverse the superscript N^* will denote the relativization of a notion to N^*. We wish to remark that $X^{N^*} = X$ and $Y^{N^*} = Y$ for any N^* such that $N \subset N^*$.

Since PI holds in N there is by [10] a linear ordering of $\bigcup_{i\epsilon\omega} Z_i$. It follows that $\bigcup_{i\epsilon\omega} Z_i$ has no countable (infinite) family of finite subsets in any N^*, $N \subset N^*$. If $\{F_i\}_{i\epsilon\omega}$ were such a family we could define $g_n \epsilon \bigcup_{i\epsilon\omega} Z_i$ as the least (in the ordering of $\bigcup_{i\epsilon\omega} Z_i$) element of the nonempty $F_i - \{g_o,\dots,g_{n-1}\}$ where i is least. $\{g_n\}$ would be a countable subset of $\bigcup_{i\epsilon\omega} Z_i$.

Let $e(Z_i) = \{\Phi:Z_i \xrightarrow{\Phi} \mathbb{R} \wedge \{z\epsilon Z_i:\Phi(z) \neq 0\}$ is finite$\}$. We claim that in N^* and in the universe both $\ell_1(Z_i)$ and $c_o(Z_i)$ reduce (as sets) to $e(Z_i)$. Hence $\ell_1(Z_i)$ and $c_o(Z_i)$ are absolute ($e(Z_i)$ is clearly absolute since $\mathbb{R} \subset N \subset N^*$). $e(Z_i) \subset \ell_1(Z_i) \subset c_o(Z_i)$ from the definitions. $c_o(Z_i) \subset e(Z_i)$ since if $\Phi \epsilon c_o(Z_i)$ let $A_n = \{z \epsilon Z_i: \frac{1}{n+1} \leq \Phi(z) < \frac{1}{n}\}$. Only finitely many of the A_n are nonempty since otherwise we would have a countable family of finite subsets of Z_i. This shows $\Phi \epsilon e(Z_i)$.

We now claim that in the universe and N^* both $\sum_{i\epsilon\omega} {}_M e(Z_i)$ and $\sum_{i\epsilon\omega} {}_{\ell_1} e(Z_i)$ reduce (as sets) to $\sum_{i\epsilon\omega} e(Z_i) = \{x \epsilon \prod_{i\epsilon\omega} (Z_i):\{i:x_i \neq 0\}$ is finite$\}$. Thus $\sum_{i\epsilon\omega} {}_M e(Z_i)$ and $\sum_{i\epsilon\omega} {}_{\ell_1} e(Z_i)$ are absolute. In fact $\prod_{i\epsilon\omega} e(Z_i) = \sum_{i\epsilon\omega} e(Z_i)$. If $x \epsilon \prod_{i\epsilon\omega} e(Z_i)$ define $F_i = \{z\epsilon Z_i:x_i(z) \neq 0\}$. F_i is finite since $x_i \epsilon e(Z_i)$. No countable collection of finite subsets of $\bigcup_{i\epsilon\omega} Z_i$ exists so only finitely many F_i are nonzero. It follows that $x \epsilon \sum_{i\epsilon\omega} e(Z_i)$.

If $N \subset N^*$ and N^* is a subuniverse the following will be absolute from N^* to the universe: $\|x\|$ in X or Y, the dual space of Y (From [4] this will always be $\sum_{i\epsilon\omega} {}_M e(Z_i) = X$), the unit ball in X, the standard basis for the weak* topology on the unit ball of X, and the nonexistence of an extreme point on the unit ball of X. The unit ball of X is weak* compact in N since PI holds there.[11] We will have a counterexample to KM once we prove it compact in the universe. Some care is required since there may be open covers in the universe which do not exist in N.

To check compactness it suffices to consider open covers from a basis for the topology.[12] Let $\mathcal{C}$ be such an open cover. For some $n \epsilon \omega$ we have $\mathcal{C} \epsilon \nabla n$. Define N* via: $N_0^* = \{\emptyset\}$, $N_{\alpha+1}^* = \{A \subset N_\alpha^* : A \epsilon \nabla_1\}$, $N_\alpha^* = \bigcup_{\beta<\alpha} N^*_\beta$ for limit α, and $N^* = \bigcup_{\alpha\epsilon O_n} N_\alpha^*$. We have, as with N, that $N \subset N^*$, N* is a sub-universe, and PI holds in N*. $\mathcal{C} \subset N \subset N^*$, $\mathcal{C} \epsilon \nabla n$, so $\mathcal{C} \epsilon N^*$. $\mathcal{C}$ has a finite subcover since the unit ball of X is weak* compact in N*. Therefore the unit ball of X is weak* compact and is a counterexample to KM.

Our proof of HB requires some lemmas.

<u>2.6 Lemma</u>. Let X be a real vector space. Let $p: X \to \mathbb{R}$ be a function satisfying $p(x+y) \leq p(x) + p(y)$ and $p(\lambda x) = \lambda p(x)$ for $x,y \epsilon X$, $\lambda \geq 0$ in $\mathbb{R}$.

There is a function $E(f,m) \epsilon \nabla n$ such that if $m \geq n$ and $f \epsilon \nabla n$ is a linear functional defined on the subspace $Y \subset X$ satisfying $f(y) \leq p(y)$ on Y then $E(f,m)$ is a linear extension of f to $(X \cap \nabla m) + Y$ satisfying $E(f,m)(x) \leq p(x)$ there.

In the statement of <u>Lemma 2.6</u> the + denotes subspace summation.

---

[11] As remarked in [12] we only use the Tychanoff theorem for Hausdorff spaces.

[12] If is an arbitrary open cover consider the cover ' of <u>all</u> basic open subsets of elements of .

$X \cap \nabla m$ is a subspace since if $x,y \in X \cap \nabla m$ and $\lambda_1,\lambda_2 \in \mathbb{R}$ then the operations of X, λ_1, and λ_2 are all in ∇m so $\lambda_1 x+\lambda_2 y \in X \cap \nabla m$. Lemma 2.6 is proved by reducing it to Lemma 2.7 below.

2.7 Lemma. Let X,p, and n be as in Lemma 2.6.

There is a function $D(f,m,Q) \in \nabla n$ such that if f and m are as in Lemma 2.6 and Q is a choice function for $\{P_j\}_{j<m}$ then $D(f,m,Q)$ is a linear extension of f to $(X \cap \nabla_m) + Y$ satisfying $D(f,m,Q)(x) \leq p(x)$ there.

Reduction of Lemma 2.6 to Lemma 2.7

Fix a choice function Q_0 for $\{P_j\}_{j<n}$. Let $\mathbb{Q} = \{Q \supset Q_0 : Q$ is a choice function for $\{P_j\}_{j<m}\}$. We see that $\mathbb{Q} \in \nabla_n$ and $\mathbb{Q}$ has 2^{m-n} elements. Set

$$E(f,m) = \frac{1}{2^{m-n}} \sum_{Q \in \mathbb{Q}} D(f,n,Q).$$

It is immediate that $E \in \nabla n$ and $E(f,m)$ is linear on $(X \cap \nabla_m) + Y$. For $x \in (X \cap \nabla_m) + Y$,

$$E(f,m)(x) = \frac{1}{2^{m-n}} \sum_{Q \in \mathbb{Q}} D(f,n,Q) \leq \frac{1}{2^{m-n}} (2^{m-n} p(x)) = p(x).$$

Proof of Lemma 2.7.

Our use of nonstandard analysis occurs here. It is desirable to avoid the use of higher order language but it will be clear that our arguement is based on a nonstandard proof of HB similar to those of [11] and [12].

Let f,m and Q be given as in the conclusion of Lemma 2.7. Our object is to explicitly define $D(f,m,Q)$ from these and other parameters in ∇n. We recall in particular that the function $T(Q,B)$ is defined from the parameter $P \in \nabla n$.

As in the arguement of Banach, [4], define for $x \in X$:

$$\mu_1(x) = \sup_{y \varepsilon Y} f(y)-p(y-x)$$
$$\mu_2(x) = \inf_{y \varepsilon Y} p(x+y)-f(y).$$

Clearly $\mu_1, \mu_2 \ \varepsilon \ \nabla n$. Banach shows that $\mu_1(x) \leq \mu_2(x)$ on X, $\mu_1(y) = f(y) = \mu_2(y)$ on Y, and for any linear extension g of f and any $x \ \varepsilon$ Domain g; $g(z) \leq p(z)$ for $z \ \varepsilon$ Span $(x) + Y$ if and only if $\mu_1(x) \leq g(x) \leq \mu_2(x)$.

Consider the first order structure

$\mathcal{M} = (\overline{X}, \mathbb{R}, \overline{\cdot}, \overline{+}, \overline{f}, \overline{\mu}_1, \overline{\mu}_2)$ where the $\overline{}$ denotes the restriction to ∇_m (e.g. $\overline{f} = f \restriction (Y \cap \nabla_m)$). Let $\mathcal{T}$ be the complete first order theory of this structure with constants for all the members. We have $\mathcal{M}$, $\mathcal{T} \ \varepsilon \ \nabla_m$ and $\mathcal{T} \subset \nabla_m$.

Let $\mathcal{T}'$ be the extension of $\mathcal{T}$ with a new function symbol $\underline{g}$ and first order axioms which formalize:

(a) Domain $\underline{g} \subset \overline{X}$, Range $\underline{g} \subset \mathbb{R}$.

(b) $\underline{g}$ is a linear functional.

(c) $\underline{g}$ extends $\overline{f}$

(d) $\overline{\mu}_1(2) \leq \underline{g}(2) \leq \overline{\mu}_2(2)$ for $z \ \varepsilon$ Domain $\underline{g}$

(e) (Schema as $\overline{x}$ ranges over $\overline{X}$).

$\overline{x} \ \varepsilon$ Domain $\underline{g}$.

$\mathcal{T}'$ is consistent since a finite set of the above axioms involving $\overline{x}_1, \ldots, \overline{x}_n$ is modeled by $\mathcal{M}, g$ where $g = h \restriction (\text{Span}(\overline{x}_1, \ldots, \overline{x}_n) + \overline{Y})$ and h is an extension of f to $\text{Span}(\overline{x}_1, \ldots, \overline{x}_n) + Y$ which satisfies $\mu_1(2) \leq h(2) \leq \mu_2(2)$ there. h exists by finite induction on Banach's results.

Let $\mathcal{T}''$ be the canonical $\exists$-closure of $\mathcal{T}'$ and let B be the Lindenbaum algebra of $\mathcal{T}''$ - the statements of $\mathcal{T}''$ modulo provable equivalence. We have $B \ \varepsilon \ \nabla_m$ and Field $B \subset \nabla_m$. So $T(Q, B)$ is defined and is a prime ideal for B.

$$\mathcal{M}^* = (X^*, \mathbb{R}^*, \overline{\cdot}^*, \overline{+}^*, \overline{f}^*, \overline{\mu}_1^*, \overline{\mu}_2^*, g^*)$$

is the structure obtained from B and $T(Q, B)$ as in Henkin's proof of the completeness theorem, [18]. In particular, $(\overline{X}^*, \mathbb{R}^*, \overline{\cdot}^*, \overline{+}^*, \overline{f}^*, \overline{\mu}_1^*, \overline{\mu}_2^*)$ is an enlargement of $\mathcal{M}$ and g^* satisfies the axioms of $\mathcal{T}'$ with respect to the elements of $\mathcal{M}^*$. Thus if $\overline{x} \varepsilon \overline{X}$ we have $\overline{x} \varepsilon$ Domain g^* and

$$\mu_1(\overline{x}) = \overline{\mu}_1(\overline{x}) = \overline{\mu}_1^*(\overline{x}) \leq g^*(\overline{x}) \leq \overline{\mu}_2^*(\overline{x}) = \mu_2(\overline{x}).$$

$g^*(\overline{x})$ is therefore finite, hence near-standard.

Define $g(\overline{x})$ as the standard part of $g^*(\overline{x})$. We immediately check that $g:\overline{X} \to \mathbb{R}$ is a linear extension of $\overline{f}$ and $\mu_1(\overline{x}) \leq g(\overline{x}) \leq \mu_2(\overline{x})$ on $\overline{X}$.

If $z \varepsilon \overline{X} + Y$ then $z = \overline{x} + y$ some $\overline{x} \varepsilon \overline{X}$, $y \varepsilon Y$, it is well defined to set

$$D(f,m,Q)(z) = g(\overline{x}) + f(y)$$

since g extends $\overline{f} = f\restriction(Y \cap \nabla_m) = f\restriction(Y \cap (X \cap \nabla_m))$. $D(f,m,Q)$ is clearly a linear extension of f to $(X \cap \nabla_m) + Y$. If $z \varepsilon (X \cap \nabla_m) + Y$ then $z = \overline{x} + y \varepsilon$ Span $(\overline{x}) + Y$. Thus $D(f,m,Q)(z) \leq p(z)$ because $\mu_1(\overline{x}) \leq D(f,m,Q)(x) = g(x) \leq \mu_2(\overline{x})$.

<u>HB is True</u>

<u>Proof</u>. Let X be a real vector space. Let P be a functional to $\mathbb{R}$ satisfying $p(x+y) \leq p(x)+p(y)$ and $p(\lambda x) = \lambda p(x)$ for $\lambda = 0$. Let f be a linear functional defined on a subspace and satisfying $f(x) \leq p(x)$ there. Let n be large enough so that $X,p,f \varepsilon \nabla_n$. Define g_m, $m \geq n$ via $g_n = f$ and $g_{m+1} = E(g_m, m+1)$. Clearly $f \subset g_k \subset g_m$ for $n \leq k \leq m$. Thus $g = \bigcup_{n \leq m} g_m$ is well defined with domain X. g is a linear extension of f satisfying $g(x) \leq p(x)$, $x \varepsilon X$, from <u>Lemma 2.6</u> by induction.

III Lower Bounds on the Hahn Banach Theorem

The result of this section was stated in Section I as "HB is independent of any mathematical special case of AC'. Some discussion is necessary. A special case of AC is seen to be of the form $(\exists x)[\phi(x) \wedge \Omega(x)]$ where $AC \rightarrow \forall x \Psi(x)$ is a theorem of ZF and $\Omega(x)$ is some additional description of x such that $\exists x \Omega(x)$ is consistent with AC. This is consolidated (and weakened) by asserting that a special case of AC has the form $\exists x \Phi(x)$ and is consistent with AC.

The following example shows that some further limitation of $\Phi(x)$ is necessary. Let X, p, and f be the vector space, sub-linear functional, and partial linear functional defined as follows.

Case 1 A counterexample to HB exists.

Let S be the set of all triples (Y,q,g), (vector space, sub-linear functional, and partial functional) of least possible rank which are counterexamples to HB. X is the direct sum (finitely many nonzero coordinates) of the $\{Y_s\}_{s\varepsilon S}$. p and f are defined:

$$p(\sum_{s\varepsilon S} y_s) = \sum_{s\varepsilon S} q_s(y_s), \quad f(\sum_{s\varepsilon S} y_s) = \sum_{s\varepsilon S} g_s(y_s).$$

f is undefined when not all $g_s(y_s)$ are defined. In this case it is clear that X, p, and f are a well defined counterexample to HB.

Case 2 HB is true.

X is the $\{0\}$ vector space. p and f are the 0 functionals.

The assertion of HB for the particular X, p, and f above is a special case of AC, consistent with AC, and implies HB. However, it is clear that this example is highly nonmathematical.

It is defined in terms of all sets. A mathematical example is generally defined only in terms of its own full structure.

We introduce some set theoretic notation. If x is a set and $\alpha \varepsilon O_n$ define $R_\alpha(x)$ inductively via $R_o(x) = x$, $R_{\alpha+1}(x) = R_\alpha(x) \cup P(R_\alpha(x))$, and $R_\alpha(x) = {}_{\beta<\alpha}R_\beta(x)$ for limit α.[1] Recall that a term or formula of set theory is <u>absolute</u> if it means the same thing with respect to any transitive submodel of ZF with all the ordinals (subuniverse). For example ω is absolute but $P(\omega)$ is not. $\Phi(\vec{x})$ would be called mathematical in the above paragraph if

$$\Phi(x) \longleftrightarrow \Phi^{R\omega(\cup\vec{x})}(\vec{x})$$

were provable.[2] We weaken this by considering boundable formulae. Jech and Sochor ([8], also see [15]) say that $\Phi(\vec{x})$ is boundable if

$$\Phi(\vec{x}) \longleftrightarrow \Phi^{R\alpha(\cup\vec{x})}(\vec{x})$$

is provable where α is an absolute term denoting an ordinal. A boundable statement is the existential closure of a boundable formula.

<u>Theorem A</u> HB is independent of any boundable statement which is consistent with AC.

<u>Example</u> HB is independent of $2VM(\omega)$. (This implies both independences of (6) in the introduction).

---

[1]The vector notation $\vec{x}$ describes a finite tuple of variables stands for an appropriate $x_1 x_2 \ldots$ etc. Similarly $\cup\vec{x}$ stands for $x_1 \cup x_2 \cup \ldots$ $P(x)$ denotes the power set of x.

[2]Superscripting a formula (or term) by A denotes relativization to the ε structure of A.

$2VM(\omega)$ can be analyzed as

$(\exists x)(\exists y)(\exists \mu)$["x is transitive, ordered under ε, nonempty closed under successor and nonzero predecessor. $y = P(x)$. $\mu : y \to \{0,1\}$ is a measure. μ is 0 on finite sets."]

A scan of the above statement shows that it can be relativized to $R_5(x \cup y \cup \mu)$.

In the above example ω might be replaced by ω_1, the least measurable ordinal etc. e.g. the existence of a measure on x can be relativized to $R_3(x)$.

A generalization of Theorem A is possible. If x is a set $|x|$ denotes its injective cardinal (or aleph) defined as the least well ordered cardinal greater than or equal to the cardinal of every well orderable subset of x.

Theorem B Let $\psi(\alpha)$ and $\Phi(x)$ be boundable formulae where $\psi(\alpha) \to \alpha \in O_n$ is provable. If

$$(\exists \alpha)(\forall \vec{x})[(|\cup\vec{x}| < \alpha \wedge \psi(\alpha)) \to \Phi(\vec{x})]$$

is consistent with AC then it does not imply HB.[3]

Example HB is independent of "For every infinite x, $P(x)$ has a measure which is 0 on finite sets."

At first sight the above example is surprising since it is shown in [12] that HB is equivalent to "For every infinite X and every ideal J of $P(x)$ there is a measure on $P(x)$ which is 0 on J." However, every infinite x has a subset y such that $|y| \leq \omega$. If μ is a measure on $P(y)$ with the desired property then μ is extended to $P(x)$ by letting $\mu(z) = \mu(z \cap y)$. It is now clear that the example can be put in the form of Theorem B with

[3]Theorem B could be stated for injectively boundable statements (see [15]).

$\phi(\alpha) \leftrightarrow \alpha = \omega_1$.

A final remark on Solvay's proof of the independence of HB from ZF.

Theorem C $M(\omega)$ is independent of $\sim$BP. (see (8) of the introduction).

Theorems A, B, and C are quickly derived from the following theorem on models of set theory.

3.1 Theorem Let L be a universe of ZF + AC and let $\varkappa$ be a regular cardinal of L.[4] There is a universe M of ZF with L as a subuniverse and the following hold in M.

3.1.1 If $x \in L$ and $|P^L(x)|^L < \varkappa$ then $P^L(x) = P(x)$.

3.1.2 HB is false. In fact $P(\varkappa)$ has no measure which is 0 on sets of cardinal $< \varkappa$ (See the equivalent to HB given following Theorem B).

3.1.3 If $\lambda < \varkappa$ is an L-cardinal (hence a cardinal by 3.1.1) and $|\underline{x}| < \lambda$ then x is well orderable.

3.1.4 If CH (the continuum hypothesis) holds in L then there is a set without the Baire property.

We deduce Theorems A, B, and C before proceeding to the proof of Theorem 3.1.

Proof of Theorem A Let $\exists \vec{x} \phi(\vec{x})$ be a boundable statement consistent with AC. Let L be a universe of ZF + AC in which it holds and let $\vec{a} \in L$ be a vector such that $\phi(\vec{a})$ holds in L. Let α be as in the definition of boundable statement and let $\varkappa$ be a regular cardinal greater than $|R^L_\alpha(\cup\vec{a})|^L$. Let M be the model of Theorem 3.1.

---

[4] $\varkappa$ is not a limit of $< \varkappa$ smaller cardinals.

Since α is absolute it has the same meaning in L and M. It follows by induction from well known absoluteness results (see [2] or [7]) and <u>3.1.1</u> that $R^L_\beta(U\vec{a}) = R_\beta(U\vec{a})$, $\beta \leq \alpha$, holds in M. Thus in M we can argue:

$$\Phi^L(\vec{a}) \to \Phi^{R^L_\alpha(U\vec{a})}(\vec{a}) \to \Phi^{R_\alpha(U\vec{a})}(\vec{a}) \to \Phi(\vec{a}) \to \quad x\Phi(\vec{x}).$$

<u>Proof of Theorem B</u> Let L be a universe in which $\exists\alpha\forall\vec{x}[(|\underline{U\vec{x}}| < \alpha \wedge \Psi(\alpha)) \to \Phi(\vec{x})]$ holds. Let α satisfy $\Psi(\alpha)$ in L. Let β be as in the boundability of $\Psi(\alpha)$ and γ be as in the boundability of $\Phi(x)$. Let λ be a cardinal greater than the product $[\alpha \times R_1(\alpha)\times\ldots.\times R_\gamma(\alpha)]^L$. Choose $\varkappa$ greater than λ and $|R_\beta(\alpha)|^L$. Let M be the model of <u>Theorem 3.1</u>.

In M suppose $|\underline{U\vec{x}}| < \alpha$. By our choice of λ certainly $|\underline{U\vec{x}}| < \lambda$ so by <u>3.1.3</u> $U\vec{x}$ is well orderable. It is now easy to see that

$$|R_\gamma(U\vec{x})| \leq |(U\vec{x})\times\ldots\times R_\gamma(U\vec{x})| \leq |\alpha \times R_1(\alpha)\times\ldots\times R_\gamma(\alpha)|^L < \gamma$$

So $R_\gamma(Ux)$ is well orderable hence is similar to a set $A \in L$ with $|A| < \varkappa$. The similarity of $U\vec{x}$ and A carries the ε relation on $U\vec{x}$ to a well founded relation on A. This relation is in L by <u>3.1.2</u>. By a familiar construction there is a set $B \in L$ whose ε relation is isomorphic to the given well founded relation on A and the nonminimal elements of B are the sets of their predecessors. We now have an ε-isomorphism $R_\gamma(U\vec{x}) \to B$. It is not hard to check that $B = R_\gamma(U\vec{b})$ where $\vec{b}$ is the image of $\vec{x}$. By our choice of $\varkappa$, $|R^L_\gamma(U\vec{b})| < \varkappa$ so $R^L_\gamma(U\vec{b}) = R_\gamma(U\vec{b})$. $\Phi^{R^L_\gamma(U\vec{b})}(\vec{b})$ is true and by the isomorphisms $\Phi^{R_\gamma(U\vec{x})}(\vec{x})$ is true so $\Phi(\vec{x})$ is true. Other points check out as in the proof of Theorem A so $\exists\alpha\forall x[(|\underline{U\vec{x}}| < \alpha \cap \Psi(\alpha)) \to \Phi(\vec{x})]$ is true in M and hence does not imply HB.

<u>Proof of Remark C</u> Let L be the constructable universe and let $\varkappa = \omega$. By <u>3.1.2</u> and <u>3.1.4</u> M is a model in which a set without the Baire property exists but $P(\omega)$ has no nontrivial real valued measure which is 0 on finite sets.

<u>Proof of Theorem 3.1</u> This proof will occupy the remainder of the paper. It begins similarly to the proof of <u>Theorem 2.1</u>. In L let $\mathbb{P}$ denote the set of functions from bounded subsets of $\varkappa$ to 2. $D \subset \mathbb{P}$ is <u>dense</u> if every function in $\mathbb{P}$ has an extension in D. $g \subset \varkappa$ is <u>generic</u> if it extends a function from every dense $D \in L$. It is consistent, [19], that a generic g exists (although not in L). We fix a generic g and form the model M as an inner model of L(g) containing L. From, [19], AC holds in L(g) and if $x \in L$ and $|P(x)|^L < \varkappa$ then $P^L(x) = P^{L(g)}(x)$.

<u>Proof of 3.1.1 in M</u> Since $L \subset M \subset L(g)$ and M is a subuniverse of L(g) it follows (even though M hasn't been defined yet) that if $x \in L$, $|P(x)|^L < \varkappa$ then

$$P^L(x) \subset P^M(x) \subset P^{L(g)}(x) \subset P^L(x)$$

so equality holds throughout.

Whether one is in L,M, or L(g) there is no ambiguity concerning $\mathbb{P}$; $|x|$ for $x \in L$, $|x| < \varkappa$; and such notions as cardinal number less than $\varkappa$. Henceforth no relativizations will be indicated when such terms appear. We now state an analogue to (2.1).

(3.1) Let $\Phi(x)$ be a formula with only the free variable x and with parameters from L. If $\Phi(g)$ is true in L(g) then there is an $f \in \mathbb{P}$ such that g extends f and $\Phi(g')$ is true in L(g) for any generic g' extending f with $L(g') = L(g)$.

One can give in L a sequence $\langle\varphi_\alpha\rangle$, $\alpha \leq k$, of 1:1 onto functions $\alpha \times \varkappa \to \varkappa$. The following definitions refer to a fixed sequence of this type. If $\langle h_\beta \rangle_{\beta<\alpha}$ is a sequence of subsets of $\varkappa$ define $\underset{\beta<\alpha}{m} h_\beta$ as follows.[5]

$$\underset{\beta<\alpha}{m} h_\beta (\gamma) = h_\delta(\varsigma) \quad \text{where} \quad \varphi_\alpha(\delta,\varsigma) = \gamma$$

Intuitively the $\langle h_\beta \rangle_{\beta<\alpha}$ are "interlaced" according to φ_α by the m operator. Clearly if $h \subset \varkappa$ and $\alpha \leq \varkappa$ the expression $h = \underset{\beta<\alpha}{m} h^\alpha_\beta$ determines a unique sequence $\langle h^\alpha_\beta \rangle_{\beta<\alpha}$. The superscript α will be omitted when $\alpha = \varkappa$. An analogue to (2.2) and (2.3) is:
(3.2) If g is generic then the g_β, $\beta < \varkappa$, are distinct. If $f \in \mathbb{P}$ is fixed there is an unbounded set of β for which g_β extends f.

The model M is defined as $L(M_o(g))$ where

$$M_o(g) = \{h \subset \varkappa : (\exists \alpha < \varkappa)[h \in L(\underset{\beta<\alpha}{m} g_\beta)\}.$$

As in Section IIA M is an increasing union of M_α, $\alpha \in O_n$ where $M_o = M_o(g)$ (notice that this set is transitive), $M_\alpha = \bigcup_{\beta<\alpha} M_\beta$ for limit α, and $M_{\alpha+1}$ is the collection of subsets of M_α defined over M_α with parameters in $M_\alpha \cup R^L_{\alpha+1}(\emptyset)$. M is, to all intents and purposes, identical with Fefferman's model of [5] in which PI fails. Fefferman considered only the case $\varkappa = \omega$.

M will be endowed with a support structure defined from the parameter $M_o(g)$. It follows that M and its support structure will be preserved if g is replaced by a generic g^* such that $L(g) = L(g^*)$ and $M_o(g) = M_o(g^*)$. A number of ways of obtaining such g^* are listed below. Only the last of these is new, the others having been considered in [5]. The methods used here for the last operation also apply, and considerably more easily, to the others.

[5] Here and elsewhere subsets of $\varkappa$ are used interchangeably with their characteristic functions.

(3.3) Let $\sigma \in L$ be a permutation of κ. g can be replaced by $\underset{\alpha<\kappa}{m}\, g_{\sigma(\alpha)}$.

(3.4) Let $\gamma < \kappa$. For each $\beta < \gamma$ let g^*_β differ from g_β on a bounded subset of κ. For $\beta \geq \alpha$ let $g^*_\beta = g_\beta$. Then g can be replaced by $\underset{\beta<\kappa}{m}\, g^*_\beta$.

(3.5) Let $\sigma < \kappa$. Let $g^*_\beta = g_\beta$ for $\beta \neq \sigma$ and let $g^*_\sigma = \kappa - g_\sigma$. g can be replaced by $\underset{\beta<\kappa}{m}\, g^*_\beta$

(3.6) Let $\gamma, \delta < \kappa$. Let $g^*_\beta = g_\beta$ for $\beta \neq \gamma, \delta$. Let $g^*_\gamma = g_\gamma = g_\delta$ and let $g^*_\delta = \underset{i<2}{m}\, h_i$ where

$h_o(\tau) = 1$ iff the τth element of $\kappa - (g_\gamma \cap g_\sigma)$ is in g_γ

$h_1(\tau) = 1$ iff the τth element of $\kappa - g_\gamma$ is in g_δ.

Then g can be replaced by $\underset{\beta<\kappa}{m}\, g^*_\beta$.

<u>Proof of the Properties of the Transformation in (3.6)</u>

By conjugating the given transformation with transformations of the type (3.3) it may be assumed that $\gamma = 0$, $\delta = 1$. We first note that $L(g_o, g_1) = L(g^*_o, g^*_1)$. $L(g_o, g_1) \supset L(g^*_o, g^*_1)$ is clear since g^*_o and g^*_1 are absolutely defined from g_o and g_1. Conversely h_o and h, are defined as $(g^*_1)^2_o$ and $(g^*_1)^2_1$, g_1 is defined from g^*_o and h_o via

$g_o(\gamma) = 1$ iff $g^*_o(\gamma) = 1$ or γ is the δth element of $\kappa - g^*_o$ and $h_o(\delta) = 1$.

g_1 is defined from g_o, g^*_o, and h_o via

$g_1(\gamma) = 1$ iff $g^*_o(\gamma) = 1$ or γ is the δth element of $\kappa - g_o$ and $h_o(\gamma) = 1$.

Therefore $L(g_o, g_1) = L(g^*_o, g^*_1)$. It quickly follows that $L(g) = L(g^*)$ and $L(\underset{\beta<\alpha}{m}\, g_\beta) = L(\underset{\beta<\alpha}{m}\, g^*_\beta)$ so $M(g) = M(g^*)$.

To prove that g^* is generic we invoke the theory of product generic sequences as in [19] or [20]. If $\alpha \leq \kappa$ let $\mathbb{P}^{\underline{\alpha}}$ denote the set of α-sequences from $\mathbb{P}$ with $<\kappa$ nonzero entries. $\mathbb{P}^{\underline{\alpha}}$ is partially ordered by saying that

$$\langle f_\beta\rangle_{\beta<\alpha} \leq \langle f'_\beta\rangle_{\beta<\alpha} \text{ iff each } f_\beta \leq f'_\beta.$$

As with subsets of κ, if $\langle f_\beta\rangle_{\beta<\alpha} \in \mathbb{P}^{\underline{\alpha}}$ define $\underset{\beta<\alpha}{m} f_\beta \in \mathbb{P}$ via

$$(\underset{\beta<\alpha}{m} f_\beta)(\gamma) = f_\delta(\zeta) \text{ where } \varphi(\delta,\zeta) = \gamma.$$

This is undefined when $f_\delta(\zeta)$ is undefined. Conversely if $f \in \mathbb{P}$ then $f = \underset{\beta<\alpha}{m} f^\alpha_\beta$ for a unique sequence $\langle f^\alpha_\beta\rangle_{\beta<\alpha} \in \mathbb{P}^{\underline{\alpha}}$. It is quickly verified that the map

$$\langle f_\beta\rangle_{\beta<\alpha} \to \underset{\beta<\alpha}{m} f_\beta$$

is an isomorphism of $\mathbb{P}^{\underline{\alpha}}$ and $\mathbb{P}$. A sequence $\langle h_\beta\rangle_{\beta<\alpha}$ of subsets of κ is said to be generic if for every dense subset $D \subset \mathbb{P}^{\underline{\alpha}}$, $D \in L$, there is an $\langle f_\beta\rangle_{\beta<\alpha} \in \mathbb{P}$ such that h_β extends f_β all $\beta < \alpha$. By the isomorphisms $\langle h_\beta\rangle_{\beta<\alpha}$ is a generic sequence iff $\underset{\beta<\alpha}{m} h_\beta$ is a generic set.

Write $\mathbb{P}^{\underline{\kappa}} = \mathbb{P}^{\underline{\kappa-2}} \times \mathbb{P}^2$. The product lemma of [19] shows that $\underset{\beta<2}{m} h_\beta$ is generic iff $\underset{2\leq\beta<k}{m} h_\beta$ is generic over $L(\underset{i<2}{m} h_i)$ (i.e. the dense subsets of $\mathbb{P}^{\underline{\kappa-2}}$ can come from $L(\underset{i<2}{m} h_i)$.) and $\underset{i<2}{m} h_i$ is generic over L. g is given to be generic so $\underset{2\leq\beta<k}{m} g_\beta = \underset{2\leq\beta<k}{m} g^*_\beta$ is generic over $L(\underset{i<2}{m} g_\beta) = L(\underset{i<2}{m} g^*_\beta)$. Also $\underset{i<2}{m} g_i$ is generic so $\langle g_0, g_1\rangle$ is a generic sequence. It remains only to deduce from this that $\langle g^*_0, g^*_1\rangle$ is a generic sequence to conclude that $\underset{i<2}{m} g^*_i$ is generic and hence by the product lemma that g^* is generic. Applying the isomorphisms again over $L(g^*_0)$ we conclude that g^*_1 is $L(g^*_0)$ generic if $\langle h_0, h_1\rangle$ is an $L(g^*_0)$ generic sequence. From the product lemma it follows that we need only show that $<g_0^*, h_1, h_2>$ is a generic sequence.

Let $D_2 \subset \mathbb{P}^{\underline{2}}$ and $D_3 \subset \mathbb{P}^{\underline{3}}$ be defined as follows.

$D_2 = \{<f_0, f_1> \in \mathbb{P}^{\underline{2}}$: Domain f_0 = Domain $f_1 = \alpha$ for some $\alpha<k\}$

$D_3 = \{<f_0, f_1, f_2> \in \mathbb{P}^{\underline{3}}$: Domain f_i is an ordinal for $i<3$.

Domain f_1 is the order type of the 0's in the domain of f_0. Domain f_2 is the order type of the 0's of f' where Domain f' = Domain f_0 and $f'(\gamma) = 1$ iff $[f_0(\gamma) = 1$ or (γ is the δth element of Domain f_0 such that $f_0(\gamma) = 0$) and $f_2(\delta) = 1.]\}$

It is immediate that D_2 is dense in $\mathbb{P}^{\underline{2}}$.
D_3 is dense in $\mathbb{P}^{\underline{3}}$ since given (f_0, f_1, f_2) in $\mathbb{P}^{\underline{3}}$ one can perform the following process.

a) Extend f_0, f_1, and f_2 arbitrarily so that their domains are ordinals.

b) Add 0 values to either f_0 or f_1 in such a way that its domain remains an ordinal and Domain f_1 is the order type of the 0's in f_0.

c) If Domain f_2 is less than the order type of the 0's in the f^1 produced from f_0 and f_1 then extend it to this order type. Otherwise extend both f_0 and f_1 by adding 0's until the order type of the 0's in f^1 is equal to Domain f_2. It will remain true that the domain of f_1 is the order type of the 0's in f_0.

Consider the following map $D_2 \to D_3$, $\varphi(f_0, f_1) = <f_0^*, f_1^*, f_2^*>$ where:

$f_0^*(\gamma) = 1$ iff $f_0(\gamma) = f_1(\gamma) = 1$. Domain f_0^* = Domain f_0.

$f_1^*(\delta) = 1$ iff γ is the δth element of Domain f_o such that $f_o^*(\gamma) = 0$ and $f_o(\gamma) = 1$. Domain f_1^* is the order type of the 0's of f_o^*.

$f_2^*(\delta) = 1$ iff γ is the δth element of Domain f_o such that $f_o(\gamma) = 0$ and $f_1(\gamma) = 1$. Domain f_2^* is the order type of the 0's of f_o.

φ is seen to be an order isomorphism $D_2 \to D_3$. φ^{-1} is defined as follows. $\varphi^{-1}(f_o,f_1,f_2) = <f',f''>$ where f' is defined from f_o and f_1 as in the definition of D_3 and f'' is defined with Domain equal to that of f' via

$f''(\gamma) = 1$ iff either $f_o(\gamma) = 1$ or γ is the δth element of Domain f'' such that $f'(\gamma) = 0$ and $f_2(\delta) = 1$.

It is also apparent from the definitions of g_o^*, h_o, and h_1 that $<g_o,g_1>$ extends $<f_o,f_1> \varepsilon D_2$ if and only if $<g_o^*,h_o,h_1>$ extends $\varphi(f_o,f_1)$.

The genericity of $<g_o^*,h_o,h_1>$ is now clear. Let $D \subset \mathbb{P}^3$ be dense and in L. Let $D^* \varepsilon L$ be a set of members of D_3 containing only extensions of members of D and at least one extension of each member of D (use AC in L). Let $D' = \{\varphi^{-1}<f_o,f_1,f_2>:<f_o,f_1,f_2> \varepsilon D^*\}_2$. Since D is dense in $\mathbb{P}^3$, D^* is dense in D_3 hence, by the isomorphism, D' is dense in D_2 but, since D_2 is dense in $\mathbb{P}^2$, it follows that D' is dense in $\mathbb{P}^2$. Therefore $<g_o,g_1>$ extends some $<f_o,f_1>$ in D' and $<g_o^*,h_o,h_1>$ extends $\varphi<f_o,f_1>$ extends some element of D and we are done.

It is now time to put a support structure on M. This is easier than in Section II in that ∇ and T are defined by in one stage process. The only parameter in the definition is M_o.

$G \subset M_o$ is called a <u>support</u> if $G = \{h_\beta\}_{\beta<\alpha<\kappa}$ and $\underset{\beta<\alpha}{m} h_\beta \in M_o$. It quickly follows, since by <u>3.1.1</u> permutations of α are in L, that $\underset{\gamma<\delta}{m} h'_\gamma \in M_o$ for any expression $G = \{h'_\gamma\}_{\gamma<\delta}$ where g is a support. Also in such instances $\{h_\beta\}_{\beta<\alpha} = \{h'_\gamma\}_{\gamma<\delta} \rightarrow L(\underset{\beta<\alpha}{m} h_\beta) = L(\underset{\gamma<\delta}{m} h'_\gamma)$.

If G is a support ∇G is defined via:

$\nabla^o G = M_o \cap L(\underset{\beta<\alpha}{m} h_\beta)$ where $G = \{h_\beta\}_{\beta<\alpha}$

$\nabla^{\alpha+1} G$ is the set of subsets of M_α defined over M_α with parameters in $\nabla^\alpha G \cup R_{\alpha+1}(\phi)$.

$\nabla^\alpha G = \bigcup_{\beta<\alpha} \nabla^\beta G$ for limit α

$\nabla G = \bigcup_{\alpha \in O_n} \nabla^\alpha G$

If $\langle h_\beta \rangle_{\beta<\alpha}$ is a well ordering of a support and $\gamma \in O_n$ then $T(\langle h_\beta \rangle_{\beta<\alpha}, \gamma)$ is the γth element in the following well ordering of $\nabla\{h_\beta\}_{\beta<\alpha}$. Fix a well ordering of L.[6] Well order $\nabla^o G$ by least definition from $\underset{\beta<\alpha}{m} h_\beta$ and well order $\nabla^{\alpha+1} G$ by least definition from an inductive well ordering of $\nabla^\alpha G \cup R^2_{\alpha+1}(\phi)$.

An intrinsic theory of the support structure on M could be set up as in <u>Theorem 2.1</u>. However there are some technical difficulties to this approach. Also the arguments inside M are quite elementary. It is therefore easier to give several of the following propositions from the external vantage point of L(g). The proofs are essentially no different from those in Section IIA.

(2.7) (In L(g)) $(\forall x \in M)(\exists G)[G \subset \{g_\beta\}_{\beta<\kappa} \wedge x \in \nabla G]$

This is stated in L(g) since the sequence $\langle g_\beta \rangle_{\beta<\kappa}$ is not in M.

---

[6] By an unpublished theorem of P.J. Cohen and others ZF + E is a conservative extension of ZF + AC where E is Godel's axiom well ordering the universe. This result is not really used here since it suffices in our arguments to well order an arbitrarily large $\nabla^\alpha(G)$.

(2.8) (In M) ∇G contains M_o, all members of L, and all well orderings of G.

(2.9) (In M) if x is defined from parameters in L(g) then $x \in L(g)$.

(3.10) (In L(g)) Let $\Phi(x)$ have parameters in $\{M_o\} \cup L$. Let $\{h_\gamma\}_{\gamma<\delta} \subset \{g_\beta\}_{\beta<k}$. If $\Phi(<h_\gamma>_{\gamma<\delta})$ holds in M then there is a sequence $<f_\gamma>_{\gamma<\delta}$ from $\mathbb{P}$ such that h_γ extends f_γ. Also $\Phi(<h'_\gamma>_{\gamma<\alpha})$ holds in M for any $<h'_\gamma>_{\gamma<\delta}$ such that $\{h'_\gamma\} \subset \{g_\beta\}_{\beta<k}$ and h'_γ extends f_γ.

(3.11) (In M) If $<h_\gamma>_{\gamma<\delta}$ is a fixed well ordering of a support then $T(<h_\gamma>_{\gamma<\delta},\zeta)$ is, as a function of ζ, 1:1 and onto $\nabla\{h_\gamma\}_{\gamma<\delta}$.

(3.7)....(3.11) Combine as in Lemma 2.5 to give:

(3.12) If $G_1, G_2 \subset \{g_\beta\}_{\beta<k}$ then $\nabla G_1 \cap \nabla G_2 = \nabla G_1 \cap G_2$.

Proof of 3.12 Let $\mu \in M$ be a measure on $P^M(k)$ which is 0 on the sets of cardinal $<\varkappa$. Using (3.7) and (3.9), $\mu \in \nabla\{g_\beta\}_{\beta<\alpha}$ for some $\alpha<\varkappa$. Thus for some $\delta \in O_n$, $\mu = T(<g_\beta>_{\beta<\alpha},\delta)$.

Claim 1. $\mu(g_\gamma) = 1/2$ for all $\gamma \geq 2$.

Proof of Claim 1. Assume that $\mu(g_\gamma) \leq 1/2$. '"$T(<g_\beta>_{\beta<\alpha},\delta)(g_\gamma) \leq 1/2$" holds in $L(M_o(g))$.' is a true sentence in L(g). It is therefore true when g is replaced by any g* as in (3.3)...(3.6) which extends some fixed $f \in \mathbb{P}$, f restricts g. Change g in the following ways. First use (3.5) to change g_γ to $\varkappa - g_\gamma$. Then use (3.4) to change the new g_γ in $<\varkappa$ places so that it agrees with f_γ. The new g* thus extends f. Therefore '"$T(<g^*_\beta>_{\beta<\alpha},\delta)(g^*_\gamma) \leq 1/2$" holds in $L(M_o(g^*))$.' is true in L(g). $<g^*_\beta>_{\beta<\alpha} = <g_\beta>_{\beta<\alpha}$ and $L(M_o(g^*)) = M$ so $\mu(g^*_\gamma) \leq 1/2$ holds in M. Since g^*_γ differs from g_γ in $<\varkappa$ places, $\mu(\varkappa - g_\gamma) \leq 1/2$ holds in M. Thus $\mu(g_\gamma) = 1/2$ holds in M. A similar argument shows $\mu(g_\gamma) = 1/2$

under the assumption $\mu(g_\gamma) \geq 1/2$ so $\mu(g_\gamma) = 1/2$.

Claim 2. $\mu(g_\alpha \cap g_{\alpha+1}) = \mu(g_\alpha - g_{\alpha+1}) = 1/2$

Proof of Claim 2. Consider the statement in $L(g)$:

$$\text{'"}T(\langle b_\beta\rangle_{\beta<\alpha},\delta)(g_\alpha) = 1/2\text{" holds in } M.\text{'}$$

Let $f \in \mathbb{P}$ be as in (3.1) for the above statement. Change g as follows. Do a transformation of type (3.6) to replace g_α by $g_\alpha \cap g_{\alpha+1}$ while leaving $\langle g_\beta\rangle_{\beta<\alpha}$ alone. Then change the new g_α and $g_{\alpha+1}$ in $<\kappa$ places by (3.4) so as to agree with f_α and $f_{\alpha+1}$. The resulting g^* extends f so re-reading the statement with g^* replacing g gives $\mu(g^*_\alpha) = 1/2$. g^*_α differs from $g_\alpha \cap g_{\alpha+1}$ in $<\kappa$ places so $\mu(g_\alpha \cap g_{\alpha+1}) = 1/2$.

To show $\mu(g_\alpha - g_{\alpha+1}) = 1/2$ use the same statement and insert a first step in the transformation of g which changes $g_{\alpha+1}$ to $\kappa - g_{\alpha+1}$.

Conclusion. Claims 1 and 2 give the contradiction

$$1/2 = \mu(g_\alpha) = \mu(g_\alpha \cap g_{\alpha+1}) + \mu(g_\alpha - g_{\alpha+1}) = 1/2 + 1/2 = 1.$$

Therefore no μ exists with the specified properties.

Proof of 3.1.3 Let $x \in M$ not be well orderable in M. Let $\lambda<\kappa$ be a cardinal, $\lambda \leq |x|$ must be shown.

$x \in \nabla\{g_\beta\}_{\beta<\alpha}$ for some $\alpha < k$. Since x is not well orderable in M, $x \notin \nabla\{g_\beta\}_{\beta<k}$. It follows that there are $\delta,\zeta \in O_n$ and a sequence $\langle h_\mu\rangle_{\mu<\sigma}$ with $\{h_\mu\}_{\mu<\sigma} \subset \{g_{\beta\alpha}\}_{\leq\beta<k}$ such that $x = T(\langle g_\beta\rangle_{\beta<\alpha},\delta)$ and

$$T(\langle g_\beta\rangle_{\beta<\alpha} * \langle h_\mu\rangle_{\mu<\sigma},\zeta) \in T(\langle g_\beta\rangle_{\beta<\alpha},\delta) - \nabla\{g_\beta\}_{\beta<\alpha}.$$

where * denotes concatonation of sequences. From (3.10) there is a sequence $\langle f_\mu\rangle_{\mu<\sigma}$ such that the above equation holds with

$\langle h_\mu \rangle_{\mu<\sigma}$ replaced by any $\langle h'_\mu \rangle_{\mu<\sigma}$ satisfying the same assumptions and with h'_μ extends f_μ.

λ and σ are $<\varkappa$ so by (3.2) there is a sequence $\langle h_{\mu r} \rangle_{\mu<\sigma, r<\lambda}$ with distinct terms from $\{g_\beta\}_{\beta<\varkappa}$ such that $h_{\mu r}$ extends f_μ. $\varkappa$ is regular so there is a $z<\varkappa$ such that $\{h_{\mu r}\}_{\mu<\sigma, r<\lambda} \subset \{g_\beta\}_{\beta<z}$. For $r<\lambda$ set

$$y_r = T(\langle g_\beta \rangle_{\beta<\alpha}{}^* \langle h_{\mu r} \rangle_{\mu<\sigma}, \zeta).$$

The sequence $\langle y_r \rangle_{r<\lambda}$ is in M since it is definable from $\underset{\beta<\alpha}{m}\, g_\beta$ using T, α, and ζ. Reading the original equation with $\langle h_\mu \rangle_{\mu<\sigma}$ replaced by $\langle h_{\mu r} \rangle_{\mu<\sigma}$ gives

$$y_r \in x - \nabla\{g_\beta\}_{\beta<\alpha}.$$

The y_r are distinct since if $y_r = y_j$ then
$y_r \in \nabla(\{g_\beta\}_{\beta<\alpha} \cup \{h_{\mu r}\}_{\mu<\sigma}) \cap \nabla(\{g_\beta\}_{\beta<\alpha} \cup \{h_{\mu J}\}_{\mu<\sigma}) = \nabla\{g_\beta\}_{\beta<\alpha}$

by (3.12) since the $h_{\mu r}$ were chosen distinct.
Therefore $\{y_r\}_{r<\lambda}$ is a subset of x with cardinal λ so $\lambda \leq |\underline{x}|$.

Proof of 3.1.4 If $\varkappa \supset \omega$ and CH holds in L then $P^L(\omega) = P^M(\omega)$ so $\mathbb{R}$ can be well ordered and a set without the Baire property exists. It remains to consider the case $\varkappa = \omega$.

The Halpern Levy model, H, of [7] is $L(H_o)$ in $L(g)$ where $H_o = \{g_n\}_{n \in \omega}$. Evidently

$$M_o = \{h \subset \omega : (g_o, \ldots, g_k \in H_o[h \in L(\underset{i\leq k}{m}\, g_i)]\}$$

It follows that $M_o \in H$ and $L \subset M \subset H \subset L(g)$.

In fact $M_o = P^M(\omega) = P^H(\omega)$. $M_o \subset P^M(\omega) \subset P^H(\omega)$ is clear. To see that $P^H(\omega) \subset M_o$ set up, as in M, a support structure in H where supports are finite subsets of H_o. If G is a support in

H then $\nabla^M G \subset \nabla^H G$ is easily seen. The relation $R(f_1,\dots,f_k,\alpha,\beta)$ which says in H

"For every $g_1,\dots,g_k \in H_o$ such that g_i extends f_i, $i = 1,\dots,k$ we have $\alpha \in T(\langle g_1,\dots,g_k\rangle,\beta)$."

is in L (see [3]). Now if $h \in P^H(\omega)$ then for some $g_1,\dots,g_k \in H_o$, $h \in \nabla^H(\{g_1,\dots,g_k\})$ and for some $\beta \in O_n$ $h = \{n \in \omega: \exists f_i \in \mathbb{P}, f_i \subset g_i, i = 1,\dots,k$ and

$$R(f_1,\dots,f_k,n,\beta)\}.$$

So $h \in L(\underset{1\leq i\leq k}{m} g_i) \subset M_o$. Therefore $P^H(\omega) \subset M_o$.

As corollaries to $P^M(\omega) = P^H(\omega)$ are; $\mathbb{R}^M = \mathbb{R}^H$, the open sets of $\mathbb{R}$ in M are identical to those of $\mathbb{R}$ in H, and the closed-nowhere dense sets of $\mathbb{R}$ in M are identical to those of $\mathbb{R}$ in H. These are all clear once it is noted that M and H have the same sequences of reals. But the sequences from $P(\omega)$ are in 1:1 correspondence with $P(\omega)$ via $\langle g_n\rangle_{n\in\omega} \to \underset{n\in\omega}{m} g_n$.

The following is a lemma in [16].

<u>Lemma</u>. (In H) Let $A \in L$ be such that $A \subset \mathbb{R}$ and A has countable intersection with every closed nowhere dense L-subset of $\mathbb{R}^L$. Then A has countable intersection with every closed nowhere dense subset of $\mathbb{R}$.

The foregoing discussion implies that the same lemma holds in M. An A satisfying the hypotheses of the lemma exists by CH in L (a standard construction). The conclusion of the lemma implies that A does not have the Baire property.

Bibliography

1. J. L. Bell and D. H. Fremlen, A geometric form of the axiom of choice, Fund. Math. (1972), to appear.

2. P. J. Cohen, Independence of the axiom of choice, mimeographed notes, Stanford University.

3. P. J. Cohen, Set Theory and the Continuum Hypothesis, Benjamin, 1966

4. M. M. Day, Normed Linear Spaces, Springer, 1962.

5. S. Fefferman, Some applications of the notions of forcing and generic set, Fund. Math. 56 (1965), pp. 325-345.

6. J. D. Halpern and H. Lauchli, A partition theorem, Trans. A.M.S. 124 (1966), pp. 360-367.

7. J. D. Halpern and A. Levy, the Boolean prime ideal theorem does not imply the axiom of choice, Proc. Symp. Pure Math. Vol. XIII Part I, pp. 83-134.

8. T. Jech and A. Sochor, Applications of the θ model, Bull. Acad. Polon. Ser. Sec. 14 (1966), pp. 351-355.

9. J. Los and C. Ryll-Nardzewski, Effectiveness of the representation theory for Boolean algebras, Fund. Math. 41 (1954), pp. 49-56.

10. J. Los and C. Ryll-Nardzewski, On the applications of Tychanov's theorem in mathematical proofs, Fund. Math. 38 (1951), pp. 233-237.

11. W. A. J. Luxemburg, 2 applications of the method of construction by ultrapowers to analysis, Bull. A.M.S. 68 (1962), pp. 416-419.

12. W. A. J. Luxemburg, Reduced powers of the real number system and equivalents of the Hahn-Banach extension theorem, Int. Sympos on the Application of Model Theory to Alg. Anal. and Probability, Holt, Rinehart and Winston 1969.

13. J. Mycielski, Two Remarks on Tychonoff's Product Theorem, Bull. Akad. Polon. Ser. Sci. 12 (1964), pp. 439-441.

14. D. Pincus, Support structures for the axiom of choice, J.S.L. 35 (1971), pp. 28-38.

15. D. Pincus, Zermelo Fraenkel consistency results by Fraenkel Mostowski Methods J.S.L. (to appear).

16. D. Pincus, Nonperfect sets, Baire diagonal sets, and well ordering the continuum. Submitted for publication.

17. D. Pincus, Independence of the Prime Ideal theorem from the Hahn Banach Theorem Bull. A.M.S. 78 (1972), to appear.

18. J. R. Shoenfield, Mathematical Logic, Addision Wesley (1967).

19. J. R. Shoenfield, Unramified Forcing, Proc. Symp. Pure Math. XIII Vol. I, pp. 357-381.

20. R. M. Solovay, A model of set theory in which every set of reals is Lesbague measurable Ann. of Math. 92 (1970), pp. 1-58.

21. W. Sierpinski, Sur un probleme conduisant a un ensemble non measurable, Fund. Math. 10 (1927), pp. 177-179.

22. W. Sierpinski, Functions additives non completement additives et functions non measurables, Fund. Math. 30 (1938), pp. 96-99.

ENLARGED SHEAVES

Abraham Robinson

Yale University, New Haven

1. The present note is related to [4], but we shall not assume familiarity with that paper. The reader may consult [1], [2] for standard results from the theory of functions of several complex variables used here and to [3] for basic concepts in nonstandard analysis.

Let $\mathcal{S} = (S,D,\pi)$ be a sheaf of rings with base space D and map $\pi : S \to D$. Let ${}^*\mathcal{S} = ({}^*S, {}^*d, \pi)$ be an enlargement of $\mathcal{S}$ where, as usual, we have not appended the star to the extended π. For any section f in $\mathcal{S}$ with domain $U \subset D$ (where U is always assumed open) and for any point $z \in U$, we denote by $\gamma_z(f)$ the germ generated by f at z. Thus, the mapping $(f,z) \to \gamma_z(f)$ is defined provided z is in the domain of f. The mapping extends to ${}^*\mathcal{S}$ where it is applicable also to nonstandard internal f and to nonstandard z in the domain of f.

Again, let z be a standard point of the base space D and let $\mu(z)$ be its monad. For any standard section f which includes z and hence also $\mu(z)$ in its domain, we denote by $\beta_z(f)$ the restriction of f to $\mu(z)$. Thus, $\beta_z(f)$ is, generally speaking, external. We may extend this definition also to nonstandard internal f, provided z is included in the domain of f. Let

$$G_z = \{y | y = \gamma_z(f) \text{ for some section } f\}$$

and let

$$M_z = \{y | y = \beta_z(f) \text{ for some section } f\}$$

where M_z is defined only for standard z and where, in both cases, the f that are to be taken into account are standard or nonstandard internal. Also, for

---

Research supported in part by the National Science Foundation, Grant No. GP - 34088.

standard z, let

$$^{o}G_z = \{y \in G_z | y = \gamma_z(f),\ \ f \ \text{standard}\}$$

and let

$$^{o}M_z = \{y \in M_z | y = \beta_z(f),\ \ f \ \text{standard}\} .$$

Then $^{o}G_z$ is simply the stalk of $\mathcal{G}$ at z and $G_z = {}^*({}^{o}G_z)$ is the stalk of ${}^*\mathcal{G}$ at z. Both $^{o}G_z$ and G_z are endowed with a ring structure such that $^{o}G_z$ is (or may be regarded as) a subring of G_z. Moreover, M_z has a ring structure which is given by pointwise addition, subtraction, and multiplication on $\mu(z)$ and includes $^{o}M_z$ as a subring.

For standard z, there is a natural map $\lambda : M_z \to G_z$ as follows. Let U be any (internal) open neighborhood of z such that $U \subset \mu(z)$. For any $g \in M_z$, $g|U$ is a section which determines an element f of the stalk of the sheaf at z, i.e. of G_z, where the particular choice of U is irrelevant. We then put $f = \lambda(g)$. λ is a ring homomorphism and the restriction $^{o}\lambda$ of λ to $^{o}M_z$ is an isomorphism onto $^{o}G_z$ since each element of the stalk of $\mathcal{G}$ at z determines standard sections which coincide on $\mu(z)$. To this extent, $^{o}M_z$ reflects the properties of the stalk of germs $^{o}G_z$ and this fact has been used in [4] for a discussion of the Rückert Nullstellensatz. At the same time, since they are genuine functions, the elements of M_z also reflect the properties of (standard) sections. In the present paper we offer some remarks concerning $^{o}M_z$ and M_z from the latter point of view, and in this context we prefer to call the elements of M_z <u>monadic sections</u> rather than <u>nonstandard germs</u> as in [4].

As usual in this area, we shall make use of the Weierstrass preparation and division theorems. As for the latter, a glance at the proof given on page 70 of [1], shows that if $h(z_1, \ldots, z_n)$ is a Weierstrass polynomial in z_n of degree $k > 0$ which is holomorphic for $|z_j| < \delta$, $j = 1, \ldots, n-1$, $\delta > 0$, then there exists a δ', $0 < \delta' > \delta$ such that for every function which is holomorphic for $|z_j| < \delta'$, $j = 1, \ldots, n$, we have a representation

$$f(z_1, \ldots, z_n) = g(z_1, \ldots, z_n)h(z_1, \ldots, z_n) + r(z_1, \ldots, z_n)$$

where $g(z_1, \ldots, z_n)$ is holomorphic for $|z_j| < \delta'$, and $r(z_1, \ldots, z_n)$ is a polynomial in z_n of degree $< k$ with coefficients which are holomorphic functions of $z_1, \ldots, z_{n-1}$ for $|z_j| < \delta'$.

2. From now on, we shall assume that $\mathcal{Y} = (S,D,\pi)$ is the sheaf of germs of holomorphic functions on an open set D in n-dimensional complex space. Let $g_1, \ldots, g_k$ be elements of ${}^oM_z$ for some $z = (z_1, \ldots, z_n) \in D$ and let f be an element of ${}^oM_z$. If $\lambda(f)$ is included in the ideal generated in ${}^oG_z$ by $\lambda(g_1), \ldots, \lambda(g_k)$, i.e. if

$$\lambda(f) = \sum_{j=1}^{k} p_j \lambda(g_j) \quad \text{with} \quad p_j \in {}^oG_z$$

then the isomorphism provided by λ ensures that a corresponding fact is true within ${}^oM_z$, that is to say, putting $p_j = \lambda(q_j)$ we have

$$f = \sum_{j=1}^{k} q_j g_j \; .$$

Thus, f belongs to the ideal generated by $g_1, \ldots, g_k$. The following theorem states that this conclusion can be extended to nonstandard internal functions f which are holomorphic on $\mu(z)$. More precisely - and more generally - we shall prove

2.1. Theorem. Let

$$G_j = (g_{j1}, g_{j2}, \ldots g_{jm}), \; m \geq 1, \; j = 1, \ldots, k, \; k \geq 1$$

where $g_{ji} \in {}^oM_z$ and let $F = (f_1, \ldots, f_m)$ where $f_j \in M_z$ (so that the $f_j(z_1, \ldots, z_n)$ are holomorphic but not necessarily standard on $\mu(z)$). Suppose that $\lambda(F)$ belongs to the module generated by $\lambda(G_1), \ldots, \lambda(G_k)$ over G_z

where

$$\lambda(F) = (\lambda(f_1), \ldots, \lambda(f_m)), \lambda(G_j) = (\lambda(g_{j1}), \ldots, \lambda(g_{jm})) .$$

Then F belongs to the module generated by $G_1, \ldots, G_k$ over M_z, i.e.

$$F = \sum_{j=1}^{k} p_j G_j$$

for certain monadic sections p_j in M_z .

For the proof, we first reformulate 2.1, reinterpreting the g_{ji} and f_j as holomorphic functions, standard or internal respectively which have domains <u>including</u> $\mu(z)$. On this assumption we have to establish -

2.1'. Suppose that there exist internal functions $p_j(z_1, \ldots, z_n)$, $j = 1, 2, \ldots, k$, which are holomorphic in some internal (but not necessarily standard) open neighborhood U of z such that

$$F = \sum_{j=1}^{k} p_j G_j$$

on U. Then there exist internal functions $q_j(z_1, \ldots, z_n)$ which are holomorphic on sets including $\mu(z)$ such that

$$F = \sum_{j=1}^{k} q_j G_j \quad .$$

<u>Proof of 2.1'</u>. We may suppose that $z = \underline{0} = (0, \ldots, 0)$. There is nothing to prove for $n = 0$ (for in that case the g_{ji}, f_j and p_j are constants, by convention). Suppose that the assertion has been proved for $n - 1$, $n > 1$, and for all positive integers m and k, and let n be the number of variables and $m = 1$. Replacing the notation g_{j1}, f_1, p_1 by g_j, f, and p, respectively we then have

$$f = \sum_{j=1}^{k} p_j g_j$$

in some internal open neighborhood U of $\underline{0}$.

Suppose first that $k = 1$. We may assume, if necessary after first carrying out a suitable standard nonsingular linear transformation of the independent variables, that g_1 is regular of order ρ in z_n. Then $g_1 = eh$ where e and h are both standard with domains including $\mu(\underline{0})$ and e is invertible on $\mu(\underline{0})$ while h is a Weierstrass polynomial,

$$h(z_1, \ldots, z_n) = z_n^{\rho} + a_1 z_n^{\rho-1} + \ldots + a_{\rho} ,$$

where the a_j are standard functions of $z_1, \ldots, z_{n-1}$ with $a_j(0, \ldots, 0) = 0$ and holomorphic in a standard neighborhood of the origin in $(z_1, \ldots, z_{n-1})$ - space. Since $f = p_1 g_1$ we have $f = p_1 eh$. At the same time, by the Weierstrass division theorem, there is a representation $f = sh + r$ where, according to a remark in section 1 above, s and h are holomorphic in domains which include $\mu(\underline{0})$. Thus, $p_1 eh = sh + r$ in some internal neighborhood of $\underline{0}$. But then, by the uniqueness of the representation $f = sh + r$, we must have $p_1 e = s$ and $r = 0$, and so $p_1 = e^{-1}s$ is holomorphic in a domain which includes $\mu(\underline{0})$. This remains true if we first had to introduce the linear transformation which made g_1 regular in z_n, for a standard nonsingular transformation maps monads on monads.

Still assuming $m = 1$, suppose that we have proved the assertion of the theorem for positive integers less than some $k \geq 2$. We may assume that the functions $g_1, \ldots, g_k$ are regular in z_n for if this is not the case from the outset we may achieve it for all g_j simultaneously by means of a linear transformation of the independent variables, as before. Furthermore, we may assume that none of the g_j vanish identically (otherwise we omit the g_j in question, unless they all vanish, in which case f also vanishes identically). We may then write

$$g_j = e_j h_j, \ j = 1, \ldots, k,$$

as before, where we may suppose that h_k is the Weierstrass polynomial of highest degree, ρ. The e_j are all holomorphic and invertible in some standard neighborhood of $\underline{0}$, while

$$h_j = z_n^{\rho_j} + a_1^j z_n^{\rho_j - 1} + \ldots + a_{\rho_j}^j, \; \rho_j \leq \rho, \; j = 1, \ldots, k, \; \rho_k = \rho,$$

where the a_i^j are standard and holomorphic in the neighborhood of the origin of $(z_1, \ldots, z_{n-1})$ - space and vanish at the origin. By the division theorem, $f = sh_k + r$ where s and r are holomorphic in $\mu(\underline{0})$. Accordingly, it only remains to be shown that r can be written as a linear combination of the h_j, thus, $r = \Sigma s_j h_j$ where the s_j are internal and holomorphic on $\mu(\underline{0})$.

Let

$$r = b_1 z_n^{\rho-1} + b_2 z_n^{\rho-2} + \ldots + b_\rho .$$

Also, let A be the ideal in ${}^{o}G_{\underline{o}}$ which is generated by $\gamma_{\underline{0}}(h_1), \ldots, \gamma_{\underline{0}}(h_j)$, and let B be the module consisting of $(\rho + 1)$ - tuples of germs of standard holomorphic functions in $(z_1, \ldots, z_{n-1})$-space, $(c_o, \ldots, c_\rho)$ such that $c_o z_n^\rho + c_1 z_n^{\rho-1} + \ldots + c_\rho \in A$. By the finite basis theorem for modules of germs, B has a basis (set of generators) $B_1, \ldots, B_\ell$, $B_j = (\gamma'(b_0^j), \gamma'(b_1^j), \ldots, \gamma'(b_\rho^j))$, where the b_i^j are standard holomorphic functions in the neighborhood of the origin of $(z_1, \ldots, z_{n-1})$ - space and the γ' indicates that the germs are taken at that point. Let

$$d^j = b_0^j z_n^\rho + b_1^j z_n^{\rho-1} + \ldots + b_\rho^j \;;$$

then $\gamma_{\underline{0}}(d_j)$ belongs to A, $j = 1, \ldots, \ell$, and so

$$d^j = \sum_j s_i^j h_i$$

where the s_i^j are standard and holomorphic on the monad $\mu(\underline{0})$.

So far, everything we have said about A and B involves standard functions only. However, by the definition of B we also have that $(0,b_1,b_2, \ldots, b_\rho)$ belongs to ${}^*B$ and so, by the assumption of our induction,

$$(0,b_1,b_2, \ldots, bo) = \sum_j k_j (b_0^j, b_1^j, \ldots, b_\rho^j) ,$$

where the k_j are internal and holomorphic on the monad of $(z_1, \ldots, z_{k-1})$-space. It follows that

$$r = \sum_j k_j (b_0^j z_n^\rho + b_1^j z_n^{\rho-1} + \ldots + b_\rho^j) = \sum_{j,i} k_j s_i^j h_i .$$

But the $k_j s_i^j$ are internal and holomorphic on $\mu(\underline{0})$ and so our assertion is proved for this case. This is still true if we first have to carry out a linear transformation on the independent variables in order to make $g_1, \ldots, g_k$ regular in z_n.

Suppose finally that the assertion has been proved up to and including some $m \geq 1$ and that

$$F = \sum_{j=1}^{k} p_j G_j$$

where

$$F = (f_1, \ldots, f_m, f_{m+1}), \; G_j = (g_{j1}, \ldots, g_{jm}, g_{j,m+1})$$

where the f_j are internal and the g_{ji} are standard, and both the f_j and the g_{ji} are holomorphic on $\mu(\underline{0})$ while the p_j are internal but holomorphic only on some internal neighborhood of the origin, to begin with. In particular,

$$f_{m+1} = \Sigma p_j g_{j,m+1} .$$

By what has been proved already we may replace the p_j in this equation by functions p_j^{m+1} which are internal and holomorphic on $\mu(\underline{0})$, thus

$$f_{m+1} = \sum_j p_j^{m+1} g_{j,m+1} \ .$$

Then the last component in

$$F' = F - \sum_j p_j^{m+1} G_j = \sum_j (p_j - p^{m+1}) g_j$$

vanishes and so, by the inductive assumption on m, there exist functions q_j which are internal and holomorphic on the monad of the origin such that

$$(f_1', \ldots, f_m', 0) = \sum_j q_j (g_{j1}, \ldots, g_{jm}, 0)$$

where

$$F' = (f_1', \ldots, f_m', 0), \ f_i' = f_i - \sum_j p_j^{m+1} g_{j,m+1} \ .$$

But

$$\sum_j (p_j - p_j^{m+1}) g_{j,m+1} = 0$$

in some internal neighborhood of $\underline{0}$ and so

$$\sum q_j g_{j,m+1} = 0$$

in the same neighborhood and, hence on the entire monad $\mu(\underline{0})$. Thus, on $\mu(\underline{0})$,

$$(f_1', \ldots, f_m', 0) = \sum_j q_j (g_{j1}, \ldots, g_{jm}, g_{j,m+1})$$

and further

$$(f_1, \ldots, f_m, f_{m+1}) = \sum_j (q_j + p_j^{m+1}) (g_{j1}, \ldots, g_{jm}, g_{j,m+1}) \ .$$

This completes the proof of 2.1' and hence, of 2.1.

Remarks. In the above proof, we adopted the convention that, for $n = 0$, sections and germs are just complex numbers. Alternatively, we might have considered the case $n = 1$ separately. This would have led to a module B of $(\rho+1)$-tuples of constants in the last part of the proof.

2.2. Corollary. Suppose that the monadic sections $f_1, \ldots, f_m$ in

Theorem 2.1. are finite on $\mu(z)$. Then we may choose the functions $p_1, \ldots, p_k$ so as to be finite on $\mu(z)$.

A standard argument of nonstandard analysis ("Robinson's lemma") shows that the assumption that an internal function is finite on $\mu(z)$ is equivalent to the condition that the function be standardly bounded on $\mu(z)$. Accordingly we may replace "finite," by standardly bounded" both in the hypothesis and in the conclusion of 2.2.

In order to prove 2.2, we only have to check through the proof of 2.1 (or, rather, 2.1'). All the standard holomorphic functions introduced in the course of the proof must be finite (or, equivalently, standardly bounded) on $\mu(\underline{0})$. In addition, we introduce internal functions s and r by means of Weierstrass' division theorem, $f = sh + r$ where h is a standard Weierstrass polynomial and f is standardly bounded on $\mu(\underline{0})$ and, hence, on some polydisc including $\mu(\underline{0})$. Another glance at the integral formula for s on page 70 of [1] (where our s is called g) shows that s and hence r, also are standardly bounded on $\mu(\underline{0})$. Moreover, an appeal to Cauchy's integral formula shows that the coefficients of r as a polynomial of z_n must then be standardly bounded on the monad of the origin of $(z_1, \ldots, z_{n-1})$-space. By means of these facts it is now easy to verify that the p_j also may be chosen so as to be standardly bounded on $\mu(\underline{0})$.

2.1 and 2.2 together provide an easy proof of the following classical result.

2.3 Closure of modules theorem. Suppose that $F = (f_1, \ldots, f_m)$ has holomorphic components on an open neighborhood U of a point z and suppose that F can be uniformly approximated on compact subsets of U by holomorphic functions $(g_1, \ldots, g_m)$ such that $(\gamma_z(g_1), \ldots, \gamma_z(g_m))$ belongs to a submodule A of the m-module of germs at z. Then $(\gamma_z(f_1), \ldots, \gamma_z(f_m))$ also belongs to A.

Proof. Suppose without loss of generality that $z = \underline{0}$. Let $G_1, \ldots, G_\ell$ be a basis of A. Choose a closed polydisc $P\colon |z_j| \le \delta$ which is a subset of U such that the functions of $G_1, \ldots, G_\ell$ are holomorphic on P.

By transferring the hypothesis of the theorem we see that there exist internal functions $h_1, \ldots, h_m$ which are holomorphic on P such that

$$(\gamma(h_1), \ldots, \gamma(h_m)) \in {}^*A$$

and such that

$$\sup|f_j - h_j| \simeq 0, \; j = 1, \ldots, m \text{ on } P.$$

Then the h_j are standardly bounded on P. Hence, by 2.1 and 2.2 there exists a standard δ', $0 < \delta' < \delta$ such that $(h_1, \ldots, h_m) = \sum_j p_j G_j$ for $|z_j| \leq \delta'$, where the p_j are holomorphic and standardly bounded for $|z_j| \leq \delta'$. Writing $G_j = (g_{j1}, \ldots, g_{jm})$ we then have

2.4. $$f_i - \sum_j p_j g_{ji} \simeq 0 \quad \text{for } |z_j| \leq \delta', \; i = 1, \ldots, m$$

But since the p_j are standardly bounded they have (as for functions of a singlecomplex variable) standard parts

$$q_j(t_1, \ldots, z_n) = {}^o p_j(z_1, \ldots, z_n) .$$

Then, by 2.4,

$$f_i - \sum_j q_j g_{ji} \simeq 0$$

for standard points such that $|z_j| < \delta'$. But $f_i - \sum_j q_j g_{ji}$ is standard and so

$f_i - \sum_j q_j g_{ji} = 0$ for $|z_j| < \delta'$. This proves that $(\gamma_{\underline{0}}(f_1), \ldots, \gamma_{\underline{0}}(f_m))$ belongs to A.

The reader may find it instructive to compare our argument with the standard proof given in [1], chapter 2, section D.

3. As might be expected by now, the question of the coherence of a given sheaf is determined entirely by the behavior of its monadic sections. In order to illustrate this fact we shall conclude this paper by stating the nonstandard version of the assertion that every locally finitely generated subsheaf of the sheaf of germs of holomorphic functions on a given domain is coherent (Oka's theorem). If f is a monadic section on $\mu(z)$ and $\zeta \in \mu(z)$ we write $\gamma_\zeta(f)$ for the uniquely determined germ at ζ of a holomorphic function g such that $\beta_z(g) = f$. In particular $\lambda(f) = \gamma_z(f)$. We also write $\gamma_z(F)$ for $(\gamma_z(f_1), \ldots, \gamma_z(f_m))$ if $F = (f_1, \ldots, f_m)$.

<u>3.1. Theorem</u>. Let $G_1, \ldots, G_k$, $k \geq 1$, be m-tuples of standard monadic sections of holomorphic functions at a standard point z,

$$G_j = (g_{j1}, \ldots, g_{jm}), \quad m \geq 1.$$

Then there exist k-tuples of standard monadic sections at z,

$$F_1, \ldots, F_\ell, F_j = (f_{j1}, \ldots, f_{jk})$$

such that

$$\sum_{i=1}^{k} f_{ji} G_i = 0 \quad \text{on} \quad \mu(z), \ j = 1, \ldots, h$$

and such that the following condition is satisfied. Let $\zeta \in \mu(z)$ and let $f = (f_1, \ldots, f_k)$ consist of internal functions which are holomorphic in an internal neighborhood of ζ such that $\sum_j \gamma_\zeta(f_j)\,\gamma_\zeta(G_j) = 0$. Then $\gamma_\zeta(f)$ belongs to the module generated by $\gamma_\zeta(F_1)$, $\gamma_\zeta(F_\iota)$ over G_ζ.

List of References

1. Gunning, R. C. and Rossi, H. Analytic Functions of Several Complex Variables, Englewood Cliffs, N. J. 1965.
2. Hörmander, L. An Introduction to Complex Analysis in Several Variables, Toronto-New York-London 1966.
3. Robinson, A. Nonstandard Analysis, Studies in Logic and the Foundations of Mathematics, Amsterdam 1966.
4. __________. Germs, Applications of Model Theory to Algebra, Analysis, and Probability (ed. W. A. J. Luxemburg) New York, etc. 1969 pp. 138-149.

A NONSTANDARD CHARACTERIZATION OF MIXED TOPOLOGIES

K. D. Stroyan
University of Wisconsin
Madison, Wisconsin 53706

1. INTRODUCTION

In a recent paper [15] we characterized the Mackey topology on L^{∞} for the dual pair $\langle L^{\infty}, L^{1} \rangle$ in terms of an infinitesimal relation on the nonstandard extension ${}^{*}L^{\infty}$. In this note we show how that example can be viewed in terms of the older standard notion of mixed topologies. We also give some other examples. The note is restricted to our characterization and some immediate applications, tho undoubtedly the nonstandard formulation is advantageous for certain other problems in mixed topologies.

The standard notions are due to Alexiewicz, Semadeni, and Wiweger; they are nicely summarized in Cooper [5]. Amongst our applications is a simple proof of the theorem of Cooper [5] and Dorroh [6] that the strict topology on the space of bounded continuous functions over a locally compact space is the mixture of the uniform norm and uniform convergence on compact sets.

The reader is referred to the references [9, 11, 12, 13] for an introduction to nonstandard analysis which we shall not give.

2. INFINITESIMAL RELATIONS, BOUNDED AND FINITE POINTS

In this section we summarize some facts from the nonstandard theory of linear spaces which we use in our characterization. The reader is referred to the papers of Luxemburg [10], Young [17], and especially Henson & Moore [8] for details (or to the notes [11]).

Let E be a linear space and u a family of subsets of $E \times E$. In the non-standard model we form the monad $\mu(u) = \bigcap[{}^*U : U \in u]$. Luxemburg proves that u is a uniformity if and only if $\mu(u)$ is an equivalence relation. In that case we write $x \overset{u}{=} y$ if and only if $(x, y) \in \mu(u)$. We use the notation $o_u[x] = \{y \in {}^*E : x \overset{u}{=} y\}$ for the infinitesimal neighborhood of x in ${}^*E$.

The uniformity u is linear if and only if the infinitesimal relation $\overset{u}{=}$ is linear, that is, if $x \overset{u}{=} x'$, $y \overset{u}{=} y'$ and $\lambda \approx \lambda'$ in the scalars. Then

$$\lambda x + y \overset{u}{=} \lambda' x' + y' \ .$$

We write $a \approx b$ for "$(a - b)$ is infinitesimal" and $a \sim b$ for "$(a - b)$ is finite" in the scalars.

The linear uniformity is locally convex if and only if the infinitesimal neighborhoods are hyperconvex sets, that is, all internal convex combinations are still in the infinitesimals.

The precise way in which the uniformity u is characterized by its infinitesimals is: $U \in u$ if and only if ${}^*U \supseteq \{(x, y) : x \overset{u}{=} y\}$, a standard set is an entourage if and only if its nonstandard extension contains all the infinitesimals (this is a weak precise formulation of Cauchy's heuristic continuity principle).

In the above references and results the fact that the infinitesimal relation is given by the intersection of the standard members of a standard uniformity, that is, the fact that it is monadic, is used. The infinitesimal relation we introduce for the mixture of two locally convex linear uniformities will not generally be monadic, in particular, the remark about convexity will not apply in that setting in general.

The finite points of ${}^*E$, $\mathrm{fin}_u({}^*E)$, for a locally convex linear uniformity u are the points for which all the standard u-seminorms are finite. Henson and Moore [8] make the important observation that the finite points characterize a locally convex uniformity, that is, u is finer than v if and only if $\mathrm{fin}_u({}^*E) \subseteq \mathrm{fin}_v({}^*E)$. A standard set $B \subseteq E$ is bounded if and only if ${}^*B$ consists only of finite points.

The bounded points of ${}^*E$ are $\mathrm{bdd}_u({}^*E) = \bigcup[{}^*B : B$ is a standard bounded set$]$, the union monad of the standard bounded sets. The bounded points and finite points are equal if and only if u is defined by a single seminorm. The fact that the bounded points are a union monad plays a role in our result. Of course, each bounded point is finite by the result quoted above.

Another way to describe the u-infinitesimals for a locally convex linear uniformity: $x \overset{u}{=} y$ if and only if $p(x - y) \approx 0$ for each standard u-continuous seminorm p. Moreover, this characterizes u-continuous standard seminorms.

If $\overset{u}{=}$ is an infinitesimal relation and $f : {}^*E \to {}^*\mathbb{C}$ is any functional we say f is <u>$\overset{u}{=}$-continuous provided</u> $x \overset{u}{=} y$ <u>implies</u> $f(x) \approx f(y)$. In the case of a standard functional $f : E \to \mathbb{C}$, the $\overset{u}{=}$-continuity will refer to ${}^*f$. In a saturated model of a uniform space an <u>internal</u> $\overset{u}{=}$-continuous function is uniformly S-continuous (this is an easy generalization of Robinson [3, Theorem 4.5.8] see [11, 8.4.22]). In particular, <u>a standard function is uniformly continuous if and only if it is</u> $\overset{u}{=}$-<u>continuous on all of</u> ${}^*E$.

Finally, <u>a uniformity</u> u <u>is finer than another</u>, v, <u>if and only if</u> $x \overset{u}{=} y$ <u>implies</u> $x \overset{v}{=} y$.

3. MIXING

Let E be a linear space with two locally convex linear uniformities u and v satisfying:

(1) u is finer than v.

(2) (E, u) is a (DF)-space, that is, there is a fundamental sequence (B_n) of bounded sets and E has the property that whenever (U_n) is a sequence of closed absolutely convex neighborhoods of zero so that $U = \bigcap_{n=1}^{\infty} U_n$ absorbs bounded sets of E, then U is also a neighborhood of zero.

(3) There is a base (B_n) of absolutely convex bounded sets such that $B_n + B_n \subseteq B_{n+1}$ and each B_n is v-closed.

<u>The mixture of</u> u <u>and</u> v, m(u, v) <u>is defined to be the finest locally convex linear uniformity on</u> E <u>which agrees with</u> v <u>on</u> u-<u>bounded sets.</u> We refer the reader to the references [1, 2, 5, 16] for properties of this uniformity.

We define an infinitesimal relation on ${}^*E$ as follows:

"$x \overset{M}{=} y$ <u>if and only if</u> $(x - y)$ <u>is</u> u-<u>bounded and</u> $x \overset{v}{=} y$."

(3.1) Theorem: The $\overset{M}{=}$-continuous standard seminorms $p = E \to \mathbb{R}$ (such that $x \overset{M}{=} y$ implies $p(x-y) \approx 0$) generate the mixed uniformity $m(u,v)$, that is, a standard seminorm is m-continuous if and only if it is $\overset{M}{=}$-continuous.

Proof: If p is a seminorm of $m(u,v)$ and $x \overset{M}{=} y$, then there is a standard u-bounded set $B \subseteq E$ so that $(x-y) \in {}^*B$. (We could assume it is one of the special B_n's.) Since p is m-continuous, it is v-continuous on B or equivalently (since p is standard) $x \overset{v}{=} y$ on ${}^*B$ implies $p(x-y) \approx 0$. This shows that p is $\overset{M}{=}$-continuous.

Conversely, if p is $\overset{M}{=}$-continuous, then it is v-continuous on any u-bounded set, because $\overset{v}{=}$ agrees with $\overset{M}{=}$ on ${}^*B$. Since $m(u,v)$ is the finest locally convex topology that agrees with v on u-bounded sets, p is m-continuous.

Remark: It does not follow that $x \overset{m}{=} y$ implies $x \overset{M}{=} y$.

We introduce another infinitesimal relation on ${}^*E$ as follows:

"$x \overset{\mathcal{M}}{=} y$ if and only if $(x-y)$ is u-finite and $x \overset{v}{=} y$."

This is coarser than $\overset{M}{=}$, and by the last result, standard $\overset{\mathcal{M}}{=}$-continuous seminorms are $\overset{m}{=}$-continuous. Moreover, $\overset{\mathcal{M}}{=}$ agrees with the mixed uniformity on the bounded points and yields the same notion of continuity when there is only one u-seminorm. Since the finite points are easier to work with in most examples, it is desirable to know that $\overset{\mathcal{M}}{=}$ also describes the mixed uniformity.

Next, we define standard part operations:

$$st_m(A) = \{x \in E : {}^*x \overset{m}{=} a \text{ for some } a \in A\}$$

$$st_M(A) = \{x \in E : {}^*x \overset{M}{=} a \text{ for some } a \in A\}$$

$$st_{\mathcal{M}}(A) = \{x \in E : {}^*x \overset{\mathcal{M}}{=} a \text{ for some } a \in A\} .$$

In the case of the conventional infinitesimal relation $\overset{m}{=}$, Luxemburg [10] shows that the closure in the mixed topology of a set $A \subseteq E$ is given by

$$cl_m(A) = st_m({}^*A) .$$

(3.2) Applications:

A1. The Mackey uniformity of $\langle L^\infty, L^1 \rangle$ on L^∞ for finite measures is the mixture of the L^∞-norm and the L^1-norm. The infinitesimal relation on ${}^*L^\infty$ is: "$f \overset{M}{=} g$ if and only if $f(x) \sim g(x)$ except on a set of measure zero and $f(x) \approx g(x)$ except on a set of infinitesimal measure." See [15].

A2. The strict uniformity on $H^\infty(\Omega)$, the space of bounded holomorphic functions (assuming Ω is a region which supports bounded holomorphic functions), is the mixture of the uniform norm and uniform convergence on compact subsets of Ω. It is also the mixture of the uniform norm and pointwise convergence by Robinson's [13] Theorem 6.2.1 which says a finite *-holomorphic function is S-continuous. The infinitesimal relation is: "$f \overset{M}{=} g$ if and only if $f(z) \sim g(z)$ for all $z \in {}^*\Omega$ and $f(z) \approx g(z)$ for z near standard." (The strict seminorms are given in the next example.) Work on H^∞ including this example, but otherwise unrelated to mixing topologies will appear elsewhere.

Consider the case of $\Omega = U = \{z : |z| < 1\}$, the unit disk. Bounded holomorphic functions have non-tangential limits almost everywhere and thus we can view $H^\infty(U)$ as a subspace of $L^\infty(T)$ where $T = \{z : |z| = 1\}$. The function $f(z) = z^\lambda$ for λ an infinite natural number is infinitesimal in the sense of A2. The map $(\ell(z))^\lambda$ is infinitesimal in the sense of A1 where ℓ maps the unit disk onto a lens:

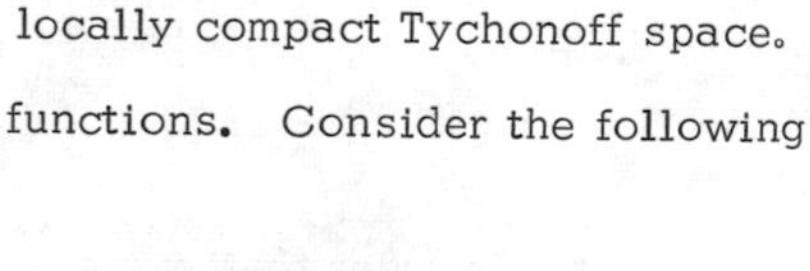

$\xrightarrow{\ell}$

Neither of these functions are norm infinitesimals since they have norm one.

A3. Let X be a locally compact Tychonoff space. Let BC(X) be the space of bounded continuous functions. Consider the following uniformities:

u given by the uniform norm $\|f\|_\infty = \sup[|f(x)| : x \in X]$,

κ given by the seminorms $|f|_K = \sup[|f(x)| : x \in K]$, for K a compact subset of X, the <u>compact convergence uniformity</u>,

β given by the seminorms $|f|_\varphi = \sup[|f(x)\varphi(x)| : x \in X]$, for $\varphi \in C_0(X)$, the <u>strict uniformity</u>.

The infinitesimal relation of compact convergence is $f \overset{\kappa}{=} g$ if and only if $f(x) \approx g(x)$ for $x \in ns({}^*X)$, the near-standard points of ${}^*X$. See [15] Lemma 2 and Luxemburg [10] Theorem 3.7.1, we have used local compactness, $ns({}^*X) = \bigcup[{}^*K : K \subset\subset X]$.

The infinitesimal relation of the mixture of u and κ is "$f \overset{M}{=} g$ if and only if $f(x) \sim g(x)$ for all $x \in {}^*X$ and $f(x) \approx g(x)$ for $x \in ns({}^*X)$."

<u>Theorem</u>: (Cooper [5] and Dorroh [6].) <u>The strict uniformity is the mixture of</u> u <u>and</u> κ.

<u>Proof</u>: Each of the β-seminorms $|\circ|_\varphi$ for $\varphi \in C_0(X)$ is $\overset{M}{=}$-continuous, because if x is remote, $\varphi(x) \approx 0$, so $f \overset{M}{=} 0$ implies $f(x)\varphi(x) \approx 0$, for all $x \in {}^*X$.

Nonstandardizing Buck's original proof that the dual of BC(X) in the strict topology is M(X) [3, Sect. 4], we see that the $\overset{\beta}{=}$-continuous functionals are exactly the *-measures which are finite in variation and nearly concentrated on the near standard points (the variation on internal measurable sets of remote points is infinitesimal). Also see [4].

Note: The *-measures or internal measures in $M({}^*X)$ are the internally countably additive (Radon) measures with internal total variation $|\mu|(X) \in {}^*\mathbb{R}^+$, but this could be infinite.

Without loss of generality we may assume our standard $\overset{M}{=}$-continuous seminorm has the form

$$p(f) = \sup[\langle f, \mu \rangle : \mu \in A]$$

where $A \subseteq M(X)$ is an absorbent set. In order that p is $\overset{M}{=}$-continuous, we must have

$$\int_X f(x)d\mu(x) \approx 0$$

for all $\mu \in {}^*A$ when $f \overset{M}{=} \varphi$. This in turn implies that each member of ${}^*A$ is concentrated on the near-standard points and has finite variation. First, if $E \subseteq \text{support}(\mu)$ is a *-measurable set such that $E \cap ns({}^*X) = \phi$, we approximate the characteristic function χ_E by *-continuous functions in $BC({}^*X)$ with $0 \leq f \leq 1$ by taking a closed subset F of E and an open neighborhood G of E with $\mu(G\setminus F) < \epsilon \approx 0$, $\chi_F \leq f \leq \chi_G$. Each such f satisfies $f \overset{M}{=} 0$ provided we take G so it does not contain near standard points, whence $\int f d\mu \approx 0$ and therefore $\int_E d\mu \approx 0$. Second, by the embedding of $M(X)$ in the uniform dual of $BC(X)$, the total variation of μ is given by

$$|\mu|(X) = \sup[\int f d\mu : \|f\|_u \leq 1,\ f \in BC({}^*X)] \ .$$

However, if $\|f\|_u \leq 1$, then for each infinitesimal scalar $\epsilon \approx 0$

$$\epsilon \int f d\mu = \int \epsilon f d\mu \approx 0$$

since $\epsilon f \overset{M}{=} 0$. Therefore $\int f d\mu$ is finite for all $f \in BC({}^*X)$ with $\|f\| \leq 1$ and the supremum is also finite (being internal and bounded by all infinite numbers).

Since ${}^*A$ consists only of $\overset{\beta}{=}$-continuous functionals, A is β-equicontinuous and p is a β-seminorm which proves the theorem. (A standard family whose *-extension contains only $\overset{u}{=}$-continuous functions is u-equicontinuous; this notion in nonstandard analysis follows from a semantic argument, see [8, 10, 11, 13, 15]. It was first used by Robinson to characterize normal families.) This completes the proof.

Unlike the case in application 2, it is easy to verify that the mixture of the uniform norm and pointwise convergence is not the strict topology. Simply take an infinitesimally good *-continuous approximation to the characteristic function of a nonstandard near-standard point x so that it is a pointwise infinitesimal with $|f|_\varphi \approx \varphi(st(x))$. ($f$ is a blip which vanishes at all standard points and is one at x. x is infinitesimally close to $st(x) \in {}^\sigma X$.)

A4. <u>A sequence</u> $(x_n) \subseteq E$ <u>converges to</u> $x_0 \in E$ <u>in</u> $m(u,v)$ <u>if and only if</u> (x_n) <u>is uniformly</u> u-<u>bounded and</u> $x_n \xrightarrow{v} x_0$. <u>Moreover</u>, $x_n \xrightarrow{m} x_0$ <u>if and only if</u> $x_\omega \overset{M}{=} x_0$ <u>for finite</u> $\omega \in {}^*\mathbb{N}$.

<u>Proof</u>: If $x_\omega \overset{M}{=} x_0$ for all infinite ω, ${}^*(x_n) \subseteq \mathrm{bdd}({}^*E)$ and therefore is a bounded set. On bounded sets m-convergence is equivalent to v-convergence, which follows from the condition $x_\omega \overset{v}{=} x_0$.

Conversely, if p is an M-continuous standard seminorm and $x_n \xrightarrow{m} x_0$ then $p(x_n - x_0) \to 0$, in particular, ${}^*(x_n)$ consists only of finite points and therefore forms a bounded set. Since m is finer than v, $x_n \xrightarrow{v} x_0$. (m and u have the same bounded sets.)

<u>Remark</u>: A question which frequently arises is whether a uniformity is characterized in some sense by its convergent sequences. In this case the convergence notion is $x_n \xrightarrow{B} x_0$ provided $(x_n - x_0)$ forms a u-bounded set and $x_n \xrightarrow{v} x_0$. In particular, when is $m(u,v)$ the finest uniformity with these convergent sequences?

A negative answer in case $\overset{M}{=}$ is not locally convex is easy to give. Precisely, if the uniformity whose entourages are the sets U so that ${}^*U \supseteq \{(x,y) : x \overset{M}{=} y\}$ is <u>not</u> locally convex, then $m(u,v)$ is not the finest. This is clear in view of application 4.

Moreover, when the basic bounded sets are metrizable in the v-uniformity, then the mixture is the finest if $\overset{M}{=}$ is locally convex. The convexity of $\overset{M}{=}$ is nearly a necessary and sufficient condition for the mixture to arise from the convergence notion.

A5. On the space of internal smooth functions ${}^*C^\infty({}^*\mathbb{R}) = {}^*(C^\infty(\mathbb{R}))$ we define the relations:

$\varphi \overset{\mathcal{D}}{=} \psi$ if and only if $\varphi(x) = \psi(x)$ for infinite $x \in {}^*\mathbb{R}$ and whenever $\kappa \in {}^\sigma\mathbb{N}$ is a standard derivative $\varphi^{(\kappa)}(x) \approx \psi^{(\kappa)}(x)$ for all $x \in {}^*\mathbb{R}$.

$\varphi \overset{\mathcal{D}}{\sim} \psi$ if and only if $\varphi(x) = \psi(x)$ for infinite x and $\varphi^{(\kappa)}(x) \sim \psi^{(\kappa)}(x)$ for $\kappa \in {}^\sigma\mathbb{N}$ and $x \in {}^*\mathbb{R}$.

$\varphi \overset{\mathcal{E}}{=} \psi$ if and only if $\varphi^{(\kappa)}(x) \approx \psi^{(\kappa)}(x)$ for finite $x \in {}^*\mathbb{R}$ and $\kappa \in {}^\sigma\mathbb{N}$.

$\varphi \overset{\mathcal{E}}{\approx} \psi$ if and only if $\varphi^{(\kappa)}(x) \sim \psi^{(\kappa)}(x)$ for finite $x \in {}^*\mathbb{R}$ and $\kappa \in {}^\sigma\mathbb{N}$.

These are <u>infinitesimal</u> and <u>finiteness relations</u> for the test spaces $\mathcal{D}$ and $\mathcal{E}$ of distribution theory. (The space $\mathcal{S}$ for tempered distributions could also be described similarly as could the tempered distributions $\mathcal{S}'$.) Note that the $\mathcal{D}$-finite elements mod the $\mathcal{D}$-infinitesimals is Schwartz' space $\mathcal{D}$ and that $\mathcal{E}$-finite mod $\mathcal{E}$-infinitesimal is the space $\mathcal{E}$. The appropriate uniformities are inherited from the infinitesimals.

An internal pairing on ${}^*C^\infty({}^*\mathbb{R})$ is given by

$$(\varphi, f) \to \langle \varphi, f \rangle = \int \varphi(x) f(x) dx \in {}^*\mathbb{R} ,$$

when $\varphi(x)f(x)$ is internally Lebesgue integrable (otherwise $\langle \varphi, f \rangle$ is undefined.)

We shall say a function $f \in {}^*C^\infty({}^*\mathbb{R})$ is a <u>finite distribution</u>, $f \in D$, provided $\langle \varphi, f \rangle$ is finite whenever $\varphi \overset{\mathcal{D}}{\sim} 0$. This is equivalent to saying $\langle \varphi, f \rangle$ is infinitesimal whenever $\varphi \overset{\mathcal{D}}{=} 0$. We call a distribution f so that $\langle \varphi, f \rangle$ is infinitesimal whenever $\varphi \overset{\mathcal{D}}{\sim} 0$ an <u>infinitesimal distribution</u> $f \in d$. The space D/d equals the Schwartz distributions $\mathcal{D}'$ by restricting $\langle \circ, f \rangle$ to standard test functions and ignoring infinitesimal differences.

We call a function $f \in {}^*C^\infty({}^*\mathbb{R})$ a <u>compact distribution</u>, $f \in E$, provided $\langle \varphi, f \rangle$ is defined and finite whenever $\varphi \overset{\mathcal{E}}{\sim} 0$, or $\langle \psi, f \rangle$ is defined and infinitesimal whenever $\varphi \overset{\mathcal{E}}{=} 0$. A function $f \in {}^*C^\infty({}^*\mathbb{R})$ is an <u>infinitesimal compact distribution</u>, $f \in e$, if $\langle \varphi, f \rangle$ is infinitesimal whenever $\varphi \overset{\mathcal{E}}{\sim} 0$. The space E/e equals the compact Schwartz distributions $\mathcal{E}'$.

Notice that our approach to distributions is slightly different from Robinson's [13]; we are allowing internal test functions so continuity is built in.

The strong uniformity on $\mathcal{D}'$ (resp. $\mathcal{E}'$) is induced by the entourages $U \subseteq C^\infty \times C^\infty$ so that ${}^*U \supseteq \{(f,g) : f - g \in d\}$ (resp. $f - g \in e$). The weak uniformity, $\sigma(\mathcal{D}', \mathcal{D})$, (resp. $\sigma(\mathcal{E}', \mathcal{E})$) is measured by standard test functions $\varphi \in {}^\sigma C_K^\infty$ (resp. $\varphi \in {}^\sigma C^\infty$), that is, ${}^*V_\varphi = \{(f,g) : |\langle \varphi, f-g \rangle| \leq 1\}$ forms a base.

Now if f is strongly finite, $f \in D$, and weakly infinitesimal so $\langle \varphi, f \rangle \approx 0$, for $\varphi \in {}^\sigma C_K^\infty$ (resp. $\varphi \in {}^\sigma C^\infty$), then we actually have $\langle \psi, f \rangle \approx 0$ for all $\psi \overset{\mathcal{D}}{\sim} 0$

since $\psi \overset{\mathcal{D}}{=} \varphi \in {}^{\sigma}C_K^{\infty}$ (resp. $\varphi \in {}^{\sigma}C^{\infty}$) where $\varphi(x) = st(\psi(x))$, $x \in {}^{\sigma}\mathbb{R}$. Therefore mixed infinitesimals are strong infinitesimals and the mixture of the weak and the strong is the strong.

In a way this is a negative example since it says you can't weaken strong infinitesimals by mixing even with the weak ones. It may be that there are concrete examples of mixing a proper DF-space (i.e., not a normed space) with a weaker topology, unfortunately the author does not know one. Most "bounded convergences" in analysis involve a single norm. (Recall that only in such examples is there a difference between finite and bounded points.) Of course, the abstract Banach-Dieudonné theorem has a nonstandard description. Such descriptions are interesting, but we expect nonstandard analysis to be most useful where it is used in conjunction with "all the properties" of a particular example.

BIBLIOGRAPHY

[1] Alexiewicz, A., On the two-norm convergence, Studia Math 14 (1953), 49-56.

[2] Alexiewicz, A., and Semadeni, Z., Some properties of two-norm spaces and a characterization of reflexivity of Banach spaces, Studia Math 19 (1960), 115-132.

[3] Buck, R. C., Bounded continuous functions on a locally compact space, Mich. Math. J., 5 (1958), 95-104.

[4] Conway, J. B., The strict topology and compactness in the space of measures II, TAMS 126 (1967), 474-486.

[5] Cooper, J. B., The strict topology and spaces with mixed topologies, PAMS 30 (1971), 583-592.

[6] Dorroh, J. R., The localization of the strict topology via bounded sets, PAMS 20 (1969), 413-414.

[7] Fichtenholz, G., Sur les fonctionnelles linearies continues au sens generalise, Recueil Math 4 (1938), 193-214.

[8] Henson, C. W., and Moore, L. C., Jr., The nonstandard theory of topological vector spaces, TAMS (to appear).

[9] Luxemburg, W. A. J., editor, Applications of Model Theory to Algebra, Analysis, and Probability, Holt, Rinehart and Winston, New York, 1969.

[10] Luxemburg, W. A. J., A general theory of monads, in [9].

[11] Luxemburg, W. A. J. and Stroyan, K. D., Lecture Notes on Nonstandard Analysis (mimeographed), Caltech bookstore, Pasadena, 1973.

[12] Machover, M., and Hirschfeld, J., Lectures on Non-Standard Analysis, Springer-Verlag lecture notes 94, Berlin, 1969.

[13] Robinson, A., Non-Standard Analysis, Studies in Logic and the Foundations of Math, North Holland, Amsterdam, 1966.

[14] Robinson, A., and Luxemburg, W. A. J., Contributions to Non-standard Analysis, North Holland, Amsterdam, 1972.

[15] Stroyan, K. D., A characterization of the Mackey uniformity $m(L^{\infty}, L^{1})$ for finite measures, to appear.

[16] Wiweger, A., Linear spaces with mixed topology, Studia Math 20 (1961), 47-68.

[17] Young, L., Functional analysis - a non-standard treatment with semifields, in [14].

ACKNOWLEDGEMENT: The author was supported by N. S. F. grant GP-24182.

APPLICATIONS OF FRACTIONAL POWERS OF DELTA FUNCTIONS

James K. Thurber* and Jose Katz**

*Purdue University

**Free University, Berlin

Introduction

By means of Non-Standard Analysis a new method is open to theoretical physicists to deal with infinite renormalization, a route which is however closely geared to orthodox physical intuition. Three interrelated problems are pursued here.

1) The self-energy of a classical electron.

2) Some formalisms in quantum scattering theory.

3) The infinite-star finite quantum many body problem.

The star-finite many body problem is of interest because potentially it is an alternative to standard quantum field theory. It is a simple matter to give a non-trivial theory for (some infinite but star finite natural number) interacting Dirac electrons which is manifestly Lorentz covariant and which is such that there is no violation of macro or micro causality except infinitesimally. Considering the efforts in the foundations of quantum field theory this assertion seems hard to believe. Here we limit the discussion to a relativistically invariant model for a classical electron. Once a reasonable theory is established here the transition to the Dirac many electron theory is clear. The interesting question then becomes what kind of calculations can be performed within the context of non-standard analysis. This will be left for a later exposition. In this paper preliminary calculations are made which indicate the feasibility of such an endeavour. One comment however - it will be apparent after the reader has finished this paper that one can make sense out of an infinite sea of electrons without quantum field theory.

Again the primary question becomes where does this lead one.

Basically the reason why one can obtain results that are thought to be a prerogative of quantum field theory, are first that the state vectors and operators are defined on an enlarged Non-Standard Hilbert space which is so big that it is bigger than the space used in the usual Fock representation of the method of second quantization. The natural numbers in the star finite interval $1 \leq j \leq \omega$ are uncountably infinite. Thus the quantum ω-body system is much larger than a standard Fock space. And second the peculiar features of enlargements a la Non-Standard Analysis where many of the essential properties of pre-enlarged systems are preserved. In the language of Non Standard Analysis the interval $1 \leq j \leq \omega$ where ω is star finite is an internal set.

Thus the basic features of these enlargements which will condition the attitude and contents of this paper is that if a class of problems is known to be well posed for the pre-enlarged system then for internal subsystems of the enlarged system the corresponding problems are automatically well posed. From the foundation point of view the finite many body problem is simpler to justify, whether relativistic or non-relativistic, and thus it follows from the theorems of Non-Standard Analysis that the star finite many body problem is easier from the foundation point of view than quantum field theory. Here some calculations that indicate that such an approach is feasible are given.

Part I.

The Classical Electron Revisited

First we consider a static electron because it is simpler to illustrate the main ideas and second a relativistically invariant formulation of an electron moving at constant speed.

As basic references for Non-Standard Analysis see (refs. 1, 2, 3, 4). In particular for the connection between Non-Standard Analysis and Quantum Mechanics see ref. 5. We will use the notation of ref. 6.

The notation

$$\Delta(x) = \left(\frac{n}{\pi}\right)^{1/2} e^{-nx^2} \tag{1}$$

where n is a star-finite positive infinite constant will be used to distinguish the Non-Standard Gaussian as a delta function from the Non-Standard extension of the delta function of L. Schwarz's theory which will be denoted by

$$^{*}\delta(x) \tag{2}$$

$$\Delta^{p}(x-a) = \left(\frac{n}{\pi}\right)^{p/2} e^{-np(x-a)^2} \tag{3}$$

For $p > 0$, we define

$$\hat{\Delta}^{p}(k) = {}^{*}\!\int_{-\infty}^{\infty} e^{ikx}\Delta^{p}(x)\,dx = \left(\frac{n}{\pi}\right)^{\frac{p-1}{2}} p^{-\frac{1}{2}} e^{\frac{-k^2}{4np}} \tag{4}$$

And in general we will denote Fourier transforms of functions by the symbol ^ placed above. By direct calculations we obtain

$$\Delta^{p}(x-a)\Delta^{q}(x-b) = \Delta^{p+q}\left\{x - \left[\frac{ap+bq}{p+q}\right]\right\} e^{\left[-n\left(\frac{pq}{p+q}\right)(a-b)^2\right]} \tag{5}$$

a useful identity.

$$\tilde{\Delta}_q(x) = \left(\frac{nq}{\pi}\right)^{1/2} e^{-nqx^2} \quad \text{for } q > o, \tag{6}$$

is a delta function of the same type as $\Delta(x)$, which will occur in our calculations.

Maxwell's equations and the Lorentz force on a charged particle will be written in the following form (see ref. 6).

$$\nabla \times H = \frac{1}{c}\left(\frac{\partial E}{\partial t}\right) + \frac{4\pi J}{c} \tag{7}$$

$$\nabla \times E = -\frac{1}{c}\left(\frac{\partial H}{\partial t}\right) \tag{8}$$

$$\nabla \cdot H = 0 \tag{9}$$

$$\nabla \cdot E = 4\pi\rho \tag{10}$$

$$\mathbf{E} = -\frac{1}{\mathbf{c}} \frac{\partial A}{\partial t} - \nabla\phi \tag{11}$$

$$H = \nabla \times A \tag{12}$$

A and ϕ are the vector and scalar potentials, which are related by

$$\nabla \cdot A = -\frac{1}{c}\left(\frac{\partial\phi}{\partial t}\right) \tag{13}$$

and are arbitrary to within a guage transformation of the form

$$A' = A + \nabla f, \ \phi' = \phi - \frac{1}{c}\frac{\partial f}{\partial t} \tag{14}$$

where f is an arbitrary function of position (x_1, x_2, x_3) and time t.

The Lorentz force per unit charge is given by

$$F = E + \frac{1}{c}\,(V\times H), \text{ where} \tag{15}$$

V is the velocity of the particle in the reference frame in which the force is being measured.

The energy density of the electromagnetic field is given by

$$U = \frac{1}{8\pi}\,{}^{*}\!\int\!\!\int\!\!\int (E\cdot E+H\cdot H)\,dx_1dx_2dx_3, \tag{16}$$

where the integral is taken for, $-\infty < x_1,x_2,x_3 < \infty$.

For the case of an electron at rest, the magnetic field $H = 0$, and the curl of the electric field $\nabla \times E = 0$. Thus we have a scalar function ϕ such that

$$E = -\nabla\phi \tag{17}$$

Then the charge density ρ and potential ϕ are related by Poissons eq.

$$\nabla^2\phi = -4\pi\rho \tag{18}$$

Further we have

$$U = \frac{1}{2}\,{}^{*}\!\int\!\!\int\!\!\int \rho\phi\,dx_1dx_2dx_3 \tag{19}$$

For our model centered at $x_1 = x_2 = x_3 = 0$ we take

$$\rho = \alpha\,\Delta^{p}(x_1)\Delta^{p}(x_2)\Delta^{p}(x_3) \tag{20}$$

where $\alpha, p > 0$ are to be determined.

Thus
$$\hat{\rho} = \alpha\left(\frac{n}{\pi}\right)^{\frac{3(p-1)}{2}} p^{-3/2}\exp\left[\frac{-k_1^2-k_2^2-k_3^2}{4np}\right] \tag{21}$$

The Non-Standard extension of Parsevals theorem has the usual form, thus

$$2U = {}^{*}\!\int \phi\rho\,dx_1dx_2dx_3 = \frac{1}{8\pi^3}\,{}^{*}\!\int \hat{\phi}\,\hat{\rho}\,dk_1dk_2dk_3 \tag{22}$$

From the Poisson equation (18) connecting ρ and ϕ we have

$$\left(k_1^2 + k_1^2 + k_3^2\right)\hat{\phi} = 4\pi\hat{\rho} \tag{23}$$

Solving for $\hat{\phi}$ in terms of $\hat{\rho}$ and using (21) and (22) we obtain

$$U = \frac{\alpha^2}{\sqrt{2}\; p^{5/2}} \; \frac{n^{3p - 5/2}}{\pi^{3p - 5/2}} \tag{24}$$

Under the assumption that α and p are finite and positive, since n is infinite-star finite, the only way to avoid either an infinitesimal or infinite self-energy U is to choose

$$p = \frac{5}{6} \; . \tag{25}$$

This gives

$$U = 4(3/5)^{5/2}\alpha^2 \; . \tag{26}$$

If we identify U with the energy of the rest mass of the electron then

$$4(3/5)^{5/2}\alpha^2 = mc^2 \quad \text{and} \tag{27}$$

$$\alpha = (5/3)^{5/4} \frac{m^{\frac{1}{2}}c}{2} \tag{28}$$

In terms of (6) we now write

$$\rho = \alpha\left(\frac{6}{5}\right)^{3/2}\left(\frac{\pi}{n}\right)^{1/4} \tilde{\Delta}_{5/6}(x_1)\tilde{\Delta}_{5/6}(x_2)\tilde{\Delta}_{5/6}(x_3) \tag{29}$$

Since the factor $\left(\frac{\pi}{n}\right)^{1/4}$ is an infinitesimal it would seem that we have a contradiction with reality since forces between particles would be infinitesimal unless the particles were infinitely close. A contradiction apparently to the long range nature of Coulomb forces. The interpretation of this apparent nonsense will be given in the last section where the renormalized effective interaction due to infinitely many interacting particles will be seen to be finite. The observed particles are the dressed or quasi-particles (see ref. 7). The infinite positive constant n has the dimensions of reciprocal square of length.

Now we construct a relativistically invariant formulation which will allow for an electron moving with finite constant velocity.

In terms of the potentials A and ϕ and the current J and density ρ Maxwell's equations take the form

$$\frac{1}{c^2}\frac{\partial^2 A}{\partial t^2} - \nabla^2 A = \frac{4\pi J}{c} \tag{30}$$

$$\frac{1}{c^2}\frac{\partial^2 \phi}{\partial t^2} - \nabla^2 \phi = 4\pi\rho \tag{31}$$

$$\nabla \cdot A = -\frac{1}{c}\frac{\partial\phi}{\partial t} \tag{32}$$

There are many possibilities. We define our model in terms of the charge distribution of the electron in its rest frame. Our analysis is based on the previous static computation. All but an infinitesimal percentage of the charge will lie within the monad of the center of charge of the electron. Hence from the standard point of view the electron will be a point electron.

The superscript L will denote the Lorentz transformation to a moving frame having velocity V. Our model is

$$\frac{1}{c^2}\frac{\partial^2 A}{\partial t^2} - \nabla^2 A = \frac{4\pi V\gamma\rho^L}{c} \tag{33}$$

$$\frac{1}{c^2}\frac{\partial^2 \phi}{\partial t^2} - \nabla^2 \phi = 4\pi\gamma\rho^L \tag{34}$$

$$\nabla \cdot A = -\frac{1}{c}\frac{\partial\phi}{\partial t} \quad , \tag{35}$$

where

$$\rho^L = \alpha(\Delta^{5/6}(x_1^L - x_1^{oL})\, \Delta^{5/6}(x_2^L - x_2^{oL})\, \Delta^{5/6}(x_3^L - x_3^{oL})\, {}^*\Theta(t^L - t^{oL}) \tag{36}$$

where
$$ {}^*\Theta(s) = \begin{cases} 1 & \text{for} \quad s > 0 \\ 0 & \text{for} \quad s < 0 \end{cases} \tag{37}$$

A and ϕ are given in terms of

$$\Psi = \frac{\alpha}{4\pi} \int dy_1 dy_2 dy_3 \; \frac{{}^*\Theta(t-t^o-|y|)}{|y|}\, \Delta^{5/6}(x_1-x_1^o-y_1)\Delta^{5/6}(x_2-x_2^o-y_2)$$

$$\text{times} \;\; \Delta^{5/6}(x_3-x_3^o-y_3) \tag{38}$$

where $\quad |y| = (y_1^2 + y_2^2 + y_3^2)^{1/2} \quad ,$

$$A = \frac{4\pi V\gamma\psi^L}{c} \quad , \quad \phi = \frac{4\pi\gamma\Psi^L}{c} \tag{39}$$

$$\gamma = \left(1 - \frac{V^2}{c^2}\right)^{-1/2} \tag{40}$$

c being the speed of light.

$$\Psi^L = \Psi(t^L, t^{oL}, x_1^L, x_1^{oL}, x_2^L, x_2^{oL}, x_3^L, x_3^{oL}) \ . \tag{41}$$

A question which naturally arises, is, to what extent is this model consistent with causality requirements. To answer this consider an electron moving with speed V in the x direction with respect to the observer's frame of reference. The corresponding Lorentz transformations take the form

$$t^L - t^{oL} = \left[t - t^o - \frac{V(x_1 - x_1^o)}{c^2} \right] \gamma$$

$$x_1^L - x_1^{oL} = \left[x_1 - x_1^o - V(t - t^o) \right] \gamma \qquad (42)$$

$$x_2^L - x_2^{oL} = x_2 - x_2^o$$

$$x_3^L - x_3^{oL} = x_3 - x_3^o$$

From the form of ρ^L we see that all but an infinitesimal percentage of the charge lies in the region where

$x_1^L - x_1^{oL}$, $x_2^L - x_2^{oL}$ and $x_3^L - x_3^{oL}$ are all in the monad of zero. But

$x_1 - x_1^o = V(t - t^o) + (x^L - x^{oL})\gamma^{-1}$, which gives

$x_1 - x_1^o = V(t - t^o)$ + an infinitesimal. This in turn implies that

$t^L - t^{oL} = (t - t^o)(1 - \frac{V^2}{c^2})\gamma$ + an infinitesimal

$= (t - t^o)\gamma^{-1}$ + an infinitesimal. (43)

Due to the factor ${}^*\theta$ (see (36) and (37)) we only have charge for

$$t^L - t^{oL} \geq 0. \qquad (44)$$

Combining (43) and (44) we conclude that all but an infinitesimal percentage of the charge lies in the region

$$t - t^o \geq \text{an infinitesimal}. \qquad (45)$$

Relations (44) and (45) are the form in which our model satisfies causality. Stated in words if the electron is born at time t^{oL} according to an observer, except for an infinitesimal percentage of its charge it will be born at t^o in its own reference frame. It is important to emphasize that only infinitesimal percentage of the charge violates causality, since later after renormalization by an infinite-star finite constant the charge distribution will be finite rather than infinitesimal, however again only an infinitesimal percentage of the charge will violate causality.

Thus our model satisfies reasonable causality requirements.

Before going onto the many body problem and an interpretation of the renormalization of infinitesimal charge, we will describe non-standard wave-packet analysis.

Part II.

Non-Standard Wave Packets and T-Matrix Formalism

(see ref.8)

To illustrate the main ideas in the simplest context we begin with the Schrodinger wave equation in one space dimension for one particle.

$$i\hbar \frac{\partial \Psi}{\partial t} = -\frac{\hbar^2}{2m} \frac{\partial^2 \Psi}{\partial x^2} + V\Psi \tag{46}$$

$$-\infty < x < \infty, \; t \geq 0$$

$\hbar = h/2$ (h = Plancks constant).

V is the potential which is assumed to be a function of the space coordinate x only.

$$\int_{-\infty}^{+\infty} |\Psi|^2 dx = 1 \quad \text{for all} \quad t \quad \text{and}$$

$|\Psi|^2$ is usually interpreted as a probability density. In calculations however one sometimes finds initial data of the following forms

$$\Psi_o = Ae^{ik_o x} \tag{47}$$

$$\Psi_o = A\delta(x) \tag{48}$$

where A is a constant $\neq 0$.

The first choice is not square integrable in the interval $-\infty < x < \infty$, and the second is not square integrable in any interval containing $x = 0$. A problem also arises because $\delta(x)\delta(x)$ is generally not well defined. A common definition is to assume that

$$\delta(x)\delta(x) = B\delta(x) \tag{49}$$

where B is some constant tailored to suit a particular calculation. However in this case both

$$\left[\delta(x)\ \delta(x)\right]\ \delta'(x) = \delta(x)\left[\delta(x)\ \delta'(x)\right] \quad \text{and}$$

$\delta(x)\ \delta'(x) = \delta'(x)\ \delta(x)$ cannot be true, for if so then on the one hand

$$\begin{aligned}\left[\delta(x)\ \delta(x)\right]\ \delta'(x) &= B\delta(x)\ \delta'(x)\\ &= B/2(\delta(x)\delta(x))'\\ &= B/2(B\delta'(x))\\ &= B^2/2\ \delta'(x), \text{ and}\end{aligned}$$

on the other hand

$$\begin{aligned}\delta(x)\left[\delta(x)\ \delta'(x)\right] &= \delta(x)/2\left[\delta(x)\ \delta(x)\right]'\\ &= B/2\ \delta(x)\ \delta'(x)\\ &= B/4\left[\delta(x)\ \delta(x)\right]'\\ &= B^2/4\ \delta'(x)\ .\end{aligned}$$

So that $B^2/2\ \delta'(x) = B^2/4\ \delta'(x)$.

This is a contradiction unless $B = 0$, nonetheless in many applications $B > 0$.

For $1 = \int_{-\infty}^{+\infty} |\Psi|^2 dx = \int_{-\infty}^{+\infty} |A\ \delta(x)|^2 dx = |A|^2 B$, we have to choose $B = 1/|A|^2 > 0$.

In quantum mechanics when physicists wish to give a more rigorous derivation of their results they work with so called wave packets (see ref. 8).

The first initial condition is replaced by

$$\Psi_o = (2\pi)^{-1/2} \int_{-\infty}^{+\infty} \left(\frac{n}{\pi}\right)^{1/4} e^{-\frac{n}{2}(k-k_o)} e^{ikx} dx \tag{50}$$

and the second by

$$\Psi_o = \left(\frac{n}{\pi}\right)^{1/4} e^{-\frac{nx^2}{2}} \tag{51}$$

However instead of choosing n to be finite as is the case in standard physics we choose n to be infinite-star finite. Using again the notation of relation (3) we have

$$\Psi_o = (2\pi)^{-1/2} \; {}^*\!\int_{+\infty}^{-\infty} \Delta^{1/2}(k-k_o)e^{ikx}\,dk, \tag{52}$$

and $\Psi_o = \Delta^{1/2}(x)$. (53)

Thus by reinterpreting the Schrodinger equation in the larger Non-Standard Hilbert-Space we have achieved the intent of the original but non-rigorous formulation of having a point particle or a plane wave, "essentially", and still remaining in a Hilbert-Space.

As in standard quantum mechanics the uncertainty relation still remains valid. For illustration we consider the case where the potential $V \equiv 0$ and the initial condition (52) is used. The solution for $t \geq 0$ is given as

$$\Psi(x,t) = \left(\frac{n}{\pi s^2}\right)^{1/4} \exp\left[i(k_o x - \omega_o t) - \frac{(x-ut)^2}{2s}\right] , \tag{54}$$

where

$$s = n + \frac{i\hbar t}{m} , \quad |s| = \left[n^2 + \frac{\hbar^2 t^2}{m^2}\right]^{1/2} .$$

$$u = \frac{k_o \hbar}{m} \quad \text{and} \quad \omega_o = \frac{\hbar k_o^2}{2m} .$$

The mean values and standard deviations of the position x and momentum operators

$$p = -i\hbar \frac{\partial}{\partial x} \tag{55}$$

$$\langle x\rangle = {}^{*}\int_{-\infty}^{+\infty} x\bar{\Psi}\Psi dx = ut \ . \tag{56}$$

$$\langle p\rangle = {}^{*}\int_{-\infty}^{+\infty} \bar{\Psi}(-i\hbar \ \frac{\partial}{\partial x})\Psi dx = mu \tag{57}$$

$$\Delta x = (\langle x^2\rangle - \langle x\rangle^2)^{1/2} = |s|(2n)^{-1/2} \tag{58}$$

$$\Delta p = (\langle p^2\rangle - \langle p\rangle^2)^{1/2} = \hbar \ (2n)^{-1/2} \tag{59}$$

and the uncertainty relation reduces to

$$\Delta x\Delta p = \frac{\hbar}{2}\left(1 + \frac{\hbar^2 t^2}{n^2 m^2}\right)^{1/2} \geq \frac{\hbar}{2} \ . \tag{60}$$

Thus our version of a plane wave which we call a quasi plane wave has the properties that

1) Its spread in momentum Δp is a constant infinitesimal.

2) Its spread in position Δx is $(n/2)^{1/2}$ + an infinitesimal for finite times t or even for infinite t such that $|t| \leq n^{1/2}$ say.

3) It has a definite mean position $\langle x\rangle$ which moves with constant velocity u.

4) Its average momentum is its mass m times its finite average velocity u.

5) For finite times or infinite times $|t| \leq n^{1/2}$ the uncertainty $\Delta x\Delta p$ is $\hbar/2$ plus a positive infintesmal.

Thus we have the best of both worlds, the convenience of plane waves and a home in a Hilbert Space.

Before going on to the T-Matrix formalism we will need the Non-Standard concept of an approximate free particle. Just temporarily and for illustration purposes only we restrict our attention to one space dimension.

In the momentum representation (Fourier transform of the space variable) the free particle Schrodinger equation takes the form,

$$\frac{\partial\Psi}{\partial t} = \frac{-i\hbar}{2m} \ k^2\Psi \ . \tag{61}$$

For $\Psi = \Delta^{1/2}(k-k_o)$ at $t = 0$ we have

$$\Psi = e^{\frac{-i\hbar k^2 t}{2m}} \Delta^{1/2}(k-k_o) \, . \tag{62}$$

We will now assume that

$$|t| \leq n^{1/8} \tag{63}$$

where the n occuring in the right hand side of relation (63) is the same n as that occuring in the definition of Δ (see (1) and (3)).

$$\text{If} \quad (k-k^o)^2 \leq n^{-1/2}, \text{ then} \tag{64}$$

$$k = k_o + O(n^{-1/4})$$

$$t^{1/2}k = t^{1/2}k_o + O(t^{1/2}n^{-1/4})$$

$$tk^2 = tk_o^2 + O(tk_o n^{-1/4}) + O(tn^{-1/2}) \quad \text{and}$$

$$tk^2 - tk_o^2 = O(n^{-1/8}) \tag{65}$$

Define

$$\tilde{\Psi} = e^{\frac{-i\hbar k_o^2 t}{2m}} \Delta^{1/2}(k-k_o), \text{ then}$$

$$\Psi/\tilde{\Psi} = 1 + \text{an infinitesimal.} \tag{66}$$

But

$|\tilde{\Psi}| = |\Psi|$ for all values of k and in particular for $|k-k_o| \geq n^{-1/4}$.

$$\underset{(k-k_o)^2 \geq n^{-1/2}}{{}^*\!\!\int} |\Psi|^2 dk = \underset{(k-k_o)^2 \geq n^{-1/2}}{{}^*\!\!\int} |\tilde{\Psi}|^2 dk$$

$$= 2\left(\frac{n}{\pi}\right)^{1/2} {}^*\!\!\int_{n^{-1/4}}^{\infty} e^{-nx^2} dx = 2\pi^{-1/2} {}^*\!\!\int_{n^{-1/4}}^{\infty} e^{-y^2} dy \, .$$

For positive infinite-star finite z

$$2ze^{z^2} \, {}^*\!\int_z^\infty e^{-y^2} dy = 1 + \text{an infinitesimal}$$

Hence

$${}^*\!\int_{(k-k_o)^2 \geq n^{-1/2}} |\Psi|^2 dk = {}^*\!\int_{(k-k_o)^2 \geq n^{-1/2}} |\tilde{\Psi}|^2 dk = \frac{e^{-n^{1/2}}}{\pi^{1/2} n^{1/4}} (1 + \text{an infinitesimal}),$$

but

$${}^*\!\int_{-\infty}^{\infty} |\Psi|^2 dk = {}^*\!\int_{-\infty}^{\infty} |\tilde{\Psi}|^2 dk = 1 .$$

Thus the region where $|k-k_o| \geq n^{-1/4}$ is a region of infinitesimal total probability. With probability $1 - \pi^{-1/2} n^{-1/4} e^{-n^{1/2}}$ (1 + an infinitesimal) ,

$|k-k_o| \leq n^{-1/4}$. This is our motivation for replacing the free particle wave function Ψ by its approximation $\tilde{\Psi}$.

Next define

$$\tilde{\tilde{\Psi}} = \exp(-t^2/4n)\Psi. \tag{67}$$

Since $|t| \leq n^{1/8}$

$$\tilde{\tilde{\Psi}} = \tilde{\Psi} \text{ (1 + an infinitesimal)} \tag{68}$$

$$= \Psi \text{ (1 + an infinitesimal), with}$$

probability 1 minus an infinitesimal. Clearly $\tilde{\tilde{\Psi}}$ is also an infinitely close approximation to Ψ for $|t| \leq n^{1/8}$, and we have

$$\tilde{\tilde{\Psi}} = \exp\left[-i\omega_o t - \frac{t^2}{4n}\right] \Delta^{1/2}(k-k_o) . \tag{69}$$

where $\omega_o = \frac{\hbar k_o^2}{2m}$. By direct calculation one sees that

$$\tilde{\tilde{\Psi}} = (2\pi)^{-1/2} \, {}^*\!\int_{-\infty}^{\infty} e^{-i\omega t} \Delta(\omega-\omega_o) \Delta^{1/2}(k-k_o) d\omega \tag{70}$$

where $\omega = \frac{\hbar k^2}{2m}$.

Thus in the $\omega - k$ representation the free particle is infinitely nearly a product of a delta function by a square root of a delta function, namely

$$\Delta(\omega\ \omega_o)\Delta^{1/2}(k-k_o)\ . \tag{71}$$

This obviously generalizes to higher space dimensions. In the case of 3 dimensional space which is of further interest, a free particle in the (ω,k_1,k_2,k_3) representation is infinitely nearly equal to

$$\Delta(\omega-\omega_o)\Delta^{1/2}(k_1-k_{10})\Delta^{1/2}(k_2-k_{20})\Delta^{1/2}(k_3-k_{30})\ , \tag{72}$$

for $|t| \leq n^{1/8}$ and where

$$\omega = \frac{\hbar(k_1^2+k_2^2+k_3^2)}{2m} \quad \text{and} \quad \omega_o = \frac{\hbar\ (k_{10}^2+k_{20}^2+k_{30}^2)}{2m} \tag{73}$$

The above considerations are now applied to the calculation of the collision cross section of a scattering process in the center of mass system with two incoming particles and N produced (see ref.8).

In the expression (see 6)

$$\tilde{\Delta}_\beta(x) = \left(\frac{n\beta}{\pi}\right)^{1/2} e^{-n\beta x^2}\ , \tag{74}$$

β will be taken as an infinite star-finite positive constant. In the following calculations three kinds of delta functions will concern us, namely, $\tilde{\Delta}_\beta$, Δ and $\Delta^{1/2}$. The $\tilde{\Delta}_\beta$ will occur as factors of the T-Matrix elements and correspond to the conservation laws of momentum and energy for an elastic collision. The reason for taking β to be infinite is therefore clear. In the limit

$$\text{as } \beta \xrightarrow{*} +\infty\ , \tilde{\Delta}_\beta(x) \xrightarrow{*} {}^*\delta(x) \tag{75}$$

the replacement of $\tilde{\Delta}_\beta$ (s) by $*\delta(x)$ would guarantee that the conservation laws are satisfied exactly. We could in fact perform the calculations with $*\delta$ in place of $\tilde{\Delta}_\beta$, but will use $\tilde{\Delta}_\beta$ instead just to show that it is possible. The results of the calculations differ only by an infinitesimal percent.

Rather than k and ω we now use the notation

P for momenta and (76)

E for energy (77)

The superscripts "I" will denote the incoming states and "F" the final or outgoing states.

$$\phi_j^I = \Delta^{1/2}(P_1^j - P_{10}^j)\Delta^{1/2}(P_2^j - P_{20}^j)\Delta^{1/2}(P_3^j - P_{30}^j)\Delta(E - E_o^j) \quad (78)$$

For j = 1, 2

$$S = (P_1, P_2, P_3, E),\ S^I \text{ for } S \text{ initial and } S^F \text{ for } S \text{ final.} \quad (79)$$

The T-Matrix element takes the form

$$\left\langle S_1^F, S_2^F, \ldots\ldots\ldots, S_N^F | T | S_I^I, S_2^I \right\rangle$$

$$= {}^*\delta(\Sigma P_1^F - \Sigma P_I^I)\,{}^*\delta(\Sigma P_2^F - \Sigma P_2^F)\,{}^*\delta(\Sigma P_3^F - \Sigma P_3^I) \quad (80)$$

$$\text{times } {}^*\delta(\Sigma E^F - \Sigma E^I) \left\langle \Omega | T | S_1^I, S_2^I \right\rangle$$

or

$$\tilde{\Delta}_\beta(\Sigma P_1^F - \Sigma P_1^I)\tilde{\Delta}_\beta(\Sigma P_2^F - \Sigma P_2^I)\tilde{\Delta}_\beta(\Sigma P_3^F - \Sigma P_3^I)$$

$$\text{times } \tilde{\Delta}_\beta(\Sigma E^F - \Sigma E^I) \left\langle \Omega | T | S_1^I, S_2^I \right\rangle \ . \quad (81)$$

The symbol $\overset{*}{\sim}$, will be (82)
used to denote Non-Standard asymptotic equivalence. That is the quotient of both sides of the relation is equal to one plus an infinitesimal.

Of the two alternative assumed forms of $<|T|>$, namely (80) or (81) we choose (81) and will merely quote the results for (80).

Ω is defined by the differential relation

$$dS_1^F dS_2^F . . . dS_N^F = d(\Sigma P_1^F) d(\Sigma P_2^F) d(\Sigma P_3^F) d(\Sigma E^F) d\Omega \ . \tag{83}$$

$$\phi_N^F = {}^*\!\!\int \left\langle S_1^F, . . . , S_N^F | T | S_1^I, S_2^I \right\rangle \phi_1^I \phi_2^I dP_1^1 dP_2^1 dP_3^1 dE^1 dP_1^2 dP_2^2 dP_3^2 dE^2 \tag{84}$$

$$dw = d\Omega \ {}^*\!\!\int |\phi_N^F|^2 d(\Sigma P_1^F) d(\Sigma P_2^F) d(\Sigma P_3^F) d(\Sigma E^F) \tag{85}$$

$$\Psi_j = (2\pi)^{-3/2} \ {}^*\!\!\int \phi_j^I \exp\left[i(P_1^j x_1 + P_2^j x_2 + P_3^j x_3 - E^j t) \right] dP_1^j dP_2^j dP_3^j dE^j \tag{86}$$

for $j = 1,2.$

$$\rho_j = |\Psi_j|^2 \quad \text{for} \quad j = 1,\ 2. \tag{87}$$

The center of mass system differential cross section is given by

$$d\sigma = \frac{dw}{(V_1 + V_2) \left[{}^*\!\!\int \rho_1 \rho_2 dx_1 dx_2 dx_3 dt \right]} \tag{88}$$

(see ref.8)

where $V_1 + V_2$ is the sum of the velocities of the incoming particles.

Performing indicated integrations we obtain after squaring

$$|\phi_N^F|^2 \ \underset{\sim}{*} \ \frac{\pi}{n} \left(\frac{4\beta}{\beta+1} \right)^{3/2} \left(\frac{2\beta+1}{4\beta+1} \right)^{3/2} \left(\frac{\beta}{4\beta+2} \right)^{1/2}$$

$$\text{times} \quad \tilde{\Delta}_{\left(\frac{2\beta}{4\beta+1}\right)} (P_{10}^2 + P_{10}^1 - \Sigma P_1^F) \ \tilde{\Delta}_{\left(\frac{2\beta}{4\beta+1}\right)} (P_{20}^2 + P_{20}^2 - \Sigma P_2^F) \tag{89}$$

$$\text{times} \quad \tilde{\Delta}_{\left(\frac{2\beta}{4\beta+1}\right)} (P_{30}^2 + P_{30}^1 - \Sigma P_3^F) \ \tilde{\Delta}_{\left(\frac{2\beta}{4\beta+1}\right)} (E_o^2 + E_o^1 - \Sigma E^F)$$

$$\text{times} \quad \left\langle \Omega | T | S_1^I,\ S_2^I \right\rangle^2 \ .$$

This gives

$$dw \overset{*}{\sim} \frac{\pi}{n}\left(\frac{4\beta}{\beta+1}\right)^{3/2}\left(\frac{4\beta+2}{4\beta+1}\right)^{3/2}\left(\frac{\beta}{4\beta+2}\right)^{1/2} \left\langle \Omega|T|S_1^I\ ,\ S_2^I\right\rangle^2 d\Omega \tag{90}$$

$$\Psi_j = \frac{\exp\left[P_{10}^j x_1 + P_{20}^j x_2 + P_{30}^j x_3 - E_o^j t - \frac{1}{2n}(x_1^2+x_2^2+x_3^2 - \frac{t^2}{4n}\right]}{n^{3/4}\pi^{3/4}} \tag{91}$$

$$j = 1,2.$$

$$\rho_j = |\Psi_j|^2 = \frac{\exp\left[-\frac{1}{n}\ (x_1^2+x_2^2+x_3^2) - \frac{t^2}{2n}\right]}{n^{3/2}\pi^{3/2}} \tag{92}$$

$$j = 1,2\ .$$

$$\rho_1\rho_2 = \frac{\exp\left[-\frac{2}{n}(x_1^2+x_2^2+x_3^2) - \frac{t^2}{2n}\right]}{\pi^3 n^3} \tag{93}$$

$${}^*\!\int \rho_1\rho_2 dx_1 dx_2 dx_3 dt = \pi^{-1}n^{-1}2^{-3/2} \tag{94}$$

Thus

$$d\sigma \overset{*}{\sim} \frac{\left(\frac{\beta}{\beta+1}\right)^{3/2}\left(\frac{4\beta+2}{4\beta+1}\right)^{3/2}\left(\frac{\beta}{\beta+1/2}\right)^{1/2} (2\pi)^2 \left\langle \Omega|T|S_1^I, S_2^I\right\rangle^2 d\Omega}{v_1+v_2} \tag{95}$$

Since β is infinite-star finite positive we have

$$d\sigma \overset{*}{\sim} \frac{(2\pi)^2 \left\langle \Omega|T|S_1^I,\ S_2^I\right\rangle^2 d\Omega}{v_1 + v_2}\quad . \tag{96}$$

Using $^*\delta(x)$ in place of $\tilde{\Delta}_\beta(x)$ we would have obtained (96) with $\overset{*}{\sim}$ replaced by $=$. For comparison with a standard derivation of (96) see ref. 8.

This can be extended to relativistic quantum scattering where normalization factors involving modified Bessel functions of the second kind occur. This latter work will be presented elsewhere.

Part III.

Infinite-Star Finite-Many Body Theory With Infinitesimal Two Body Interaction Forces

As we saw in Part I, in order to get a finite self energy for an electron we were led to construct a model with a charge distribution of the form

$$\rho = \alpha\Delta^{5/6}(x_1)\Delta^{5/6}(x_2)\Delta^{5/6}(x_3), \quad \text{or} \tag{97}$$

equivalently

$$\rho = \varepsilon\alpha\tilde{\Delta}_{5/6}(x_1)\tilde{\Delta}_{5/6}(x_2)\tilde{\Delta}_{5/6}(x_3) \quad , \tag{98}$$

where

$$\varepsilon = \pi^{1/4}(6/5)^{3/2}n^{-1/4} \quad \text{is a} \tag{99}$$

positive infinitesimal. (97) or (98) gives rise to a corresponding potential of the form

$$\varepsilon\phi, \text{ where} \tag{100}$$

$$\phi(x_1,x_2,x_3) = \alpha \; {}^*\!\!\int \frac{\tilde{\Delta}_{5/6}(x_1-y_1)\tilde{\Delta}_{5/6}(x_2-y_2)\tilde{\Delta}_{5/6}(x_3-y_3)dy_1dy_2dy_3}{\left[(x_1-y_1)^2 + (x_2-y_2)^2 + (x_3-y_3)^2\right]^{1/2}} \tag{101}$$

Because of the infinitesimal factor ε the corresponding interparticle forces are infinitesimal. Therefore unless two particles interacting with such a force are an infinitesimal distance apart they effectively do not interact. A finite system of such particles is therefore not likely to lead to much of interest. On the other hand if we have an infinite star-finite system of such particles the situation in general is different. As will be shown we may indeed have a non-trivial and interesting interaction.

Let ω denote an infinite-star finite positive integer, and consider the following star-Schrodinger equation .

$$i\hbar \frac{\partial \psi}{\partial t} = - \frac{\hbar^2}{2m} (\nabla_1^2 + \nabla_2^2 + \ldots + \nabla_\omega^2)\Psi + V\Psi \quad , \tag{102}$$

where

$$\nabla_j^2 = \frac{\partial^2}{\partial x_{1j}^2} + \frac{\partial^2}{\partial x_{2j}^2} + \frac{\partial^2}{\partial x_{3j}^2} \quad , \quad 1 \leq j \leq \omega \quad . \tag{103}$$

$$\Psi = \Psi(t; x_{11}, x_{12}, x_{13}; \ \ldots \ ; x_{\omega 1}, x_{\omega 2}, x_{\omega 3}) \quad , \tag{104}$$

Ψ is a *function of $3\omega + 1$ (infinitely many) variables

$$V = \underset{\omega \geq r > s \geq \omega}{\varepsilon^* \Sigma \phi_{sr}} \quad , \quad \text{and} \tag{105}$$

$$\phi_{sr} = \phi(x_{1s} - x_{1r}, x_{2s} - x_{2r}, x_{3s} - x_{3r}) . \tag{106}$$

Define

$$\Psi_{pqv} = {}^*\!\!\int \Psi dx_{1b_1} dx_{2b_3} dx_{3b_3} \ . \ . \ . \ dx_{1b_{\omega-3}} dx_{2b_{\omega-3}} dx_{3b_{\omega-3}} \quad , \tag{107}$$

where the indices b_j include all integers between 1 and ω except the three distinct integers p, q, and v. Similarly one can define Ψ_{pq} and Ψ_p. The Ψ_{pqv}, Ψ_{pq} , Ψ_p might be referred to as marginal probability amplitudes in analogy with marginal probability densities (see ref.9). However

$${}^*\!\!\int |\Psi|^2 d\tau = 1, \text{ where the } \tau \tag{108}$$

star-integration is over all the 3ω space variables, and in general it is not true that

$${}^*\!\!\int |\Psi_{pqv}|^2 d^9 \ . \ . \ . = 1 \quad \text{or} \quad {}^*\!\!\int |\Psi_{pq}|^2 d^6 \ . \ . \ . = 1 \quad \text{or}$$

$${}^*\!\!\int |\psi_p|^2 d^3 \ . \ . \ . = 1 \ .$$

Again in analogy with statistics we might say that our samples corresponding respectively to three, two or one, out of an infinite population of ω objects is biased, and as in statistics we renormalize to obtain unbiased estimators of the actual marginal probability density (see ref. 9).

As an illustration of this consider the following choice of initial value for Ψ. Namely

let Ψ at $t = o$ be

$$\Psi^{o} = \left[\frac{a}{\pi}\right]^{3/4\omega} e^{-a/2 \overset{*\omega}{\sum\limits_{j=1}} [x_{1j}^2+x_{2j}^2+x_{3j}^2]} \tag{109}$$

where a is a positive constant, possibly infinite-star finite or infinitesimal.

One obtains

$$\Psi^{o}_{pqv} = \left[\frac{4\pi}{a}\right]^{\frac{3\omega-9}{4}} \left(\frac{a}{\pi}\right)^{9/4} e^{\frac{-a}{2}(x_{1p}^2 + \ldots + x_{3v}^2)} \tag{110}$$

Unless by chance $a = 4\pi$, we would have to consider instead of Ψ^{o}_{pqv}, the renormalized or "unbiased" estimator

$$\tilde{\Psi}^{o}_{pqv} = \left(\frac{a}{4\pi}\right)^{\frac{3\omega-9}{4}} \Psi^{o}_{pqv} = \left(\frac{a}{\pi}\right)^{9/4} e^{\frac{-a}{2}(x_{1p}^2 + \ldots + x_{3v}^2)} \tag{111}$$

If the renormalization of the wave function Ψ_{pqv} happens to be independent of time then

$$\tilde{\Psi}_{pqv} = \left(\frac{a}{4\pi}\right)^{\frac{3\omega-9}{4}} \Psi_{pqv} . \tag{112}$$

Similarly

$$\tilde{\Psi}_{pq} = \left[\frac{a}{4\pi}\right]^{\frac{3\omega-6}{4}} \Psi_{pq} \quad \text{and} \tag{113}$$

$$\tilde{\Psi}_{p} = \left[\frac{a}{4\pi}\right]^{\frac{3\omega-3}{4}} \Psi_{p} . \tag{114}$$

In the illustration note that because of the infinite constant ω in the exponent, in general the renormalization factor will not be finite, but rather infinite or infinitesimal depending on the ratio $a/4\pi$.

Clearly in general the renormalization is dependent on the initial data, the time t and the sample, namely the number and set of indices p, q, v etc. which are selected.

This type of renormalization will be referred to simply as amplitude renormalization of the star wave function. In addition two different types of renormalization will also be considered here.

The star wave function Ψ and all its first partial derivatives are assumed to $*\to 0$ as the space coordinates $*\to\pm\infty$. Thus performing an obvious infinite-star finite-fold integration we obtain:

$$i\hbar\frac{\partial\Psi_1}{\partial t} = -\frac{\hbar^2}{2m}\nabla_1^2\Psi_1$$

$$+\ \varepsilon\ {}^{*}\sum_{r=2}^{\omega}\ {}^{*}\!\int\phi_{1r}\Psi_{1r}dx_{1r}dx_{2r}dx_{3r} \tag{115}$$

$$+\ \varepsilon\ {}^{*}\!\!\!\sum_{\omega\geq r>s\geq 2}{}^{*}\!\int\phi_{sr}\Psi_{1sr}dx_{1s}dx_{2s}dx_{3s}dx_{1r}dx_{2r}dx_{3r}\ ,$$

As typically occurs in statistical physics, whether in quantum mechanics or in the process of obtaining the Boltzmann equation from the Liouville equation, this contracted description leaves us with more unknown functions than we have equations. In order to close the system different approximations are introduced depending upon the physical problem. A typical approximation for some circumstances is one which we now introduce primarily for illustrative purposes.

As approximate assumptions we take

$$\Psi_{1r} = \Psi_1\Psi_r\ ,\quad \Psi_{1sr} = \Psi_1\Psi_s\Psi_r\ , \tag{116}$$

$^*\int \phi_{1r}\Psi_r \, dx_{1r} dx_{2r} dx_{3r}$ is assumed to be independent of the index $r \neq 1$, and

$^*\int \phi_{sr}\Psi_r\Psi_s dx_{1s} dx_{2s} dx_{3s} dx_{1r} dx_{2r} dx_{3r}$ is assumed to be independent of the indices s and r.

Ψ_1 then satisfies approximately the following equation

$$i\hbar \frac{\partial \Psi_1}{\partial t} = -\frac{\hbar^2}{2m} \nabla_1^2 \Psi_1 + \left[\varepsilon(\omega-1) \, ^*\!\int \phi_{12}\Psi_2^o d^3 \ldots \right] \Psi_1 \tag{117}$$

$$\left[\frac{\varepsilon(\omega-1)(\omega-2)}{2} \, ^*\!\int \phi_{23}\Psi_2^o\Psi_3^o d^6 \ldots . \right] \Psi_1 \, ,$$

where an additional approximation is made by replacing Ψ_2 and Ψ_3 by the initial value of Ψ_1, namely Ψ_1^o, and the subscripts 2 and 3 now only have the significance of dummy variables of integration. Note (117) is a linear homogeneous equation.

The next step will involve three renormalizations. The first is the amplitude renormalization of the type previously illustrated. Define

$$\tilde{\Psi}_1 = A\Psi_1, \text{ so that} \tag{118}$$

$$^*\!\int |\tilde{\psi}_1|^2 dx_{11} dx_{21} dx_{31} = 1 . \tag{119}$$

Without loss of generality we can choose A to be a positive constant. That A can be chosen to be time independent is a consequence of the form of the approximate equation (117) satisfied by Ψ_1. As we emphasize again it is a linear homogeneous partial differential equation (star of course). A in general will not be finite but infinite or infinitesimal.

Correspondingly we write

$$\tilde{\Psi}_2^o = A\Psi_2^o \text{ and } \tilde{\Psi}_3^o = A\Psi_3^o \, . \tag{120}$$

Defining $U(x_{11},x_{21},x_{31}) = \, ^*\!\int \phi_{12}\tilde{\Psi}_2^o d^3 \ldots$ (121)

and $$\lambda = \frac{\varepsilon^2(\omega-1)(\omega-2)}{2A^2} \; {}^*\!\!\int \phi_{23}\tilde{\Psi}_2^o\tilde{\Psi}_3^o \, d^6 \; . \; . \; . \tag{122}$$

we obtain

$$i\hbar \frac{\partial\tilde{\Psi}_1}{\partial t} = \frac{-\hbar^2}{2m}\nabla_1^2\tilde{\Psi}_1 + \frac{\varepsilon(\omega - 1)}{A} U\tilde{\Psi}_1 + \frac{\lambda}{\varepsilon}\tilde{\Psi}_1 \, . \tag{123}$$

ω is an infinite integer and if A is such that $\frac{\omega-1}{A}$ is also infinite, then since ε is infinitesimal we can choose ε so that

$$\frac{\varepsilon(\omega-1)}{A} = D > 0 \tag{124}$$

is a finite non zero constant. Then λ will be positive and finite and $\tilde{U}$ is a finite effective potential, where

$$\tilde{U} = DU. \tag{125}$$

This is the second renormalization, the charge or interaction renormalization and

$$i\hbar \frac{\partial\tilde{\Psi}_1}{\partial t} = \frac{-\hbar^2}{2m}\nabla_1^2\tilde{\Psi}_1 + \tilde{U}\Psi_1 + \frac{\lambda}{\varepsilon}\left[\tilde{\Psi}_1\right] . \tag{126}$$

λ/ε is a positive infinite-star finite constant energy which is added onto the effective potential $\tilde{U}$.

Next we define $\tilde{\tilde{\Psi}}_1$ by the relation

$$\tilde{\Psi}_1 = e^{\frac{-i\lambda t}{\varepsilon\hbar}}\left[\tilde{\tilde{\Psi}}_1\right] . \tag{127}$$

Clearly

$${}^*\!\!\int |\tilde{\tilde{\Psi}}_1|^2 \, dx_{11}dx_{21}dx_{31} = 1 \; . \tag{128}$$

$$\tilde{\tilde{\Psi}}_1^o = \tilde{\Psi}_1^o \quad \text{and} \tag{129}$$

$$i\hbar \frac{\partial\tilde{\tilde{\Psi}}_1}{\partial t} = -\frac{\hbar^2}{2m}\nabla_1^2\tilde{\tilde{\Psi}}_1 + \tilde{U}\tilde{\tilde{\Psi}}_1 \; . \tag{130}$$

This third renormalization gives an infinite-star finite energy shift in these star-spectrum of the star-Schrodinger operator and has no physically observable consequence.

Thus we see that an infinite number of particles interacting with infinitesimal forces can lead to finite effective potentials or forces. Although the idea of renormalization is a carry over from standard many body quantum mechanics and field theory, there is an interesting twist in point of view here. In quantum electrodynamics the bare electrons have infinite mass and charge and renormalization subtracts off infinite mass and charge to give the finite observed masses and charges. In contrast we add up infinitely many infinitesimal effects to give a finite observable effect.

One may paraphrase in the language of Non-Standard analysis various standard approximations used in many body theory. For example paraphrasing the standard random phase approximation (see ref. 7) one can obtain finite many body Schrodinger equations where the Non-Standard quasi-particles interact with a finite effective Yukawa potential of the form

$$\frac{e^{-\mu r}}{r} , \tag{131}$$

where μ can be chosen to be a positive infinitesimal. Effectively one therefore has a finite Coulomb interaction, and the quasi particles will behave like ordinary finitely charged particles.

These ideas are presently being carried over to the relativistic quantum many body problem (ref.10) where interestingly one can make sense of Dirac's infinite sea of electrons without any reference whatsoever to quantum field theory. The infinite population of electrons being star-finite of course. These investigations will be presented elsewhere.

Conclusion

A. Robinson's theory of Non-Standard Analysis offers a new view of infinite renormalization. Difficulties arising in the L. Schwarz theory of distributions when one introduces the notion of pointwise multiplication of generalized functions can be overcome by using Non-Standard generalized functions instead. Thereby one is led naturally to the notion of fractional powers of delta functions, which in effect constitutes a form of infinite renormalization. One obtains a model of a point electron in classical physics, with finite self energy, satisfying reasonable causality requirements. The price one pays is to have Coulomb interaction between electrons of infinitesimal strength. However when one looks at an infinite-star finite system of such particles one can recover finite effective interactions which are between Non-Standard quasi-particles.

Also one can reformulate quantum scattering theory in terms of Non-Standard wave packet analysis. Again fractional powers of delta functions arise naturally.

In addition star-finite body quantum mechanics may offer a feasible alternative to quantum field theory, relativistic as well as non-relativistic. This point will be elaborated on in future publications.

Acknowledgement

We thank Professor P. J. Kelemen for comments which have greatly increased the clarity of our exposition.

Note Added in Proof

ψ_p^o the initial data used in (117) bears no relation to (109), which is just used to illustrate amplitude renormalization for a case where the computations are easy to make. The charge distributions in Section III are therefore not in general point charges.

Bibliography

1. W. A. J., Luxemburg, Lectures on A. Robinson's Theory of Infinitesimals and Infinitely Large Numbers, Cal. Tech. Pasadena, 1962.

2. A. Robinson, Non-Standard Analysis, Amsterdam, North-Holland, 1966.

3. M. Machover and J. Hirschfeld, Lectures on Non-Standard Analysis, 1969, Springer - Verlag.

4. H. B. Enderton, A Mathematical Introduction to Logic, 1972, Acad. Press.

5. P. J. Kelemen and A. Robinson, The Non-Standard $\lambda:\phi_2^4(x)$: model: I, (Third Symposium on Non-Standard Analysis) Pub. J. M. P., December, 1972.

6. L. Landau and E. Lifschitz, The Classical Theory of Fields, 1951, Addison - Wesley.

7. R. D. Mattuck, A Guide to Feynman Diagrams in The Many Body Problem, 1967, McGraw-Hill.

8. Marc Ross, Editor, Quantum Scattering Theory, 1963, Indiana Univ. Press, Bloomington.

9. W. C. Hamilton, Statistics in Physical Science, 1964, The Ronald Press Co. N. Y.

10. J. D. Bjorken and S. D. Drell, Relativistic Quantum Mechanics, 1964, McGraw Hill.

TWO TOPOLOGIES WITH THE SAME MONADS

Frank Wattenberg

University of Massachusetts, Amherst

I. Introduction

Consider the set $F_C(R)$ of continuous functions $f\colon R \to R$ with compact support. If this set is given the compact-open topology then certain pathological situations can arise. In particular, there will be continuous paths $g\colon [0,1] \to F_C(R)$ which do not have compact support (i.e. for which there is no compact set $K \subseteq R$ containing the support of every $g(t)$, $0 \le t \le 1$.) An example of such a function is the "traveling bump" function,

$$\begin{aligned} g(t)(x) &= \sin(x - 1/t) \quad t \neq 0,\ |x - 1/t| \le \pi \\ &= 0 \qquad\qquad\qquad \text{otherwise.} \end{aligned}$$

From the point of view of Nonstandard Analysis the compact-open topology on $F_C(R)$ can be described by saying a function g in a nonstandard model $^*F_C(R)$ of $F_C(R)$ is "infinitely close" to a standard function $f \in F_C(R)$ provided for every nearstandard (i.e. finite) point $x \in {}^*R$, $^*f(x)$ is infinitely close to $g(x)$ [17]. Hence, if α is a positive infinitesimal the function,

$$\begin{aligned} g(x) &= \sin(x - 1/\alpha) \quad |x - 1/\alpha| \le \pi \\ &= 0 \qquad\qquad\qquad \text{otherwise} \end{aligned}$$

is infinitely close to the zero function despite the fact that these two functions differ greatly on the interval $[1/\alpha - \pi, 1/\alpha + \pi]$. This is the underlying reason why pathological situations like the "traveling bump" can occur with the compact-open topology. Thus, from this perspective the way to avoid such pathologies is to "tie down" the values of functions infinitely close to a given function on infinite (i.e. non-nearstandard) points as well as on finite points.

Classically, at least two topologies, the direct limit (or inductive) topology and the δ-topology, have been put on $F_C(R)$ to avoid such pathologies (see Section II for definitions.) The following example shows that these two topologies are distinct.

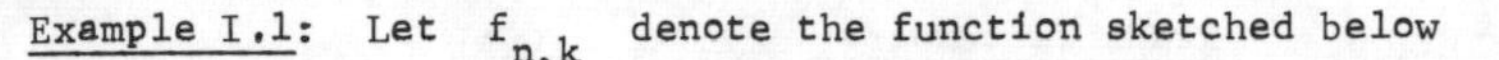

Example I.1: Let $f_{n,k}$ denote the function sketched below

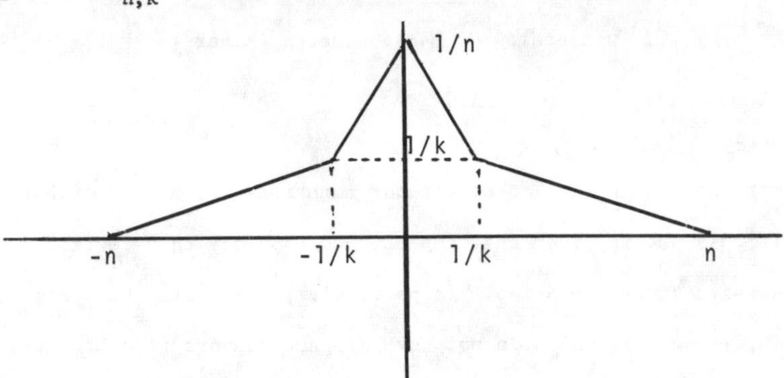

Let $F_n(R) = \{f \in F_C(R) \mid \text{support}(f) \subseteq [-n,n]\}$.

Let $A = \{f_{n,k} \mid n,k = 1,2,3,\ldots\}$. Notice, $A \cap F_n(R) = \{f_{i,k} \mid i = 1,2,\ldots,n;$ $k = 1,2,\ldots\}$ is closed for each n and, hence, A is closed in the direct limit topology. However, it is straightforward to verify that the zero function is a limit point of A in the δ-topology.

Since every open set in the δ-topology is clearly open in the direct limit topology the direct limit topology is strictly finer than the δ-topology. Nevertheless, these two topologies have strikingly similar properties and have on occasion been confused with each other. In the next section of this paper we will show that these topologies correspond to two different ways of "tying down" functional values on infinite points. The main result of this section is that in certain nonstandard models these two distinct ways become identical. Hence, the two topologies are indistinguishable with respect to those properties which can be appropriately characterized in such models [3].

Throughout this paper we will work only with Hausdorff Spaces. When we are working with several spaces X, Y, and Z, their extensions will all be taken in a single nonstandard model ${}^*M$. That is, we let M be the complete higher order structure on $X \cup Y \cup Z$ and let ${}^*M$ be a higher order elementary extension of M, [12], [13]. The sets ${}^*X$, ${}^*Y$, and ${}^*Z$ will then all live in ${}^*M$. If P denotes an object in M the corresponding object in ${}^*M$ will be denoted by ${}^*P$. In particular, an internal set $K \subseteq {}^*X$ will be said to be *compact if it is in ${}^*\mathcal{K}$ where $\mathcal{K}$ is the (standard) set of all compact subsets of X.

II. Monads of the Inductive and δ-topologies

For convenience we recall the following definitions from [13].

Definition II.1: Suppose that X is any topological space and ${}^*X$ is a non-standard model of X. If x is any standard point in X we define the monad of x, $\mu(x)$, by

$$\mu(x) = \bigcap_{x \in U,\ U \text{ standard, open}} {}^*U$$

Since X is Hausdorff, if x and y are distinct points of X then $\mu(x) \cap \mu(y) = \emptyset$. If $y \in {}^*X$ and there is a standard point x such that $y \in \mu(x)$, then we say y is nearstandard and call x the standard part of y, denoted $x = St(y)$. Two nearstandard points x and y are said to be infinitely close, written $x \approx y$, provided $St(x) = St(y)$. Many of the topological properties of X are reflected in the properties of the monads. For example, if $f: X \to Y$ is a (standard) continuous function, then for each $x \in X$, ${}^*f(\mu(x)) \subset \mu(f(x))$. In fact, if ${}^*X$ is a sufficiently rich (i.e. an enlargement) extension of X then the topology on X is completely characterized by the monads [13].

Throughout the remainder of this section X will denote a σ-compact, locally compact, Hausdorff space and $F_C(X)$ will denote the space of continuous functions $f: X \to R$ with compact support.

Definition II.2: For each compact set $K \subset X$ let $F_K(X) = \{f \in F_C(X) \mid \text{support}(f) \subseteq K\}$. Let $F_K(X)$ have the compact-open topology (Hence, the topology of uniform convergence). A set $U \subset F_C(X)$ is said to be open in the direct limit or inductive topology provided $U \cap F_K(X)$ is open in $F_K(X)$ for every compact set $K \subseteq X$.

Since X is σ-compact and locally compact we can fix a sequence of compact subsets of X; $K_1, K_2, \dots$ such that $X = \bigcup K_n$ and for each n, $K_n \subseteq \text{Int}(K_{n+1})$.

Let $F_n(X) = F_{K_n}(X)$. The proof of the following lemma is completely straightforward.

Lemma II.3: Suppose $U \subseteq F_C(X)$ then U is open in the direct limit topology if and only if $U \cap F_n(X)$ is open for every n.

Definition II.4: Suppose $f \in F_C(X)$ and $h: X \to (0,1)$ is a strictly positive continuous function. Let $N(h,f) = \{g \in F_C(X) \mid \forall x \in X \; |f(x) - g(x)| < h(x)\}$. The sets $N(h,f)$ form a basis for the δ-topology on $F_C(X)$.

The following will be useful in our investigations of these two topologies.

Definition II.5: Suppose $^*(X \times R)$ is a nonstandard model of $X \times R$. If (x,s) and (x,t) are two points in $^*(X \times R)$ we say (x,s) is infinitely close to (x,t), written $(x,s) \approx (x,t)$, provided for every standard strictly positive continuous function $f: X \to (0,1)$, $|s-t| < {}^*f(x)$. Clearly, this is an equivalence relation. (See [17], [18].) We let $k(x,s)$ denote the set $\{(x,t) \mid (x,s) \approx (x,t)\}$. Notice, if x is a standard point $k(x,s) = \{(x,t) \mid |t-s| \approx 0\}$.

A straightforward enlargement argument gives the following proposition.

Proposition II.6: Suppose $^*(X \times R)$ is an enlargement of $X \times R$ then there is an internal, strictly positive, *continuous function $f: {}^*X \to {}^*(0,1)$ such that for each $(x,s) \in {}^*(X \times R)$, $\{(x,t) \mid |t-s| < {}^*f(x)\} \subseteq k(x,s)$. In particular, $k(x,s) \neq \{(x,s)\}$ for each $(x,s) \in {}^*(X \times R)$.

The preceeding proposition says that in a sufficiently rich nonstandard model the sets $k(x,s)$ are big. However, if the model $^*(X \times R)$ is not so rich they may be only single points.

Definition II.7: An ultrafilter D on the index set $\omega = \{0,1,2..\}$ is said to be a P-point provided for every function $f: \omega \to \omega$ there is a set $A \in D$ such that $f|_A$ is either constant or finite-to-one. (See [4], [5], [6], [7], [8], [15], and [16].) If either the Continuum Hypothesis [15] or Martin's Axiom([4], [5], [6])holds there are lots of P-points.

Proposition II.8: Suppose D is an ultrafilter on ω which is a P-point and that

${}^*(X \times R)$ is the ultrapower of $(X \times R)$ with respect to D. Then if x is any infinite point of ${}^*X$, $k(x,s) = \{(x,s)\}$.

Proof: Suppose $t \neq s$. Let k be any integer such that $1/k < |t-s|$. Let ν be the least integer such that $x \in K_\nu$. Since x is infinite, ν is an infinite integer. Let ν be represented in the ultrapower by $f: \omega \to \omega$. Since D is a P-point and ν is infinite there is a set $A \in D$ such that $f|_A$ is finite-to-one. We may assume $f(p) = p$ for each $p \notin A$ and thus that f is finite-to-one on all of ω. Let $g: \omega \to \omega$ be defined by

$$g(p) = \text{Max}\,\{q\,|f(q) = p\}.$$

Then for each $q \in \omega$ $gf(q) \geq q$. Now let $h: \omega \to \omega$ represent k in the ultrapower. We may assume that h is monotonically increasing. Then for each q, $(hg)(f(q)) \geq h(q)$ so ${}^*(hg)(\nu) \geq k$. Let $p: X \to (0,1)$ be any continuous function such that for each $y \in K_n - K_{n-1}$, $p(y) \leq 1/(hg)(n)$. Then ${}^*p(x) \leq 1/{}^*(hg)(\nu) \leq 1/k < |t-s|$. Hence, $(x,t) \notin k(x,s)$.

The following proposition is immediate from Definitions II.4 and II.5.

<u>Proposition II.9</u>. Suppose $f \in F_C(X)$ then the monad of f in the δ-topology, denoted $\mu(f)$ is given by:

$$\mu(f) = \{g \in {}^*F_C(X) \mid \text{for every } x \in {}^*X,\ (x, {}^*f(x)) \approx (x, g(x))\}$$

The monads of the direct limit topology appear to be the more difficult to describe. However we can at least describe part of them as follows.

<u>Definition II.10</u>: Define $\sigma: F_C(X) \to \omega$ by $\sigma(f) = \text{Min}\,\{n\,|f \in F_n\}$. Now if $f \in F_C(X)$ we define the <u>core of f</u>, denoted core (f), by:

$$\text{core}\,(f) = \{g \in {}^*F_C(X) \mid \text{for each } x \in {}^*X,\ ({}^*\sigma(g), {}^*f(x)) \approx ({}^*\sigma(g), g(x))\}$$

<u>Proposition II.11</u>. If $f \in F_C(X)$ then core (f) is contained in the monad of f in the direct limit topology.

Proof: Suppose U is any open set in the direct limit topology and $f \in U$ we must show core $(f) \subset {}^*U$.

Since U is open in the direct limit topology $U \cap F_n(X)$ is open for each n, and hence there are positive numbers ε_n such that if $h \in F_n$ and $|h(x) - f(x)| < \varepsilon_n$ for all x then $h \in U$.

But now if $h \in$ core (f) then for all $x \in {}^*X$, $|h(x) - {}^*f(x)| < \varepsilon_{\sigma(h)}$ and, hence, $h \in {}^*U$.

Now, Proposition II.8 gives us our main theorem.

<u>Theorem II.12</u>: Suppose D is an ultrafilter on ω which is a P-point and ${}^*F_C(X)$ is the ultrapower of $F_C(X)$ with respect to D. Then if $f \in F_C(X)$,

(i) $\text{core}(f) = \mu(f) = M(f)$ where $M(f) = \{g \in {}^*F_C(X) \mid {}^*\sigma(g)$ is finite and for each $x \in {}^*X$, $|g(x) - {}^*f(x)| \approx 0\}$

(ii) $\mu(f)$ = monad of f in the direct limit topology.

Proof: (ii) Follows from (i) by II.11 and the observation that the direct limit topology is finer than the δ-topology.

(i) Clearly, core $(f) \subset \mu(f)$. Now, if $g \in \mu(f)$, by II.8 for every infinite $x \in {}^*X$, $g(x) = {}^*f(x) = 0$. Thus for each infinite integer ν, $g \in F_\nu(X)$. Hence, $\sigma(g)$ is finite and since $g \in \mu(f)$, for each $x \in {}^*X$, $|g(x) - {}^*f(x)| \approx 0$. Thus $\mu(f) \subset M(f)$. Since it is immediate that $M(f) \subset$ Core (f) this completes the proof.

Before examining a few consequences of this theorem we will look at some of the differences between these two topologies.

<u>Proposition II.13</u>: The δ-topology on $F_C(X)$ is a locally convex vector space topology. In fact it is the finest locally convex vector space topology on $F_C(X)$ for which the inclusion maps $F_n(X) \to F_C(X)$ are continuous.

Proof: The proof of the first assertion is straightforward. Before proving the second assertion we remark that a topology is locally convex if and only if each monad is <u>Q-convex</u>. A set A is Q-convex provided for every internal *finite sequence $a_1, a_2, \ldots a_\nu \in A$ and every internal *finite sequence $\lambda_1, \lambda_2, \ldots \lambda_\nu$ in ${}^*[0,1]$ such that $\Sigma\lambda_i = 1$, we have $\Sigma\lambda_i a_i \in A$. (Notice, if A is external this is not equivalent to saying A is convex in the sense that $a, b \in A$, $0 \leq \lambda \leq 1$ implies $\lambda a + (1-\lambda)b \in A$. However, for internal sets the two are, of course, equivalent.)

Now, in order to prove the second assertion it is sufficient to show that any Q-convex set A which contains core(0) also contains $\mu(0)$, where 0 is the zero function in $F_C(X)$. Let $\{\phi_n\}$ be a sequence of standard continuous functions such that

$$\text{(i)}\quad \phi_n(x) = 0 \quad \text{if } x \in K_{n-2}$$
$$= 0 \quad \text{if } x \notin K_{n+1}$$
$$\text{(ii)}\quad \Sigma\, \phi_n(x) = 1$$

Now, if $f \in \mu(0)$ then for each $\nu = 1,2,\ldots, k = \sigma(f) + 2$, $f\cdot \phi_\nu \in$ core(0) and, clearly $2^\nu \cdot f \cdot \phi_\nu \in$ core(0). So if A is any Q-convex set containing core(0), $2^\nu \cdot f \cdot \phi_\nu \in A$ and so $\sum_{\nu=1}^{k} 1/2^\nu \cdot (2^\nu \cdot f \cdot \phi_\nu + (1/2^k \cdot 0) = f \in A$ which completes the proof.

It is immediate from the definitions that the translation mappings $T_f\colon F_C(X) \to F_C(X)$ given by $T_f(g) = f + g$ are continuous with the direct limit topology on $F_C(X)$. However, a second look at Example I.1 reveals that $F_C(R)$ with the direct limit topology is not a topological vector space.

Proposition II.14: $F_C(R)$ with the direct limit topology is not a topological vector space.

Proof: Suppose it is a topological vector space. Let ${}^*F_C(R)$ be an enlargement of $F_C(R)$. We will show the set A of Example I.1 is not closed. Let ν be any infinite positive integer. By Proposition II.6 there is an infinite integer k such that $(\nu, 1/k) \approx (\nu, 0)$. Consider the function $f_{\nu,k}$ (using the notation of Example I.1). Let g be the function:

$$g(x) = 1/\nu + (1-k/\nu)\,|x| \qquad |x| \le 1/(k-\nu)$$
$$= 0 \qquad |x| > 1/(k-\nu)$$

and let $h = f_{\nu,k} - g$

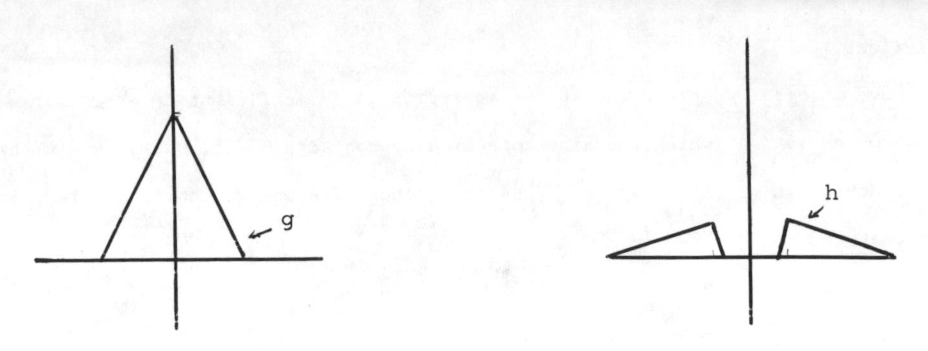

Then $h, g \in \text{core}(0)$ so under our assumption that $F_C(X)$ is a topological vector space $f_{\nu,k} = h + g$ is also in the monad of 0, contradicting the fact that A is closed. This completes the proof.

III. Some Consequences

We recall [3], [13] that the topology of a topological space X is characterized by its monads in an enlargement *X of X, or if X satisfies the First Axiom of Countability, by its monads in any nonstandard model *X. In particular, if X is any topological space and *X is any nonstandard model of X and if $\{x_n\}$ is any sequence in X then $\lim_{n\to\infty} x_n = L$ if and only if for every $\nu \in {}^*\omega - \omega$, $x_\nu \in \mu(L)$. Thus, we have:

Proposition III.1:

Suppose $\{f_n\}$ is a sequence in $F_C(X)$. Then $\{f_n\}$ converges in the δ-topology if and only if $\{f_n\}$ converges in the direct limit topology.

Proof: Immediate

Hence, in particular, the countable set A of Example I.1 has no sequence which converges to 0 in the δ-topology. Finally, the following theorem shows that the δ and direct limit topologies avoid pathologies like the "traveling" bump function.

Theorem III.2: Suppose $F_C(X)$ has either the δ or direct limit topology and that K is a compact subset of $F_C(X)$. Then there is some (standard) positive integer n such that $K \subseteq F_n(X)$.

Proof: Let D be an ultrafilter on ω which is a P-point and let ${}^*F_C(X)$ be the ultrapower of $F_C(X)$ with respect to D. Let ν be any infinite positive integer. We will show ${}^*K \subseteq F_\nu(X)$. Let $f \in {}^*K$. Since K is compact there is a unique standard function g such that $f \in \mu(g)$. (See [12], [13]. Note that the proof that every point of ${}^*K$ is nearstandard if K is compact does not require an enlargement although the converse does.) Hence, by Theorem II.12 $\sigma(f)$ is finite and $f \in F_\nu(X)$.

Now, let $T = \{\nu | {}^*K \subseteq F_\nu(X)\}$ since T is an internal set which contains all infinite positive integers, it contains some finite positive integer, n. This completes the proof.

References

1. Bell and Slomson, Models and Ultraproducts, North-Holland, Amsterdam (1969).
2. A. Bernstein, A New Kind of Compactness for Topological Spaces.
3. A. Bernstein and F. Wattenberg, Nonstandard Topology in Countable Ultrapowers and the Baire Property, these Proceedings.
4. A. Blass, Orderings of Ultrafilters, Thesis, Harvard University (1970).
5. ______________, The Rudin-Keisler Ordering of P-points, to appear.
6. D. Booth, Ultrafilters on a Countable Set, Ann. Math. Logic, 2 (1970) pp. 1-24.
7. G. Choquet, Construction d'Ultrafiltres sur N, Bull. Sci. Math. 92 (1968) pp. 41-48.
8. ______________, Deux Classes Remarquables d'Ultrafiltres sur N, Bull. Sci. Math. 92 (1968) 143-153.
9. S. Kakutani and V. Klee, The Finite Topology of a Linear Space, Arch. Math. 14 (1963) pp. 55-58.
10. W. A. J. Luxemburg, A New Approach to the Theory of Monads, California Institute of Technology, Pasadena (1967).
11. ______________, (editor), Applications of Model Theory to Algebra, Analysis and Probability, Holt, Rinehart and Winston, New York (1969).
12. M. Machover and J. Hirschfeld, Lectures on Non-Standard Analysis, Lecture Notes in Mathematics 94, Springer, Berlin (1969).
13. A. Robinson, Non-Standard Analysis, North-Holland, Amsterdam (1966)
14. ______________, Nonstandard Arithmetic, Bull. Amer. Math. Soc. 73 (1967) pp. 818-843.

15. M. Rudin, Types of Ultrafilters, Topology Seminar, Wisconsin: Edited by R. H. Bing and R. J. Bean, Princeton University Press, Princeton (1966).

16. W. Rudin, Homogeneity Problems in the Theory of Cech Compactifications, Duke Math. J. 23 No. 3 pp. 409-420.

17. F. Wattenberg, Nonstandard Topology and Extensions of Monad Systems to Infinite Points, J. Sym. Logic, 36 (1971), pp. 463-476.

18. ______________, Monads of Infinite Points and Finite Product Spaces, to appear in Trans. AMS.

A NEW VARIANT OF NON-STANDARD ANALYSIS

Elias Zakon

University of Windsor, Canada

I N T R O D U C T I O N

In [6], A. Robinson developed Non-Standard Analysis within the framework of higher order logic based on type theory. This remarkable achievement had, of necessity, to rely on rather sophisticated results of model theory and logic, as explained in Chapter II of Robinson's book.

In [8], a comparatively simpler, purely set theoretical approach to enlargements was achieved, utilizing certain injective maps ("monomorphisms") of one model of set theory ("superstructure") into another. It eliminated type theory and higher order logic, and simplified the notion of "internal elements". However, the definition of a "monomorphism" still required a rather involved set of postulated properties. Independently, Machover and Hirschfeld [5] presented their variant of Robinson's theory which, along with certain advantages, still does not completely overcome the difficulties of the original theory and thus is not entirely accessible to rank and file mathematicians.

Our present paper further develops [8]. We shall define "monomorphisms" in very simple terms, requiring only rudiments of first order logic. A novelty feature is the introduction of "non-strict" monomorphisms which are more general than Robinson's "enlargements". As in [8], we use the language of naive set theory, but a formalization in axiomatic set theory presents no difficulties, even though we admit "individuals" (non-sets).[1]

---

[1]) For an axiomatic treatment of "individuals", see [9] .

It is our aim to give a self-contained presentation of all basic notions of Non-Standard Analysis as they evolve from the newly defined "monomorphisms". At this stage, we limit ourselves to the first four sections (presented at the 1972 Victoria Symposium) of the intended work that will probably take the form of a book, as we certainly hope. Along with new results, we also give some known theorems "translated" into the language of "monomorphisms", mainly to show how the proposed method works and what simplifications it achieves.

In §1 below we briefly summarize those notions of [8] that will be necessary, and give the new definition of "monomorphisms".

§1. SUPERSTRUCTURES AND MONOMORPHISMS

The superstructure on a set $A = A_o$ of "individuals" ("Ur-elements") is, by definition, the set

$$\hat{A} = \bigcup_{n=0}^{\infty} A_n$$

where A_{n+1} is the set of all subsets of $\bigcup_{k=0}^{n} A_k$, $n = 0,1,2,\dots$. Clearly, $A_n \in A_{n+1}$ for $n = 0,1,2,\dots$, and $A_n \subset A_{n+1}$ for $n > 0$. Elements of A_n are said to be of type n (our "types" are cumulative). As noted above, elements of type 0 (those of A_o) are non-sets. Thus if $a \in A_o$, then $x \notin a$ for all x (but the formulae $x \in a$ and $x \notin a$ are always meaningful). It also follows that $A_o \cap A_n = \emptyset$ for $n > 0$, but otherwise the A_n increase with n . Therefore $A_o \cup A_n = \bigcup_{k=0}^{n} A_k$, and so $y \in x \in A_{n+1}$ implies $y \in A_o \cup A_n$. We have $\emptyset \in A_n$ for $n > 0$, and $A_n \in \hat{A}$ always, since $A_n \in A_{n+1} \subset \hat{A}$. We write C^n for $C \times C \times \dots \times C$ (n times).

If $a,b \in A_n$, then the ordered pair $(a,b) = \{\{a,b\},\{b\}\}$ is in A_{n+2} .

Thus $(a,b) \in \hat{A}$ whenever $a,b \in \hat{A}$. Similarly for ordered n-tuples, defined inductively: $(x_1)=x_1$, $(x_1,\ldots,x_n) = ((x_1,\ldots,x_{n-1}),x_n)$. A set of such n-tuples ("n-ary relation") is in $\hat{A}$ if all its n-tuples are of 'bounded type', i.e. belong to <u>one</u> A_n. In particular, a binary relation R is in $\hat{A}$ iff its domain $D_R = D(R) = \{x \mid (\exists y)\ (x,y) \in R\}$, and its range $D'_R = D'(R) = D(R^{-1})$, are in $\hat{A}$ [here $R^{-1} = \{(y,x) \mid (x,y) \in R\}$]. Also, $Y \subseteq X \in \hat{A}$ implies $Y \in \hat{A}$ if Y is a <u>set</u> (for individuals, $y \subseteq x$ holds <u>always</u>, and $x \cup y = x \cap y = x \times y = \emptyset$, by definition). Clearly, $X,Y \in \hat{A}$ implies $X \cup Y \in \hat{A}$, $X \cap Y \in \hat{A}$, $X-Y \in \hat{A}$, $X \times Y \in \hat{A}$. For any R,X, we define: $R[X] = \{y \mid (\exists x \in X)\ (x,y) \in R\}$, the "R-image of X". Obviously, $R,X \in \hat{A}$ implies $R[X] \in \hat{A}$.

It should be stressed that we do not treat $\hat{A}$ as a "universe" but only as a set in some larger universe, with an "unlimited" supply of individuals. The admission of "individuals" could be avoided, but turns out to be convenient.

We now adopt a first-order logical language L with identity $=$, writing $\wedge, \vee, \Rightarrow, \Leftrightarrow$ and $\neg$ for "and" , "or", "implies", "iff" and "not", respectively. For simplicity, we assume that all constants of L are in one-one correspondence with all elements of $\hat{A}$, and identify the constants with the corresponding elements, so that these become a part of L and denote themselves. Atomic formulae in L are those of the form $x \in y$ or $(x_1,\ldots,x_n) = y$ where x,y and the x_k are variables or constants (elements of $\hat{A}$)[2]. Well-formed formulae (WFF) and sentences (WFS) are defined as usual, with the restriction that <u>all quantifiers must have the form "$(\forall x \in C)$" or $(\exists x \in C)$"</u>, with C a <u>constant</u>

---

[2]) Actually, it suffices to adopt only $x=y$ and $x \in y$ as atomic formulae (see §3). We introduce $(x_1,\ldots,x_n)=y$ for convenience only. Observe that the language L here introduced is somewhat different from that of [8]. In particular, it automatically ensures the <u>normality</u> of the monomorphism, since $x=y$ is <u>atomic</u>.

$(C \in \hat{A})$. We denote by K the set of all WFS which are true in $\hat{A}$, with the usual set theoretical interpretation of the symbol $\in$ and the n-tuple $(x_1,\ldots,x_n)$ (see above). Sentences which belong to K (i.e. hold in $\hat{A}$) are briefly called <u>K-sentences</u>. Thus $\hat{A}$ is a model of K, by definition.

Now, let A,B be two sets of individuals, with superstructures $\hat{A}$, $\hat{B}$ respectively and let $\varphi : \hat{A} \to \hat{B}$ be a map of $\hat{A}$ into $\hat{B}$. We write $*p$ for $\varphi(p)$ (not to be confused with $\varphi[p] = \{*x \mid x \in p\}$) and put $*\hat{A} = \bigcup_{n=0}^{\infty} *A_n$ (recall that $A_n \in \hat{A}$). Given a WFF α, we denote by $*\alpha$ the formula obtained from α by replacing in it each <u>constant</u> $c \in \hat{A}$ by $*c$, without changing the variables or anything else; $*\alpha$ is called the <u>φ-transform</u> of α. E.g. the φ-transform of "$(\forall x \in C)\ x \in D$" is "$(\forall x \in *C)\ x \in *D$". Elements of the form $*c$ $(c \in \hat{A})$ are called φ-<u>standard</u> (briefly <u>standard</u>) members (of $\hat{B}$); their elements, in turn, are called φ-<u>internal</u> (briefly <u>internal</u>) members of $\hat{B}$; in particular, $*A_n$ is φ-standard; hence all its elements are φ-internal, and so are all elements of $*\hat{A} = \bigcup_{n=0}^{\infty} *A_n$. Elements of $\hat{B}$ which are not internal are called <u>external</u>.

1.1. DEFINITION. A one-to-one (i.e. injective) map $\varphi : \hat{A} \to \hat{B}$ is called a <u>monomorphism</u> of $\hat{A}$ into $\hat{B}$ iff $*\emptyset = \varphi(\emptyset) = \emptyset$ and, furthermore, for any WFS α,

<u>α holds in $\hat{A}$ iff $*\alpha$ holds in $\hat{B}$.</u>

We shall always identify $*x$ with x if $x \in A_o$. Thus we also have:

1.2. <u>$(\forall x \in A_o)\ x = *x$.</u> <u>Hence $X \subseteq *X$ whenever $X \subseteq A_o$.</u>

In the following propositions, φ is always supposed to be a monomorphism.

1.3. <u>For any constants $a, b, a_1, a_2, \ldots, a_n \in A$, we have:</u>

(i) $a \in b$ iff $*a \in *b$; (ii) $(a_1,\ldots,a_n) \in b$ iff $(*a_1,\ldots,*a_n) \in *b$;

(iii) $a \subseteq b$ iff $*a \subseteq *b$; (iv) $a=b$ iff $*a=*b$; (v) $a \in A_o$ iff $*a \in B_o$.

The proof is immediate from Defn. 1.1. E.g. (iii) is true since the WFS "$a \subseteq b$", i.e. "$(\forall x \in a)\ x \in b$", is equivalent to its φ-transform "$(\forall x \in *a)\ x \in *b$. Formula (ii) follows from (i) because $*(a_1,\ldots,a_n) = (*a_1,\ldots,*a_n)$, as will be shown in 1.6. For (v), use the fact that $a \in A_o \Leftrightarrow (a \subseteq \emptyset \wedge a \neq \emptyset) \Leftrightarrow$ $(*a \subseteq \emptyset \wedge *a \neq \emptyset) \Leftrightarrow *a \in B_o$, as follows from (iii), since $*\emptyset = \emptyset$ by definition.

NOTE 1. Thus φ preserves individuals and carries sets into sets. Also note that $A_n \subset A_{n+1}$ and $A_{n+1} \subseteq A_{n+2}$ implies $*A_n \in *A_{n+1}$ and $*A_{n+1} \subseteq *A_{n+2}$, respectively, by 1.3.

1.4 For any constants $a,b \in \hat{A}$, we have $*(a \times b) = *a \times *b$.

Proof. Use the K-sentences: "$(\forall z \in (a \times b))(\exists x \in a)(\exists y \in b)\ (x,y) = z$" and "$(\forall x \in a)(\forall y \in b)(\exists z \in (a \times b))\ (x,y) = z$". On passage to φ-transforms, this yields, in ordinary notation: $*(a \times b) \subseteq *a \times *b$ and $*a \times *b \subseteq *(a \times b)$.

NOTE 2. By induction, this applies to cartesian products of any finite number of members of $\hat{A}$. In particular, $*(C^n) = (*C)^n$ for $C \in \hat{A}$. Hence, by 1.2, $C \subseteq (A_o)^n$ implies $C \subseteq *C$.

1.5. Let $\alpha = \alpha(x_1,\ldots,x_n)$ be a WFF with $x_1,\ldots,x_n$ its only free variables. If $E = \{(x_1,\ldots,x_n) \in C^n \mid \alpha\}$, then $*E = \{(x_1,\ldots,x_n) \in *C^n \mid *\alpha\}$; i.e. $*E$ consists of all members of $*C^n$ satisfying $*\alpha$, the φ-transform of α.

Proof. For simplicity, let $n = 2$, $\alpha = \alpha(x,y)$; the general case is similar. Use the K-sentences "$(\forall z \in E)(\exists x,y \in C)\ [(x,y) = z \wedge \alpha(x,y)]$" and "$(\forall x,y \in C)\ [\alpha(x,y) \Rightarrow (\exists z \in E)\ (x,y) = z]$", to obtain $*E \subseteq \{(x,y) \in *C^2 \mid *\alpha(x,y)\} \subseteq *E$.

1.6. <u>For any constants $a, b, a_1, \dots, a_m \in \hat{A}$, we have:</u>

(i) <u>$*a - *b = *(a - b)$</u>; (ii) <u>$*a \cap *b = *(a \cap b)$</u>; (iii) <u>$*a \cup *b = *(a \cup b)$</u>;

(iv) <u>$\{*a\} = *\{a\}$</u>; (v) <u>$*\{a_1, \dots, a_m\} = \{*a_1, \dots, *a_m\}$</u>; and, finally,

(vi) <u>$*(a_1, \dots, a_m) = (*a_1, \dots, *a_m)$</u>.

<u>Proof</u>. (i) Let $E = a-b = \{x \in a \mid x \notin b\}$, and use 1.5, with $n = 1$, $C = a$, taking "$x \notin b$" for "$\alpha(x)$". Then $*E = \{x \in *a \mid x \notin *b\} = *a - *b$, as claimed. Formulae (ii) and (iii) follow from (i) on noting that $a \cap b = a - (a-b)$ and $a \cup b = c - [(c-a) - b)]$, with $c = a \cup b \in \hat{A}$. (iv) Let $a \in A_n$. Then 1.5, with $E = \{a\} = \{x \in A_n \mid x = a\}$, yields: $*E = *\{a\} = \{x \in *A_n \mid x = *a\} = \{*a\}$ [since $*a \in *A_n$; and so the restriction "$x \in *A_n$" may be dropped]. Thus $*\{a\} = \{*a\}$, as claimed. This combined with (iii) yields (v): $*\{a_1, \dots, a_m\} = \bigcup_{k=1}^{m} \{*a_k\} = \{*a_1, \dots, *a_m\}$. Hence also (vi) easily follows by the definition of an ordered pair and an ordered n-tuple. Thus all is proved.

1.7 (i) <u>If $R \in \hat{A}$ is an n-ary relation, so is $*R$ (in $*\hat{A}$)</u>;

(ii) <u>For any binary relation $R \in \hat{A}$ and any $Q \in \hat{A}$, $*(R[Q]) = (*R)[*Q]$; moreover, $*(R^{-1}) = (*R)^{-1}$, $*(D_R) = D(*R)$ and $*(D'_R) = D'(*R)$.</u>

<u>Proof</u>. (i) R is an n-ary relation in $\hat{A}$ iff $R \subseteq (A_o \cup A_m)^n$ for some m; i.e. R is a set of n-tuples of elements from $A_o \cup A_m$; but then, by 1.3 (iii) and 1.6 (iii), we have $*R \subseteq (*A_o \cup *A_m)^n$, so that $*R$, too, is a set of n-tuples.

(ii) If R is a <u>binary</u> relation in $\hat{A}$, then $R \subseteq (A_o \cup A_m)^2$; so $*R \subseteq *C^2$ where $C = A_o \cup A_m \in \hat{A}$. Hence $(x,y) \in R$ implies $(x,y) \in C^2$, i.e. $x,y \in C$; similarly, $(x,y) \in *R$ implies $x,y \in *C$, by 1.4. Hence, by their definition, the image sets $R[Q]$ and $*R[*Q]$ can be formally written as follows: $*R[*Q] = \{y \in *C \mid (\exists x \in *Q) \ (\exists z \in *R) \ (x,y) = z\}$ and

$R[Q] = \{y \in C \mid (\exists x \in Q)(\exists z \in R)(x,y) = z\}$. Then 1.5 yields $*(R[Q]) = \{y \in *C \mid (\exists x \in *Q)(\exists z \in *R)\ (x,y) = z\} = (*R)[*Q]$, as claimed. The remaining assertions in (ii) likewise easily follow by 1.5.

NOTE 3. By 1.6(v), we have $*C = \varphi(C) = \varphi[C]$ if C is <u>finite</u> (but not otherwise).

1.8. <u>If $f: C \to D$ is a mapping of C into D $(C, D \in \hat{A})$, then $*f$ is a mapping of $*C$ into $*D$, and $*[f(a)] = *f(*a)$ for each (fixed) $a \in C$.</u> <u>Moreover, f is one-one (or onto D) iff $*f$ is one-one (onto $*D$, respectively).</u>

<u>Proof</u>. By 1.7(i), $*f$ is certainly a binary relation (we identify f with its graph, thus treating it as a set of ordered pairs). The fact that f is a mapping (function) is expressed by the K-sentence: $(\forall x \in C)\{[(x,y) \in f \wedge (x,z) \in f] \Rightarrow z=y\}$ or formally, $(\forall x \in C)(\forall z,y \in D)\{(\exists u,v \in f)\ [(x,z)=u \wedge (x,y)=v] \Rightarrow z = y\}$. Passage to the φ-transform shows that $*f$ is a mapping of $*A$ into $*D$, on noting that $D(*f) = *D_f = *C$ and $D'(*f) = *D'_f = *D$, by 1.7(ii). Also, $*\{f(a)\} = *\{f[*\{a\}]\} = *f[\{*a\}] = \{*f(*a)\}$, by previous propositions 1.7(ii) and 1.6(v). The rest of 1.8 easily follows by using appropriate K-sentences. We omit the details here as well as in the next propositions, since they follow the same pattern.

NOTE 4. It follows that, if f is a binary operation in C, i.e. $f: C^2 \to C$, then $*f$ is a binary operation in $*C$. Moreover, one easily shows that if f satisfies the commutative or associative law in C, so does $*f$ in $*C$. Similarly for distributivity, with respect to some other operation $g: C^2 \to C$. In particular, if (C,f,g) is a ring or field [with $f(x,y) = x+y$ and $g(x,y) = xy$], so also is $(*C,*f,*g)$. A similar remark applies to <u>ordered</u> algebraic systems (the ordering being some binary relation $R \in \hat{A}$), as follows from appropriate K-sentences.

§2. INTERNAL AND STANDARD ELEMENTS. STRICT MONOMORPHISMS

We recall that an element $b \in \hat{B}$ is said to be φ-internal, under a monomorphism $\varphi:\hat{A} \to \hat{B}$, iff $b \in {}^*a = \varphi(a)$ for some $a \in \hat{A}$, i.e. b is an element of some φ-standard member $^*a$ of $\hat{B}$. We now prove:

2.1. The set of all φ-internal elements of $\hat{B}$ is exactly $^*\hat{A} = \bigcup_{n=1}^{\infty} {}^*A_n$.

Proof. If $b \in {}^*\hat{A}$, then $(\exists n)\ b \in {}^*A_n$; so b is internal by definition (for $^*A_n$ is φ-standard). Conversely, if b is internal, then $b \in {}^*a$ for some $a \in A_{n+1}$; so $a \subseteq A_n \cup A_o$; hence $^*a \subseteq {}^*A_n \cup {}^*A_o$. As $b \in {}^*a$, $b \in {}^*A_n \cup {}^*A_o$; so $b \in {}^*\hat{A}$.

2.2. All φ-standard members of $\hat{B}$ are also φ-internal.

For $a \in \hat{A}$ implies $(\exists n)\ a \in A_n$; hence $^*a \in {}^*A_n$; so $^*a$ is internal.

2.3. No φ-internal element can belong to any $y \in {}^*A_o$.

Proof. As A_o consists of individuals, we have for every n: $(\forall x \in A_n)(\forall y \in A_o)\ x \notin y$, and so $(\forall x \in {}^*A_n)(\forall y \in {}^*A_o)\ x \notin y$. Now, if x is internal then, by 2.1, $x \in {}^*A_n$ for some n. Thus $(\forall y \in {}^*A_o)\ x \notin y$, and 2.3 is proved.

NOTE 1. Members of $^*A_o$ need not be genuine individuals (they may have external elements, i.e. those outside $^*\hat{A}$). However, inside $^*\hat{A}$ they behave like individuals, by 2.3. Thus we shall call them the "individuals" of $^*\hat{A}$.

NOTE 2. The K-sentence "$(\forall y \in A_{n+1})[y \neq \emptyset \Rightarrow (\exists x \in A_n \cup A_o)\ x \in y]$" shows that each non-empty set $y \in {}^*A_{n+1}$ does have internal elements $x \in {}^*A_n \cup {}^*A_o$. However, y may also have external elements. If this never occurs, the monomorphism is said to be strict. Thus we have:

2.4. DEFINITION. A monomorphism $\varphi:\hat{A} \to \hat{B}$ is strict iff $(\forall y \in {}^*\hat{A})\ y \subseteq {}^*\hat{A}$; that is: φ is strict iff every member of $^*\hat{A}$ has internal elements only (if any).

Propositions in which φ is supposed to be strict will be marked by inserting the symbol (S) after the serial number of the proposition; e.g. 2.5(S).

2.5. (S). If $x \in y \in {}^*A_{n+1}$, then $x \in {}^*A_n \cup {}^*A_o$. Thus $(\forall y \in {}^*A_{n+1})\ y \subseteq {}^*A_n \cup {}^*A_o$.

Proof. For each m, we have the K-sentence: $(\forall y \in A_{n+1})(\forall x \in A_m)[x \in y \Rightarrow x \in A_n \cup A_o]$. Hence $(\forall y \in {}^*A_{n+1})(\forall x \in {}^*A_m)[x \in y \Rightarrow x \in {}^*A_n \cup {}^*A_o]$. Now, if $x \in y \in {}^*A_{n+1}$, then y is internal; hence so is x, by Definition 2.4. Thus $x \in {}^*A_m$ for some m, and so, as noted above, $x \in y \Rightarrow x \in {}^*A_n \cup {}^*A_o$, as required.

NOTE 3. The converse to 2.5 fails; i.e., $y \subseteq {}^*A_n$ does not imply that y is internal. Thus an internal (even a standard) set may have external subsets.

2.6. (S). The union, difference and intersection of any finite number of members of ${}^*\hat{A}$ is an internal set. So also are the union and intersection of any internal (even infinite) collection Q of sets.[3] Similarly for cartesian products.

Proof. For any n, we have the K-sentence: $(\forall X,Y \in A_{n+1} \cup A_o)(\exists Z \in A_{n+1})(\forall x \in A_n \cup A_o)[x \in Z \Leftrightarrow x \in X \cup Y]$. It follows that for any $X,Y \in {}^*A_{n+1} \cup {}^*A_o$ there is a $Z \in {}^*A_{n+1}$ (hence $Z \in {}^*\hat{A}$) possessing exactly the same elements $x \in {}^*A_n \cup {}^*A_o$ as does $X \cup Y$. But, by 2.5, all elements of Z,X and Y are in ${}^*A_n \cup {}^*A_o$. Thus $Z = X \cup Y \in {}^*\hat{A}$. Similarly for X-Y and $X \cap Y$. Since any $X,Y \in {}^*\hat{A}$ are in one ${}^*A_{n+1} \cup {}^*A_o$ for a large n, all is proved for two sets; hence, by induction, for finitely many sets. Next, if $Q \in {}^*A_{n+2}$, we use the K-sentence: $(\forall Q \in A_{n+2})(\exists Z \in A_{n+1})(\forall x \in A_n \cup A_o)[x \in Z \Leftrightarrow (\exists y \in A_{n+1})\ x \in y \in Q$; the rest follows as before. The proof for cartesian products is quite analogous.

NOTE 4. If φ were not strict, the same proof would show that the set Z occurring in it differs from $X \cup Y$ (X-Y, etc.) by external elements at most. We could call Z the internal quasi-union (difference, etc.) of X and Y. As far as monomorphisms are concerned, Z is a good substitute for $X \cup Y$ (X-Y, etc.). A similar remark applies to 2.7, 2.8 and other analogous propositions. Thus, actually, strictness is an expendable property.

---

[3]) We use the terms "collection of sets" and "set family" interchangeably.

2.7.(S). If a binary relation R is internal ($R \in {}^*\hat{A}$), so are D_R and D'_R.[4]

The proof is analogous to that of 2.6. (We omit such proofs henceforth.)

2.8.(S). If two binary relations R and S are internal ($R,S \in {}^*\hat{A}$), so is the composite relation $R \circ S = \{(x,y) \mid (\exists z)\ (x,z) \in S \wedge (z,y) \in R\}$.

Even if the monomorphism is not strict, we have the following:

2.9. For any binary relations $R,S \in \hat{A}$, ${}^*(R \circ S) = {}^*R \circ {}^*S$.

Proof. Choose n such that $R,S \in A_n$ and put $D = A_n \cup A_o$. Then $(x,y) \in R \cup S$ implies $x,y \in D$; $(x,y) \in {}^*R \cup {}^*S$ implies $x,y \in {}^*D$. Therefore the definitions of $R \circ S$ and ${}^*R \circ {}^*S$ can formally be written as $R \circ S = \{(x,y) \in D^2 \mid (\exists z \in D)\ (x,z) \in S \wedge (z,y) \in R\}$ and ${}^*R \circ {}^*S = \{(x,y) \in {}^*D^2 \mid (x,z) \in {}^*S \wedge (z,y) \in {}^*R\}$.[5] By using 1.5, with $E = R \circ S$, we immediately obtain ${}^*(R \circ S) = {}^*R \circ {}^*S$, as claimed.

2.10. Two internal sets are equal iff they have the same internal elements.

Proof. Use the K-sentences $(\forall U,V \in A_{n+1})\{[(\forall x \in A_n \cup A_o)\ x \in U \Leftrightarrow x \in V] \Rightarrow U = V\}$, $n = 0,1,2,\ldots$.

In the same manner, using suitable K-sentences, we obtain:

2.11.(S). (i) If the relation R and the set Q are in ${}^*\hat{A}$, so are $R[Q]$ and R^{-1}. Similarly, if a function f is internal ($f \in {}^*\hat{A}$) and if $a \in D_f$, then $f(a) \in {}^*\hat{A}$.

(ii) If $a_1, a_2, \ldots, a_m \in {}^*A_n \cup A_o$, then $\{a_1, \ldots, a_m\} \in {}^*A_{n+1}$. Thus if the a_k are internal, so is the set $\{a_1, \ldots, a_m\}$, and so is the m-tuple $(a_1, \ldots, a_m)$.

2.12.(S). For each n, ${}^*A_n \subseteq B_n$. In particular, ${}^*A_o \subseteq B_o$ (individuals).

---

4) However the converse fails since $R \subseteq D_R \times D'_R$ does not imply $R \in {}^*\hat{A}$.

5) More formally, "$(x,z) \in S$" should be replaced by "$(\exists u \in S)\ (x,z) = u$," etc.

Proof. If $q \in *A_o$, q is internal; so q has no external elements, by Definition 2.4; and no internal ones, by 2.3. Thus q has no elements at all. Also, $q \neq \emptyset$, by the K-sentence $(\forall x \in A_o)\ x \neq \emptyset$. Thus q is an individual in $\hat{B}$, i.e. $q \in B_o$. We see that $*A_o \subseteq B_o$. Induction based on 2.5 shows that $*A_n \subseteq B_n$, $n = 1,2,\ldots$

So far we have only permitted WFF's in which all quantifiers (if any) were of the form "$(\forall x \in C)$" or "$(\exists x \in C)$," with C a constant ($C \in \hat{A}$). However, when dealing with strict monomorphisms, we may also safely admit quantifiers of the form "$(\forall y \in x)$" and "$(\exists y \in x)$," provided that each such quantifier is preceded by "$(\forall x \in C)$" or "$(\exists x \in C)$," i.e. x is a bound variable occurring in a quantifier of the kind specified above (with C a constant). Indeed, if $C \in A_{n+2}$, say, then $y \in x \wedge x \in C$ implies $y \in A_n \cup A_o$. Thus a WFF of the form "$(\forall x \in C)(\forall y \in x)\alpha$ " is equivalent to "$(\forall x \in C)(\forall y \in A_n \cup A_o)[y \in x \Rightarrow \alpha]$," while "$(\forall x \in C)(\exists y \in x)\alpha$" may be written as "$(\forall x \in C)(\exists y \in A_n \cup A_o)[y \in x \wedge \alpha]$." The φ-transform of the last formula can be written as "$(\forall x \in *C)(\exists y \in *A_n \cup *A_o)[y \in x \wedge *\alpha]$" or simply "$(\forall x \in *C)(\exists y \in x)*\alpha$"; for, by 2.5, $y \in x \in *C \in *A_{n+2}$ anyway implies $y \in *A_n \cup *A_o$, if φ is strict. Similarly in other cases of this kind.

By the same argument, we may safely admit WFF's with several quantifiers, such as $(Qx \in C)(Qy \in x)(Qz \in y)(Qu \in z)\ \alpha$, where $C \in \hat{A}$ and "Q" stands for "$\forall$" or "$\exists$"; we shall call such formulas "relaxed WFF's." Under a strict monomorphism, such a formula is equivalent to $(Qx \in *C)(Qy \in x)(Qz \in y)(Qu \in z)\ *\alpha$, with "Q" again replaced by the corresponding quantifier. As an example, we prove:

2.13. (S). $*A_{n+1}$ is exactly the set of all internal subsets of $*A_n \cup *A_o$.

Proof. For all m,n, we have the relaxed K-sentences: $(\forall x \in A_{n+1})(\forall y \in x)\ y \in A_n \cup A_o$, and $(\forall x \in A_{m+1})\{[(\forall y \in x)\ y \in A_n \cup A_o] \Rightarrow x \in A_{n+1}\}$. The result now easily follows.

Summing up, we obtain:

2.14. (Alternative definition of strict monomorphisms). A map $\varphi:\hat{A}\to\hat{B}$ of one superstructure into another is a strict monomorphism iff, writing, as usual, $*x$ for $\varphi(x)$, we have: (i) $*\emptyset=\emptyset$, and (ii) $\alpha \Longleftrightarrow *\alpha$, for each "relaxed" WFS (see above), where $*\alpha$ is obtained from α by replacing in it each constant $c\in\hat{A}$ by $*c$ (as before, $*\alpha$ is called the φ-transform of α).

Proof. If φ satisfies (i) and (ii), then φ is necessarily one-one, since $(\forall c,d\in\hat{A})\ c=d \Longleftrightarrow *c=*d$, by (ii). Also, φ is a monomorphism in the sense of Definition 1.1, since we have $\alpha\Longleftrightarrow *\alpha$ even for each "non-relaxed" WFS (the latter being a special case of a "relaxed" WFF). Finally, φ is strict; for, the "relaxed" K-sentence $(\forall y\in A_n)(\forall x\in y)\ x\in A_n\cup A_o$ yields $(\forall y\in *A_n)(\forall x\in y)\ x\in *A_n\cup *A_o$, for each n; hence $y\in *A_n\cup *A_n\subseteq *\hat{A}$ for $y\in *A_n$, $n=0,1,..$, implying $(\forall y\in *\hat{A})\ y\subseteq *A$, as required in Definition 2.4. Conversely, if φ is a strict monomorphism, then (i) and (ii) hold, as was shown above.

NOTE 5. Every monomorphism $\varphi:\hat{A}\to\hat{B}$ can be transformed into a strict one.

Proof. By 2.3, we may safely replace all members of $*A_o$ by individuals, without changing any internal elements outside $*A_o$. Next, we replace each $y\in *A_n$ $(n\geq 1)$ by $y\cap *\hat{A}$ (i.e., remove from y all its external elements, if any). We carry out this process step by step, for $n=1,2,..$. This preserves the validity of all φ-transforms of "relaxed" K-sentences, since they do not assert anything about external elements. Thus φ becomes a strict monomorphism.

§3. EXISTENCE OF MONOMORPHISMS. ULTRAPOWERS

For our purposes, we must somewhat modify the usual definition of an ultrapower. First we recall some well known notions and facts.

By a filter in a set $J\neq\emptyset$ we mean a non-empty family $\mathcal{F}$ of subsets of J such that: (a) $\emptyset\notin\mathcal{F}$, (b) $(\forall X,Y\in\mathcal{F})\ X\cap Y\in\mathcal{F}$, and (c) $(\forall X\in\mathcal{F})(\forall Y\subseteq J)[X\subseteq Y\Rightarrow Y\in\mathcal{F}]$. If $\mathcal{F}$ also satisfies (d) $(\forall X\subseteq J)[X\in\mathcal{F} \Longleftrightarrow J-X\notin\mathcal{F}]$, it is called an ultra-filter.

By Zorn's lemma, any set family $\mathcal{F}$ satisfying (a) and (b) can be extended to an ultrafilter.

3.1. DEFINITION. Given an ultrafilter $\mathcal{F}$ in a set J, and a superstructure $\hat{A}$, let M be the set of all maps of the form $f:J \to D$, $D \in \hat{A}$, i.e. mappings of J into various members of $\hat{A}$. For any such maps f,g, we write $f \dot{\in} g$ iff the set $\{i \in J \mid f(i) \in g(i)\}$ belongs to $\mathcal{F}$. Similarly, we put $f \doteq g$ iff $\{i \in J \mid f(i) = g(i)\} \in \mathcal{F}$ (this is an equivalence relation in M). The set M with "$\dot{\in}$" and "$\doteq$" so defined is called the <u>$\mathcal{F}$-ultrapower of $\hat{A}$</u> (over J).

For each $c \in A$, we denote by $\bar{c}$ the constant function on J with value c; i.e. $\bar{c}(i) = c$, $i \in J$. Thus $\bar{c} \in M$. In particular, $\bar{\emptyset} \in M$ and $\bar{A}_n \in M$, $n = 0,1,2,\ldots$

3.2. <u>With the above notation, we have, for any $a,b \in \hat{A}$ and $f,g,h \in M$:</u>
(i) $a \in b \Leftrightarrow \bar{a} \dot{\in} \bar{b}$; (ii) $\bar{a} \doteq \bar{b} \Leftrightarrow a = b \Leftrightarrow \bar{a} = \bar{b}$; (iii) $f \dot{\in} g \doteq h \Rightarrow f \dot{\in} h$;
(iv) $(\forall f \in M)(\exists n)\ f \dot{\in} \bar{A}_n$; (v) If $f \dot{\in} g \dot{\in} \bar{A}_{n+1}$, then $f \dot{\in} \bar{A}_n$ or $f \dot{\in} \bar{A}_o$;
(vi) $(g \dot{\in} \bar{A}_o$ or $g \doteq \bar{\emptyset}) \Leftrightarrow (\forall f \in M)\ f \dot{\notin} g$; (vii) $f \dot{\in} \bar{A}_n \Rightarrow f \dot{\in} \bar{A}_{n+1}$ $(n \geq 1)$.

<u>Proof</u>. (i) $a \in b \Leftrightarrow \bar{a}(i) \in \bar{b}(i) \Leftrightarrow \{i \mid \bar{a}(i) \in \bar{b}(i)\} = J \Leftrightarrow \bar{a} \dot{\in} \bar{b}$, since $J \in \mathcal{F}$ for <u>any</u> filter $\mathcal{F}$ in J. Similarly for the rest. E.g., we prove (iii): If $f \dot{\in} g$ and $g \doteq h$, then the sets $I' = \{i \mid f(i) \in g(i)\}$ and $I'' = \{i \mid g(i) = h(i)\}$ are in $\mathcal{F}$; hence so is $I' \cap I'' = \{i \mid f(i) \in g(i) = h(i)\}$. By filter properties, so also is the <u>larger</u> set $\{i \mid f(i) \in h(i)\}$, whence $f \dot{\in} h$.

We see that the $\bar{A}_n$ and the relation "$\dot{\in}$" in M behave like the $*A_n$ and "$\in$" in $*\hat{A}$. In particular, by 3.2(vi), each map g with $g \dot{\in} \bar{A}_o$ behaves like an "individual" in $*\hat{A}$ (it has no "elements"). In order to convert "$\dot{\in}$" into a genuine "$\in$", we now modify M as follows, step by step.

First of all, if $f \dot{\in} \bar{A}_o$, we replace f by some genuine individual f' (called the <u>fiber</u> of f).[6] We choose these "fibers" in such a manner that $f' = g'$ iff

---

[6]) f' need not belong to $\hat{B}$. As we have noted, we axiomatically assume that our universe contains "enough" individuals to carry out such replacements.

$f \doteq g \in \bar{A}_o$; i.e., equivalent maps get <u>the same</u> fiber. Moreover, if $f = \bar{a} \dot{\in} \bar{A}_o$, we choose $f' = a$; in particular, <u>$\bar{a}' = a$ if $a \in A_o$</u>.

Now, by 3.2(v), $g \dot{\in} f \dot{\in} \bar{A}_1 \Rightarrow g \dot{\in} \bar{A}_o$; so the fiber g' is defined. Thus, if $f \dot{\in} \bar{A}_1$, we can (and do) form the set of all fibers g' ($g \dot{\in} f$) and call it the <u>fiber</u> f' of f; so $f' = \{g' \mid g \dot{\in} f\}$. Proceeding by induction, once f' has been defined for $f \dot{\in} \bar{A}_n$, we define it for each $f \dot{\in} \bar{A}_{n+1}$ by $f' = \{g' \mid g \dot{\in} f\}$. Since each $f \in M$ satisfies $f \dot{\in} \bar{A}_n$ for some n (by 3.2(iv)), we can achieve that <u>each</u> $f \in M$ will be replaced by its fiber f', and so "$\dot{\in}$" will become the ordinary "$\in$" (for, by definition, $g' \in f' \Leftrightarrow g \dot{\in} f$). The set $M' = \{f' \mid f \in M\}$ of all such fibers will be called the <u>modified $\mathcal{F}$-ultrapower of $\hat{A}$</u>. We also define a map $\varphi : \hat{A} \to M'$, setting:

(3.3) <u>$\varphi(a) = {}^*a = \bar{a}'$</u> (= fiber of the map $\bar{a}$), for each $a \in \hat{A}$. In particular:

(3.4) <u>$\varphi(a) = {}^*a = a$ if $a \in A_o$</u>; <u>$\varphi(\emptyset) = {}^*\emptyset = \emptyset$</u>; <u>$\varphi(A_n) = {}^*A_n = (\bar{A}_n)'$</u>.

The map φ is one-one, as follows from 3.5(ii) below.

3.5. <u>For any $a, b \in \hat{A}$ and $f, g \in M$, we have</u>: (i) <u>$a \in b \Leftrightarrow {}^*a \in {}^*b$</u>;
(ii) <u>$a = b \Leftrightarrow {}^*a = {}^*b$</u>; (iii) <u>$f \doteq g \Leftrightarrow f' = g'$</u>; (iv) <u>$f \dot{\in} g \Leftrightarrow f' \in g'$</u>;
(v) <u>$g' \in {}^*A_o \cup \{\emptyset\} \Leftrightarrow (\forall f \in M)\ f' \notin g'$</u>; (vi) <u>$f' \in g' \in {}^*A_{n+1} \Rightarrow f' \in {}^*A_n \cup {}^*A_o$</u>.

Indeed, (iv) was already noted above, and the rest follows from the corresponding formulas of 3.2.

3.6. (i) $M' = \bigcup_{n=1}^{\infty} {}^*A_n = {}^*\hat{A}$; (ii) <u>${}^*A_n \in {}^*A_{n+1}$</u>; (iii) <u>${}^*A_n \subseteq {}^*A_{n+1}$ for $n \geq 1$.</u>

<u>Proof</u>. (i) Use 3.2(iv); (ii) Use 3.5(i); (iii) Use 3.2(vii).

Our next aim is to show that φ is actually a monomorphism. For this purpose, we again adopt for $\hat{A}$ the logical language L of §1, with a slight modification: only formulas of the form $x \in y$ and $x = y$, but not those of the form $(x_1, \ldots, x_m) = y$, will be treated as <u>atomic</u>. The language L so modified (called L') will be presupposed throughout this section.

NOTE 1. Even so, a sentence of the form $(a_1, \ldots, a_m) = b$ still is a <u>WFS</u>.

For we always have $a_1,..,a_m \in A_n \cup A_o$ for large n; so "$\{a_1,..,a_m\} = b$" is equivalent to the WFS "$(\forall x \in A_n \cup A_o)[x \in b \Leftrightarrow (x = a_1 \vee x = a_2 \vee \dots \vee x = a_m)]$. Hence "$\{\{a_1,a_2\},\{a_2\}\} = b$", i.e. "$(a_1,a_2) = b$", can be written as a WFS. Similarly for "$(a_1,..,a_m) = b$", by induction. The same also applies if the a_k and b are <u>bound</u> variables in L'. Thus every WFS in L is also a WFS in L', and conversely.

The notion of the <u>φ-transform</u> $*\alpha$ of a WFF α is now defined in L', in the same fashion as in L (cf. §1) [replace in α each <u>constant</u> $c \in \hat{A}$ by $*c$].

Now we show that our "modified" ultrapowers behave like the ordinary ultrapowers of model theory.

3.7. (Ultrapower theorem). <u>Let $\alpha = \alpha(x_1,..,x_m)$ be a WFF in L', with $x_1,..,x_m$ its only free variables, and let $f'_1,...,f'_m \in M'$. Then the sentence $*\alpha(f'_1,...,f'_m)$ holds in M' iff $\{i \in J \mid \alpha(f_1(i),..,f_m(i))\} \in \mathcal{F}$.</u>[7]

<u>Proof</u>. Let us call α "good" if 3.7 applies to it. We then obtain:

I. <u>Each atomic WFF (i.e. one of the form $x = y$, $x \in y$, $x = c$, $x \in c$ or $c \in x$) is "good."</u> [Follows directly by 3.5(ii,iv) and definition of $\dot{\in}$ and $\dot{=}$.]

II. <u>If α is "good," so is $\neg\alpha$</u>. For, by property (d) of ultrafilters and the "goodness" of α, $*(\neg\alpha) \Leftrightarrow \neg *\alpha \Leftrightarrow \{i \in J \mid \alpha(f_1(i),...,f_m(i))\} \notin \mathcal{F} \Leftrightarrow \{i \in J \mid \neg\alpha(f_1(i),..,f_m(i))\} \in \mathcal{F}$; so $\neg\alpha$ is "good."

III. <u>If α and β are "good," so is $\alpha \wedge \beta$</u>. [Proof as in 3.2(iii).]

IV. <u>If $\beta = \beta(x_1,..,x_m,y)$ is "good," so is $\alpha = (\exists y \in C)\beta$, $C \in \hat{A}$.</u> For, suppose $*\alpha(f'_1,..,f'_m)$ holds in M'; so $(\exists y \in *C)$ $*\beta(f'_1,..,f'_m,y)$. Then we can fix $y = g' \in M'$ such that $(g' \in *C \wedge *\beta(f'_1,..,f'_m,g'))$. By the "goodness" of β and (I),(II), this implies that $\{i \in J \mid g(i) \in C \wedge \beta(f_1(i),..,f_m(i),g(i))\} \in \mathcal{F}$. Hence also the <u>larger</u> set $\{i \in J \mid (\exists y \in C)\, \beta(f_1(i),..,f_m(i),y)\} = I$ is in $\mathcal{F}$.

---

[7]) Here, as before, M' is the modified $\mathcal{F}$-ultrapower over J. Note that all $f_k(i)$ are constants in $\hat{A}$. Thus "$\alpha(f_1(i),..,f_m(i))$" makes sense in $\hat{A}$.

Conversely, let $I \in \mathcal{F}$. We have: $(\forall i \in I)(\exists y \in C)\, \beta(f_1(i),..f_m(i),y)$. Hence, by the axiom of choice, we can fix for each $i \in I$ some $c_i \in C$, and $c_o \in C$, such that $\beta(f_1(i),..,f_m(i),c_i)$ for $i \in I$. We now define $g:J \to C$ by $g(i) = c_i$ for $i \in I$ and $g(i) = c_o$ if $i \in J-I$. Then $J \in \mathcal{F}$ yields $g \dot{\in} \bar{C}$ (by the definition of $\dot{\in}$), whence $g' \in {}^*C$. Also, $(\forall i \in I)\, \beta(f_1(i),..,f_m(i),g(i))$. As $I \in \mathcal{F}$, the assumed "goodness" of β implies ${}^*\beta(f_1',..,f_m',g')$ in M'. Thus $(\exists y \in {}^*C)^*\beta(f_1',..,f_m',y)$, i.e. ${}^*\alpha(f_1',..,f_m')$, certainly follows. We see that $I = \{i \in J \mid \alpha(f_1(i),..,f_m(i))\} \in \mathcal{F} \Leftrightarrow {}^*\alpha(f_1'..,f_m')$, and so α is "good," as claimed.

Now induction over the length of α completes the proof.

3.8. (Monomorphism theorem). <u>If M' is the modified $\mathcal{F}$-ultrapower of $\hat{A}$ (over J), then φ, as defined in (3.3), is a strict monomorphism of $\hat{A}$ into $\hat{B}$, where $\hat{B} = \bigcup_{n=1}^{\infty} B_n$ is the superstructure on $B_o = {}^*A_o = (\bar{A}_o)' \in M'$.</u>

<u>Proof</u>. First of all, $\varphi:\hat{A} \to M'$ is also a map of $\hat{A}$ into $\hat{B}$, as defined in the theorem, because $M' \subset \hat{B}$. Indeed, by construction, ${}^*A_o = B_o$. Moreover, by 3.5(vi), $g' \in {}^*A_{n+1} \Rightarrow g' \subseteq {}^*A_n \cup {}^*A_o \subseteq \bigcup_{k=0}^{n} {}^*A_k$. Hence it follows by induction that ${}^*A_n \subseteq B_n$, $n = 0,1,2,..$, and so $M' \subseteq \bigcup_{n=1}^{\infty} B_n = \hat{B}$, by 3.6(i).

We have to show that $\alpha \Leftrightarrow {}^*\alpha$ for every WFS α. Proceeding as in 3.7, we call, this time, a WFF $\alpha(x_1,..,x_m)$ "good" iff $\alpha(a_1,..,a_m) \Leftrightarrow {}^*\alpha({}^*a_1,..,{}^*a_m)$ for all constants $a_k \in A$. Then assertions (I) - (III) in the proof of 3.7 are obvious from 3.5(i,ii) and from the fact that ${}^*(\neg\alpha) \Leftrightarrow \neg\, {}^*\alpha$, and ${}^*(\alpha \wedge \beta) \Leftrightarrow ({}^*\alpha \wedge {}^*\beta)$. To complete the proof, we only have to verify (IV), i.e. to show that $(\exists y \in C)\,\beta \Leftrightarrow (\exists y \in {}^*C)^*\beta$, if β is "good" $[\beta = \beta(x_1,..,x_m,y),\ C \in \hat{A}]$.

Now, if $(\exists y \in C)\beta$ holds for $x_k = a_k \in \hat{A}$, we can fix $c \in C$ such that $\beta(a_1,... a_m,c)$ is true. Then so is ${}^*\beta({}^*a_1,..,{}^*a_m,{}^*c)$, ${}^*c \in {}^*C$, by the "goodness" of β. Hence, certainly, it is true that $(\exists y \in {}^*C)^*\beta({}^*a_1,..,{}^*a_m,y)$.

Conversely, assume the latter. Then we can fix $y = g \in {}^*C$ such that the sentence ${}^*\beta({}^*a_1,..,{}^*a_m,g')$ holds in M'. Noting that ${}^*a = \bar{a}'$ (by definition), we obtain by 3.7 that $I=\{i\in J \mid \beta(a_1,\ldots,a_m,g(i))\}\in \mathcal{F}$ (recall that $\bar{a}(i)=a$). As $I\in\mathcal{F}$, $I\neq\emptyset$ by filter properties. Thus there is an element $y=g(i)$ such that $\beta(a_1,\ldots,a_m,y)$ is true. We see that $(\exists y\in {}^*C)\,{}^*\beta \iff (\exists y\in C)\beta$, indeed.

Thus φ is a monomorphism. It is strict by 3.5(vi). Thus all is proved.

Theorem 3.8 shows that modified ultrapowers yield a general method for the formation of monomorphisms. We now "translate" a classical example into the language of monomorphisms.

EXAMPLE. Let $A=A_o$ be the set of all real numbers (treated as individuals). Let J consist of all positive integers. Let $\mathcal{D}$ be the family of all complements of finite subsets of J, i.e. $\mathcal{D}=\{X\subseteq J \mid J-X \text{ finite}\}$. Then $\mathcal{D}$ has the properties (a) and (b) of filters and thus can be extended to an ultrafilter $\mathcal{F}\supset\mathcal{D}$. Fixing the sets A,J and $\mathcal{F}$, we now let M' be the modified $\mathcal{F}$-ultrapower of $\hat{A}$ over J, and obtain the monomorphism $\varphi:\hat{A}\to\hat{B}$, as in 3.8. Note that here $A_o\subset {}^*A_o$ (properly). Indeed, define $f:J\to A_o$ by $(\forall n\in J)\ f(n)=n$. Then $f\,\dot{\in}\,\bar{A}_o$, and so $f'\in {}^*A_o$. However, for no $a\in A_o$ is $f\doteq\bar{a}$ (otherwise, $f(i)=\bar{a}(i)=a$ would hold for all but finitely many values of i, contrary to the choice of f). Hence $(\forall a\in A_o)\ f'\neq {}^*a=a$, and so $f'\notin A_o$. Thus ${}^*A_o\nsubseteq A_o$, but $A_o\subseteq {}^*A_o$ by (3.4).

NOTE 2. In any monomorphism constructed as in 3.8 (from an ultrapower), M' is exactly the set ${}^*\hat{A}$ of all φ-internal elements; cf. 3.6(i). We also have:

3.9. (Comprehensive property). Each monomorphism φ constructed from an ultrapower (as in 3.8.) has the following property, called comprehensiveness:

For any sets $C,D\in\hat{A}$ and any map $h:C\to{}^*D$, there is an internal mapping $g:{}^*C\to{}^*D$ ($g\in{}^*\hat{A}$) such that $g({}^*a)=h(a)$ for every $a\in C$.

Proof. Each $f'\in{}^*C$ is the fiber of some $f\,\dot{\in}\,\bar{C}$; so $\{i\in J \mid f(i)\in C\}\in\mathcal{F}$. We may (and shall) assume that $f(i)\in C$ for all $i\in J$ otherwise, replace f by a map $g\doteq f$, $g:J\to C$, setting $g(i)=f(i)$ if $f(i)\in C$, and $g(i)=$ arbitrary $c\in C$ if $f(i)\notin C$.

Thus, indeed, <u>each $f' \in {}^*C$ is the fiber of some map $f: J \to C$</u>.

Noting this, fix some $i_o \in J$ and consider the given mapping $h: C \to {}^*D$. For each $f: J \to C$, $f(i_o) \in C$; so $h(f(i_o)) \in {}^*D$. Thus $h(f(i_o))$ is the fiber of some map of J into D; we denote this map by hf; so $h(f(i_o)) = (hf)'$, $hf: J \to D$.

Now define a mapping $k: J \to \hat{A}$ ($k \in M$) as follows. For each $i \in J$, let k(i) be the set of all ordered pairs (f(i), hf(i)), with f ranging over all maps $f: J \to C$, so that $f' \in {}^*C$, $(hf)' \in {}^*D$. Then, for each $f' \in {}^*C$, $(f(i), hf(i)) \in k(i)$, for $i \in J$. As $J \in \mathcal{F}$, 3.7 yields $(f', (hf)') \in k'$ for each $f' \in {}^*C$; it also follows that k' is <u>exactly</u> the set of all ordered pairs $(f', (hf)')$ with $f' \in {}^*C$.[8] In other words, k' is a <u>mapping</u>, $k': {}^*C \to {}^*D$, with $k'(f') = (hf)' = h(f(i_o)) \in {}^*D$ (see above). In particular, if $f' = {}^*a \in {}^*C$, ($a \in C$), we may put $f = \bar{a}$ to obtain: $k'({}^*a) = h(\bar{a}(i_o)) = h(a)$. Finally, $k' \in M' = {}^*\hat{A}$ (since $k \in M$, by construction); so k' is <u>internal</u>. Thus k' is the required map g of the theorem.

NOTE 3. A monomorphism $\varphi: \hat{A} \to \hat{B}$ is said to be <u>comprehensive</u> iff it has the property specified in 3.9. In this case, ${}^*\hat{A}$ is called a <u>comprehensive model of $\hat{A}$</u>. (This notion and Theorem 3.9 are due to Robinson [6]).

That much of ultrapower theory will suffice for our purposes. For various generalizations, see [1]-[5]. One should note, however, that there are monomorphisms which do not originate from ultrapowers. The notion of monomorphism is much more general. It yields Robinson's "enlargements" as a special case, to be studied next.

§4. CONCURRENT RELATIONS. ENLARGEMENTS. SATURATION

A binary relation R is said to be <u>concurrent</u> iff, for any finite number of

---

8) For if $z' \in k'$ ($z \in M$), then again 3.7 yields $\{i \in J \mid z(i) \in k(i)\} \in \mathcal{F}$ and, as before, we may assume that $z(i) \in k(i)$ for <u>all</u> $i \in J$. By the definition of k(i), this means that, for some $f: J \to C$, $z(i) = (f(i), hf(i))$, $i \in J$; so, by 3.7, $z' = (f', (hf)')$. Thus <u>each</u> element $z' \in k'$ is such a pair.

elements $a_1,..,a_m$ of its domain D_R, there is some b such that $(a_k,b) \in R$, $k=1,2,..,m$. E.g. the inequality relation between real numbers is concurrent since, for any real $a_1,..,a_m$, there is a real b with $a_k < b$, $k=1,..,m$.

4.1. DEFINITION. A monomorphism $\varphi:\hat{A} \to \hat{B}$ is said to be enlarging (and $*\hat{A}$ is called an enlargement of $\hat{A}$) iff, for each concurrent relation $R \in \hat{A}$, there is some $b \in *\hat{A}$ such that $(*a,b,) \in *R$ for all $a \in D_R$ simultaneously. We then also say that φ bounds concurrent relations. The enlargement is strict if φ is.

All these notions are due to Robinson [5,6,7], as is the next theorem.

4.2. (Enlargement theorem). For every superstructure $\hat{A}$, there is a superstructure $\hat{B}$ and a monomorphism $\varphi:\hat{A} \to \hat{B}$ which is strict and enlarging.

Proof. We use 3.8 with a special choice of J and $\mathcal{F}$, as follows.

Let $\mathcal{C}$ be the set of all concurrent relations R in $\hat{A}$. Let J be the set of all maps $i: \mathcal{C} \to \hat{A}$ such that, for each $R \in \mathcal{C}$, $i(R)$ is a finite subset of D_R [$i(R)$ is not to be confused with the image set $i[R]$]. We partially order J by setting: $i \leq j$ in J iff $(\forall R \in \mathcal{C})\ i(R) \subseteq j(R)$. We also define: $[j....) = \{i \in J \mid i \geq j\}$ for $j \in J$. Clearly, $j \in [j.....)$; so $[j....) \neq \emptyset$.

Next, let $\mathcal{D}$ be the family of all subsets of J of the form $[j....)$; so $\emptyset \notin \mathcal{D}$. $\mathcal{D}$ is closed under finite intersections; for $[i....) \cap [j....) = [k....)$ where $k \in J$ is defined on $\mathcal{C}$ by $k(R) = i(R) \cup j(R)$. Thus, as noted in §3, $\mathcal{D}$ can be extended to an ultrafilter $\mathcal{F}$ in J. With J and $\mathcal{F}$ so chosen, let M' be the modified $\mathcal{F}$-ultrapower of $\hat{A}$ over J. Then $\varphi:\hat{A} \to M'$ is a strict monomorphism (by 3.8), with $M' \subseteq \hat{B}$, and $M' = *\hat{A}$ (by 3.6). We now show that φ is enlarging.

Indeed, fix any concurrent relation $R \in \hat{A}$, i.e. $R \in \mathcal{C}$. Then, by construction, $i(R)$ is a finite subset of D_R, $i \in J$. Thus, by concurrence, we can fix (by the axiom of choice) some b_i for each $i \in J$, so that $(a,b_i) \in R$ for all $a \in i(R)$. Then we define a map $b:J \to D_R$ ($b \in M$) by setting $b(i) = b_i$, $i \in J$; so the fiber b' of b is in M' (cf §3). We complete the proof by showing that b' is the required element $b \in *\hat{A}$ of Defn. 4.1, i.e. that $(\forall a \in D_R)\ (*a,b') \in *R$.

In fact, let $a \in D_R$ and put $I = \{i \in J \mid a \in i(R)\}$; so $(\forall i \in I)\ (a, b_i) \in R$, by construction. Also define $i_a \in J$ by $i_a(S) = \emptyset$ if $S \neq R$, and $I_a(R) = \{a\}$, $(S \in \mathcal{C})$. Then as is easily seen, $i \in I$ iff $i \geq i_a$; so $I = [i_a \ldots) \in \mathcal{D} \subseteq \mathcal{F}$, whence $I \in \mathcal{F}$. Also, by what was said above, $I \subseteq \{i \in J \mid (a, b, (i)) \in R\}$; so this set is in the filter $\mathcal{F}$, along with I. Hence, by 3.7, $(*a, b') \in *R$, as claimed.

By 3.9, M' constructed above is also comprehensive. Thus, summing up:

4.3. <u>Each superstructure $\hat{A}$ has a strict comprehensive enlargement.</u>

Henceforth, we shall always assume that A_o is infinite. Thus it has a countable subset N which shall be identified with the natural numbers $\{0,1,2,..\}$. The set $\varphi(N) = *N$ then will be called the <u>φ-extended</u> natural number system. The ordering "<" of N is a binary relation $R \subseteq (A_o)^2$; it extends to a total ordering *R of *N (cf. §1, Note 4). By 1.2, $N \subseteq *N$ and $R \subseteq *R$; so <u>*R coincides with R (the usual order of naturals)</u> <u>when restricted to N</u>; we shall simply write "$x < y$" for "$(x,y) \in *R$." Clearly, R is concurrent. Thus, if φ is enlarging, there is $b \in *N$ such that <u>$(\forall a \in N)\ a = *a < b$</u>. Such elements b (called <u>infinite</u> naturals) may also exist if φ is not enlarging (for instance, such is f' in the Example following 3.8). It easily follows that $*N - N = \{n \in *N \mid n \text{ infinite}\}$.[9]

4.4. DEFINITIONS. A monomorphism $\varphi : \hat{A} \to \hat{B}$ is <u>non-standard</u>, and $*\hat{A}$ is a <u>non-standard model of $\hat{A}$</u>, iff $*N - N \neq \emptyset$, i.e. iff *N has infinite elements.[10]

A set $D \in *\hat{A}$ is <u>star-finite</u> (*finite) iff $D = \emptyset$ or there is some $f \in *\hat{A}$ such that $f \cap *\hat{A}$ is a bijective map of $D \cap *\hat{A}$ onto an interval $[0,n] = \{x \in *N \mid x \leq n\}$ of *N.

Propositions in which φ is assumed enlarging will be marked by "ENL".

4.5. (ENL). <u>For any $P \in \hat{A}$, there is a *finite set Q, with $\varphi[P] \subseteq Q \cap *\hat{A} \subseteq *P$.</u> <u>In particular, if $P \subseteq (A_o)^m$, then $\varphi[P] = P \subseteq Q \cap *\hat{A} \subseteq *P$</u> (by 1.2).

---

9) Fix $n \in N$ and use the K-sentence $(\forall x \in N)[(x \neq 0 \wedge x \neq 1 \wedge \ldots \wedge x \neq n) \Rightarrow x > n]$ to show that $(\forall x \in *N - N)\ \ x > n$. As $n \in N$ is arbitrary, each $x \in *N - N$ is infinite.

10) Such is every enlargement of $\hat{A}$, by what was noted above.

Proof. This is trivial if P is finite (put $Q = {}^*P = \varphi[P]$, by Note 3, §1).

If however P is infinite, the relation $R = \{(x,Y) \mid x \in Y \subseteq P,\ Y \text{ finite}\}$ is concurrent. As is easily seen, $D_R = P$, and ${}^*R = \{(x,Y) \mid x \in Y \cap {}^*\hat{A} \subseteq {}^*P,\ Y\ {}^*\text{finite}\}$. As φ is enlarging, there is $Q \in {}^*\hat{A}$ such that $(\forall a \in P)\ ({}^*a, Q) \in {}^*R$, i.e. ${}^*a \in Q \cap {}^*\hat{A} \subseteq {}^*P$ with Q *finite. Hence $\varphi[P] = \{{}^*a \mid a \in P\} \subseteq Q \cap {}^*\hat{A} \subseteq {}^*P$, and all is proved.

4.6. (ENL). For each $n \in N$, there are *finite sets (and intervals in *N) of power $\geqslant 2^{|A_n|} > |A_n|$. Also, $|{}^*N| \geqslant |\hat{A}|$. ($|X|$ denotes the cardinality of a set X).

Proof. Taking $P = A_{n+1}$ in 4.5, we get a *finite set Q, with $Q \cap {}^*\hat{A} \supseteq \varphi[A_{n+1}]$. Hence, as φ is one-one, $|Q \cap {}^*\hat{A}| \geqslant |\varphi[A_{n+1}]| = |A_{n+1}| \geqslant 2^{|A_n|}$ (for A_{n+1} contains the power set of A_n). Also, by *finiteness, $Q \cap {}^*\hat{A}$ is equipollent with some interval of *N. This proves the first clause of 4.6. It follows that the power of all of *N exceeds all $|A_n|$. Since each interval [m,n] is equipollent with [m+n,2n] (by the map $x \leftrightarrow x + n$),[11] one can inductively construct a sequence of mutually disjoint intervals $[p_n, q_n]$ in *N, with $|[p_n, q_n]| > |A_n|$, $n = 0,1,2,\ldots$. Then $\bigcup_{n=0}^{\infty} [p_n, q_n] \subseteq {}^*N$ and $\left|\bigcup_{n=0}^{\infty} [p_n, q_n]\right| \geqslant \left|\bigcup_{n=0}^{\infty} A_n\right| = |\hat{A}|$. Thus $|{}^*N| \geqslant |\hat{A}|$, and all is proved.

NOTE 1. Theorem 4.6 may fail if φ is not enlarging. Thus, in the Example of §3 (quoted above), ${}^*N = 2^{\aleph_0} = |A_0| < |\hat{A}|$; also, $|{}^*A_0| = (2^{\aleph_0})^{\aleph_0} = 2^{\aleph_0}$.

4.7. (ENL). If a binary relation $R \in \hat{A}$ is concurrent, there is $b \in {}^*\hat{A}$ and a *finite (hence internal) set $D \supseteq \varphi[D_R]$ such that $(\forall x \in D \cap {}^*\hat{A})\ (x,b) \in {}^*R$.

Proof. By Definition 4.1, there is $b \in {}^*\hat{A}$ such that $(\forall a \in D_R)\ ({}^*a, b) \in {}^*R$. Also, by 4.5, there is a *finite set $Q \in {}^*\hat{A}$ with ${}^*D_R \supseteq Q \cap {}^*\hat{A} \supseteq \varphi[D_R]$.

Now proceeding as in the proof of 2.6 and Note 4, §2, one easily obtains a set $D \in {}^*\hat{A}$ whose internal elements coincide with those of $\{x \in Q \mid (x,b) \in {}^*R\}$; that is, $D \cap {}^*\hat{A} = \{x \in Q \cap {}^*\hat{A} \mid (x,b) \in {}^*R\}$. Thus $(\forall x \in D \cap {}^*\hat{A})\ (x,b) \in {}^*R$. Moreover, D is *finite, since $D \in {}^*\hat{A}$, and $D \cap {}^*\hat{A}$ is contained in the *finite set Q. (This follows from the fact that "each subset of a finite set in $\hat{A}$ is finite itself," which can be written as a K-sentence.) This completes the proof.

[11]) Observe that all operations defined in N carry over to *N; cf. Note 4, §1. Thus addition is defined in *N and has similar properties as in N.

We now generalize the notion of an enlargement, following Luxemburg, [4].

4.8. DEFINITION. Given an infinite cardinal $\aleph$, we say that a strict monomorphism $\varphi:\hat{A}\to\hat{B}$ is $\aleph$-saturated (and $*\hat{A}$ is an $\aleph$-saturated model of $\hat{A}$) iff, whenever a relation $R\in *\hat{A}$ is concurrent on a set $D\subseteq D_R$ with $|D|<\aleph$, there is some $b\in *\hat{A}$ such that $(\forall x\in D)\ (x,b)\in R$ [here D need not be internal; the concurrence of R on D means that $(\forall a_1,..,a_m\in D)(\exists b\in *\hat{A})\ (a_k,b)\in R,\ k=1,2,..,m$]. If φ is not strict, we call it $\aleph$-saturated iff it becomes so when transformed as in Note 6, §2.

NOTE 2. In an enlargement, no cardinality restriction was imposed on D; however, the required property was assumed only for standard relations *R and for standard elements *a of D. Thus, if φ is $\aleph$-saturated ($\aleph>|\hat{A}|$), it is also enlarging, but the converse fails. We now obtain:

4.9. Let $*\hat{A}$ be $\aleph$-saturated. Then: (i) Any set $D\in *\hat{A}$, with $|D|<\aleph$, is *finite. (ii) If a relation $R\in *\hat{A}$ is concurrent on C, there is some $b\in *\hat{A}$ and a *finite Q such that $C\cap *\hat{A}\subseteq Q\cap *\hat{A}$ and $(\forall x\in Q\cap *\hat{A})\ (x,b)\in R$. (iii) If $P\in\hat{A}$ is infinite, $|*P|\geq\aleph$.

Proof. (i) First let φ be strict. By using suitable K-sentences, one easily shows that the union of any two (hence of finitely many) *finite sets is *finite itself, and hence the relation $R=\{(x,Y)\mid x\in Y\subseteq D,\ Y\ \text{*finite}\}$ (which is internal if φ is strict) is concurrent on C whenever $C\subseteq D$. Thus, if $C\subseteq D$, if $*\hat{A}$ is $\aleph$-saturated and $|C|<\aleph$, there is $Q\in *\hat{A}$ such that $(\forall x\in C)\ (x,Q)\in R$; i.e. $C\subseteq Q\subseteq D$ and Q is starfinite. Now, if $|D|<\aleph$, put $C=D$ here to obtain clause (i).

If however φ is not strict, we convert it into a strict monomorphism φ' by the process of Note 6, §2. By Definition 4.8, φ' is still $\aleph$-saturated. The set D loses its external elements, and turns into a set D', with $|D'|\leq|D|<\aleph$. By what was shown above, D' is *finite under φ'. Hence D is *finite under φ. [For, reversing the process, we add external elements only, and this does not affect *finiteness, by Defn. 4.8.] Thus (i) is proved in full.

(ii) is proved in the same manner as 4.7, using Definition 4.8 and (i).

(iii) Let $P\in A_{n+1}$, so $P\subseteq A_n\cup A_o$ and $*P\subseteq *A_n\cup *A_o$, $*P\in *\hat{A}$. Now, if we had

$|{*P}| < \aleph$, then (i) would show that *P is *finite and <u>hence P is finite</u> (this can be written as a K-sentence: "there is a bijective map $f: P \leftrightarrow [0,q] \subset N$, for some $q \in N$"). Since P is infinite (by assumption), this is impossible; so $|{*P}| \geq \aleph$.

4.10. <u>(S). Let $*\hat{A}$ be $\aleph$-saturated; let $C \subseteq {*A_m} \cup {*A_o}$, $D \subseteq {*A_n} \cup {*A_o}$. If $D \in *\hat{A}$ and $|C| < \aleph$ then, for each map $f_o: C \to D$, there is an internal map $g: Q \to D$ such that Q is *finite, $C \subseteq Q \subseteq {*A_m} \cup {*A_o}$ and $g(x) = f_o(x)$ for all $x \in C$.</u>

<u>Proof</u>. Let $R = \{(f,g) \mid f,g$ are internal maps, with $f \subseteq g \subseteq ({*A_m} \cup {*A_o}) \times D\}$. Let P be the set of all "singleton maps" $f_a: \{a\} \to D$ with $f_a(a) = f_o(a)$, $a \in C$; so $f_a = \{(a, f_o(a))\} \subseteq C \times D \subseteq ({*A_m} \cup {*A_o}) \times D$, and $f_a \in *\hat{A}$, by 2.10. As is readily seen, R is concurrent on $P = \{f_a \mid a \in C\}$; also, $R \in *\hat{A}$ and $|P| = |C| < \aleph$. Thus, by the $\aleph$-saturation of $*\hat{A}$, there is an internal g such that $(\forall x \in C)(f_x, g) \in R$, so that g is a mapping, with $f_x \subseteq g$ (i.e., $f_x(x) = f_o(x) = g(x)$) for each $x \in C$, $g: D_g \to D$, $C \subseteq D_g \subseteq {*A_m} \cup {*A_o}$. As in 4.7, we can restrict g to a *finite domain Q, $C \subseteq Q \subseteq D_g$, to obtain the required internal map $g: Q \to D$ with $g(x) = f_o(x)$, $x \in C$.

NOTE 3. The process described in Note 6, §2, never affects the cardinality or order type of *D (or $D \in *\hat{A}$). Thus, in many proofs (e.g. in 4.12-4.14) we may (and do) assume φ <u>strict</u>, with no loss of generality.

The existence of saturated monomorphisms will be proved in a later section (to appear). Now we prove some properties of non-standard models in general.

4.11. <u>If $*\hat{A}$ is a non-standard model of $\hat{A}$ ($A_o \supseteq N$), e.g. an enlargement, then:</u>

(i) <u>The set *N-N of all infinite naturals has no least or largest element.</u>

(ii) <u>*N-N and N are external sets, and so is $\varphi[D]$ for each infinite $D \in \hat{A}$.</u>

(iii) <u>$\varphi[D] \subset *D$ (properly) for each infinite set $D \in \hat{A}$. Hence each infinite set $D \subseteq (A_o)^m$ is external, with $D = \varphi[D] \subset *D$ (properly), by 1.2</u>

<u>Proof</u>. Part (i) is obvious; for if n is infinite, so are n-1 and n+1.

(ii) The fact that N is well-ordered can be expressed in terms of K-sentences whose φ-transforms imply that each <u>internal</u> set $X \subseteq *N$ ($X \neq \emptyset$) has a <u>least</u> element; and if bounded above in *N, it also has a <u>largest</u> element. But

$*N-N$ has none, by (i); and N has no largest element, though N is bounded in $*N$; thus <u>neither set is internal</u>. Now transform φ into a strict monomorphism φ', as in Note 6, §2. If $D \in \hat{A}$ and $|D| \geq \aleph_0$, there is a mapping $f: D \xrightarrow[\text{onto}]{} N$. Fix any $a \in D$ and let $f(a) = n$, $n \in N \subseteq A_o$; so $*f(*a) = *n = n$, i.e. $*f(\varphi'(a)) = n \in N$. Conversely, fixing any $n \in N$, we get $n = *f(\varphi'(a))$ for some $a \in D$. Thus $*f[\varphi[D]] = N$. As N is φ'-external, so is $*f[\varphi'[D]]$, <u>and hence so is $\varphi'[D]$ as well</u> (for, otherwise, $*f[\varphi'[D]]$ would be φ'-internal, by 2.11). It follows that $\varphi[D]$ <u>cannot be φ-internal</u>; otherwise the process of Note 6, §2, would transform it into a <u>φ'-internal set</u>, whereas $\varphi'[D]$ is φ'-external.

(iii) We always have $\varphi[D] = \{*a \mid a \in D\} \subseteq *D$ since $a \in D$ implies $*a \in *D$. Also, if $|D| \geq \aleph_0$, $\varphi[D] \neq *D$ since $\varphi[D] \notin *\hat{A}$ while $*D \in *\hat{A}$. Thus $\varphi[D] \subsetneqq *D$, as claimed.

4.12. <u>If $*\hat{A}$ is non-standard and comprehensive, then each countable set $S \subseteq *N-N$ $(S \neq \emptyset)$ has an upper and a lower bound in $*N-N$.</u>

<u>Proof</u>. Let $S = \{n_o, n_1, \ldots, n_m, \ldots\}$, $m \in N$.[12] By comprehensiveness, there is an <u>internal</u> map $g: *N \to *N$, with $g(m) = n_m$ for all $m \in N$. Let $D = \{k \in *N \mid (\forall m \leq k)\ g(m) > k\}$. As $g \in *\hat{A}$, it easily follows that $D \in *\hat{A}$, too; so $D \neq N$, by 4.11(ii). Also, by assumption, $S \subseteq *N-N$; thus, if $m \leq k \in N$, then $g(m) = n_m$ is <u>infinite</u> and hence $> k$. It follows that each $k \in N$ is in D, and so $N \subseteq D$. As $D \neq N$, D must contain some <u>infinite</u> $k_o \in *N$. Then $k_o \in D$ yields $(\forall m \leq k_o)\ g(m) > k_o$; hence, in particular, $(\forall m \in N)\ g(m) = n_m > k_o$. Thus k_o is a lower bound of $S = \{n_m \mid m \in N\}$.

To prove the existence of an <u>upper</u> bound, we use the K-sentence: "For each $m_o \in N$, every sequence $S = \{n_m \mid m < m_o\}$, $n_m \in N$, has a largest term" (being a <u>finite</u> sequence in N). Hence, for <u>internal</u> sequences, this holds with N replaced by $*N$, and our assertion follows (by taking an <u>infinite</u> m_o). If however $S = \{n_m \mid m \in N\}$

---

[12]) We treat this sequence as a <u>mapping</u> of N into $*N$. Without loss of generality, we assume that φ is strict (see Note 3).

($n_m \in$ *N-N) is external, then comprehensiveness again yields an internal map $g: {}^*N \to {}^*N$ such that $g(m) = n_m$ for all $m \in N$. Fix an <u>infinite</u> $m_o \in {}^*N$. Then, by what was said above, the (internal) subsequence $\{g(m) \mid m < m_o\}$ has a largest term $g(m)$, and this is certainly also an upper bound for the smaller set $S = \{g(m) \mid m \in N\} \subseteq \{g(m) \mid m < m_o\}$ (for m_o is <u>infinite</u>). Thus all is proved.

4.13. <u>In a nonstandard model $^*\hat{A}$, each interval of *N (hence any *finite set) is either finite or of power $\geq 2^{\aleph_0}$. Thus certainly $|{}^*A_o| \geq |{}^*N| \geq 2^{\aleph_0}$.</u>

<u>Proof</u>. As $A_1 = 2^{A_o} \supseteq 2^N$, [14] $|A_1| \geq 2^{\aleph_0}$; so A_1 has a subset E, $|E| = 2^{\aleph_0}$. We identify E with the real field and denote by N' its set of integers ≥ 0, isomorphic to N. This implies (via K-sentences) that $^*N' \cong {}^*N$; so *N' has infinite elements, and it suffices to prove 4.13 <u>for *N'</u>. For brevity, we write N for N' and E for $\varphi[E]$; thus E consists of all <u>standard</u> elements of *E (<u>itself an ordered field</u>).

Now fix an <u>infinite</u> $m \in {}^*N$. Then subdivide the interval $[0,1)$ in *E into m subintervals $[\frac{k}{m}, \frac{k+1}{m})$ ($k \in {}^*N$, $k < m$) of "infinitely small" length $\frac{1}{m}$, less than any positive $x \in E$. Each $x \in [0,1)$ is in one such interval, but no two <u>standard</u> points $x, y \in E$ can be in <u>one and the same</u> $[\frac{k}{m}, \frac{k+1}{m})$ since their distance is not infinitely small ($> {}^1/m$). Thus there must be at least as many intervals $[\frac{k}{m}, \frac{k+1}{m})$ as there are <u>standard</u> reals in $[0,1)$, i.e. at least $2^{\aleph_0}$. But the intervals $[\frac{k}{m}, \frac{k+1}{m})$ are in one-one correspondence with the values of $k \in {}^*N$, $0 \leq k < m$; so their number equals the power of the interval $[0,m)$ in *N. Hence this power is $\geq 2^{\aleph_0}$. By translation, this applies to any interval $[n, m+n)$, $m \in {}^*N - N$. This proves the first clause of 4.13, and the 2nd is immediate.

4.14. <u>A non-trivial monomorphism $\varphi: \hat{A} \to \hat{B}$ (i.e. other than the identity map on A) yields a non-standard model $^*\hat{A}$, if $|A_o| < |\hat{N}|$ [i.e. $|A_o| \leq 2^{\aleph_0}$ or $\leq 2^{2^{\aleph_0}}$</u>

[14]) We write 2^N for the power set of N, i.e. the set of all subsets of N.

or $2^{2^{2^{\aleph_0}}}$, etc.]

<u>Proof</u>. By Note 6, §2, we can make φ <u>strict</u>, without changing $|A_o|$ or any essentials. Then, as φ is non-trivial, <u>we cannot have $A_o = {}^*A_o$</u>; for, otherwise, induction yields $(\forall n)\ A_n = {}^*A_n$ and $(\forall a \in A_n)\ {}^*a = \varphi(a) = a$, i.e. φ is trivial.[15]

Thus $A_o \neq {}^*A_o$. Also, by assumption, $|A_o| \leqslant |N_m|$ for some m, with $N_m \in \hat{N}$. Thus there is a mapping f of N_m onto A_o which yields a map ${}^*f\colon {}^*N_m \to {}^*A_o$ <u>(onto ${}^*A_o$)</u>, such that $(\forall a \in N_m)\ {}^*f({}^*a) = {}^*(f(a)) = f(a) \in A_o$ (by 1.2). It follows that ${}^*f[\varphi[N_m]] = A_o \neq {}^*A_o = {}^*f[{}^*N_m]$, and hence $\varphi[N_m] \neq {}^*N_m$. This, in turn, implies that $N = N_o \neq {}^*N_o$; for if $N_o = {}^*N_o$, then again induction easily yields $(\forall m)\ \varphi[N_m] = N_m = {}^*N_m$ (contradiction!).

Thus $N \neq {}^*N$, i.e. ${}^*N - N \neq \emptyset$, and ${}^*\hat{A}$ is non-standard, by Defn. 4.4. This completes the proof.

In conclusion of this section, we would like to pose two research problems:

1. Does Proposition 4.14 hold also if $|A_o| \geqslant |\hat{N}|$? In other words, is it true that a monomorphism yields a non-standard model ${}^*\hat{A}$ <u>if and only if</u> it is not the identity map?

2. In §3, it was shown that the axiom of choice (or Zorn's lemma) implies the existence of non-standard monomorphisms. Is the converse true?

---

[15]) Indeed, suppose that $A_o = {}^*A_o$ and that $A_m = {}^*A_m$ for <u>some</u> m. Then, by 2.5, $X \in {}^*A_{m+1} \Leftrightarrow X \subseteq {}^*A_m \cup {}^*A_o \Leftrightarrow X \subseteq A_m \cup A_o \Leftrightarrow X \in A_{m+1}$; so ${}^*A_{m+1} = A_{m+1}$. Similarly, assuming that $(\forall x \in A_m)\ {}^*x = \varphi(x) = x$, one easily shows that $(\forall X \in A_{m+1})\ {}^*X \subseteq X \subseteq {}^*X$, i.e. ${}^*X = \varphi(x) = X$, completing the induction. It also follows that $(\forall C \in \hat{A})\ C = \varphi[C]$. Thus $A_o = {}^*A$ implies that φ is trivial (if strict), even without the assumption that $|A_o| < |N|$.

REFERENCES

1. T. Frayne, D. C. Morel and D. S. Scott, Reduced direct products, Fund. Math. 51(1962), 195-227.

2. H. J. Keisler, A survey of ultraproducts, Logic, Methodology and Philosophy of Science, Proc. 1964 Intern. Congress, North Holland, Amsterdam, 1965, 112-124.

3. S. Kochen, Ultraproducts in the theory of models, Ann. Math. Ser. 2, 79 (1961), 221-261.

4. W. A. J. Luxemburg (editor), Applications of Model Theory to Algebra, Analysis and Probability (Proc. Pasadena Symposium, 1967), Holt, Reinhart & Winston, New York, 1969.

5. M. Machover and J. Hirschfeld, Lectures on Non-Standard Analysis, Springer Verlag, New York, 1969.

6. A. Robinson, Nonstandard Analysis, North Holland, Amsterdam, 1966.

7. - - - - Nonstandard theory of Dedekind rings, Proc. Neth. Acad. Sci., Amsterdam A 70(1967), 442-452.

8. A. Robinson and E. Zakon, A set theoretical characterization of enlargements, Proc. Pasadena Symposium (cf. [4]), 109-122.

9. P. Suppes, Axiomatic Set Theory, Van Nostrand, New York, 1965.

10. E. Zakon, Remarks on the nonstandard real axis, [4], 195-227.

Vol. 215: P. Antonelli, D. Burghelea and P. J. Kahn, The Concordance-Homotopy Groups of Geometric Automorphism Groups. X, 140 pages. 1971. DM 16,–

Vol. 216: H. Maaß, Siegel's Modular Forms and Dirichlet Series. VII, 328 pages. 1971. DM 20,–

Vol. 217: T. J. Jech, Lectures in Set Theory with Particular Emphasis on the Method of Forcing. V, 137 pages. 1971. DM 16,–

Vol. 218: C. P. Schnorr, Zufälligkeit und Wahrscheinlichkeit. IV, 212 Seiten. 1971. DM 20,–

Vol. 219: N. L. Alling and N. Greenleaf, Foundations of the Theory of Klein Surfaces. IX, 117 pages. 1971. DM 16,–

Vol. 220: W. A. Coppel, Disconjugacy. V, 148 pages. 1971. DM 16,–

Vol. 221: P. Gabriel und F. Ulmer, Lokal präsentierbare Kategorien. V, 200 Seiten. 1971. DM 18,–

Vol. 222: C. Meghea, Compactification des Espaces Harmoniques. III, 108 pages. 1971. DM 16,–

Vol. 223: U. Felgner, Models of ZF-Set Theory. VI, 173 pages. 1971. DM 16,–

Vol. 224: Revètements Etales et Groupe Fondamental. (SGA 1). Dirigé par A. Grothendieck XXII, 447 pages. 1971. DM 30,–

Vol. 225: Théorie des Intersections et Théorème de Riemann-Roch. (SGA 6). Dirigé par P. Berthelot, A. Grothendieck et L. Illusie. XII, 700 pages. 1971. DM 40,–

Vol. 226: Seminar on Potential Theory, II. Edited by H. Bauer. IV, 170 pages. 1971. DM 18,–

Vol. 227: H. L. Montgomery, Topics in Multiplicative Number Theory. IX, 178 pages. 1971. DM 18,–

Vol. 228: Conference on Applications of Numerical Analysis. Edited by J. Ll. Morris. X, 358 pages. 1971. DM 26,–

Vol. 229: J. Väisälä, Lectures on n-Dimensional Quasiconformal Mappings. XIV, 144 pages. 1971. DM 16,–

Vol. 230: L. Waelbroeck, Topological Vector Spaces and Algebras. VII, 158 pages. 1971. DM 16,–

Vol. 231: H. Reiter, L^1-Algebras and Segal Algebras. XI, 113 pages. 1971. DM 16,–

Vol. 232: T. H. Ganelius, Tauberian Remainder Theorems. VI, 75 pages. 1971. DM 16,–

Vol. 233: C. P. Tsokos and W. J. Padgett. Random Integral Equations with Applications to stochastic Systems. VII, 174 pages. 1971. DM 18,–

Vol. 234: A. Andreotti and W. Stoll. Analytic and Algebraic Dependence of Meromorphic Functions. III, 390 pages. 1971. DM 26,–

Vol. 235: Global Differentiable Dynamics. Edited by O. Hájek, A. J. Lohwater, and R. McCann. X, 140 pages. 1971. DM 16,–

Vol. 236: M. Barr, P. A. Grillet, and D. H. van Osdol. Exact Categories and Categories of Sheaves. VII, 239 pages. 1971. DM 20,–

Vol. 237: B. Stenström, Rings and Modules of Quotients. VII, 136 pages. 1971. DM 16,–

Vol. 238: Der kanonische Modul eines Cohen-Macaulay-Rings. Herausgegeben von Jürgen Herzog und Ernst Kunz. VI, 103 Seiten. 1971. DM 16,–

Vol. 239: L. Illusie, Complexe Cotangent et Déformations I. XV, 355 pages. 1971. DM 26,–

Vol. 240: A. Kerber, Representations of Permutation Groups I. VII, 192 pages. 1971. DM 18,–

Vol. 241: S. Kaneyuki, Homogeneous Bounded Domains and Siegel Domains. V, 89 pages. 1971. DM 16,–

Vol. 242: R. R. Coifman et G. Weiss, Analyse Harmonique Non-Commutative sur Certains Espaces. V, 160 pages. 1971. DM 16,–

Vol. 243: Japan-United States Seminar on Ordinary Differential and Functional Equations. Edited by M. Urabe. VIII, 332 pages. 1971. DM 26,–

Vol. 244: Séminaire Bourbaki – vol. 1970/71. Exposés 382–399. IV, 356 pages. 1971. DM 26,–

Vol. 245: D. E. Cohen, Groups of Cohomological Dimension One. V, 99 pages. 1972. DM 16,–

Vol. 246: Lectures on Rings and Modules. Tulane University Ring and Operator Theory Year, 1970–1971. Volume I. X, 661 pages. 1972. DM 40,–

Vol. 247: Lectures on Operator Algebras. Tulane University Ring and Operator Theory Year, 1970–1971. Volume II. XI, 786 pages. 1972. DM 40,–

Vol. 248: Lectures on the Applications of Sheaves to Ring Theory. Tulane University Ring and Operator Theory Year, 1970–1971. Volume III. VIII, 315 pages. 1971. DM 26,–

Vol. 249: Symposium on Algebraic Topology. Edited by P. J. Hilton. VII, 111 pages. 1971. DM 16,–

Vol. 250: B. Jónsson, Topics in Universal Algebra. VI, 220 pages. 1972. DM 20,–

Vol. 251: The Theory of Arithmetic Functions. Edited by A. A. Gioia and D. L. Goldsmith VI, 287 pages. 1972. DM 24,–

Vol. 252: D. A. Stone, Stratified Polyhedra. IX, 193 pages. 1972. DM 18,–

Vol. 253: V. Komkov, Optimal Control Theory for the Damping of Vibrations of Simple Elastic Systems. V, 240 pages. 1972. DM 20,–

Vol. 254: C. U. Jensen, Les Foncteurs Dérivés de $\varprojlim$ et leurs Applications en Théorie des Modules. V, 103 pages. 1972. DM 16,–

Vol. 255: Conference in Mathematical Logic – London '70. Edited by W. Hodges. VIII, 351 pages. 1972. DM 26,–

Vol. 256: C. A. Berenstein and M. A. Dostal, Analytically Uniform Spaces and their Applications to Convolution Equations. VII, 130 pages. 1972. DM 16,–

Vol. 257: R. B. Holmes, A Course on Optimization and Best Approximation. VIII, 233 pages. 1972. DM 20,–

Vol. 258: Séminaire de Probabilités VI. Edited by P. A. Meyer. VI, 253 pages. 1972. DM 22,–

Vol. 259: N. Moulis, Structures de Fredholm sur les Variétés Hilbertiennes. V, 123 pages. 1972. DM 16,–

Vol. 260: R. Godement and H. Jacquet, Zeta Functions of Simple Algebras. IX, 188 pages. 1972. DM 18,–

Vol. 261: A. Guichardet, Symmetric Hilbert Spaces and Related Topics. V, 197 pages. 1972. DM 18,–

Vol. 262: H. G. Zimmer, Computational Problems, Methods, and Results in Algebraic Number Theory. V, 103 pages. 1972. DM 16,–

Vol. 263: T. Parthasarathy, Selection Theorems and their Applications. VII, 101 pages. 1972. DM 16,–

Vol. 264: W. Messing, The Crystals Associated to Barsotti-Tate Groups: With Applications to Abelian Schemes. III, 190 pages. 1972. DM 18,–

Vol. 265: N. Saavedra Rivano, Catégories Tannakiennes. II, 418 pages. 1972. DM 26,–

Vol. 266: Conference on Harmonic Analysis. Edited by D. Gulick and R. L. Lipsman. VI, 323 pages. 1972. DM 24,–

Vol. 267: Numerische Lösung nichtlinearer partieller Differential- und Integro-Differentialgleichungen. Herausgegeben von R. Ansorge und W. Törnig, VI, 339 Seiten. 1972. DM 26,–

Vol. 268: C. G. Simader, On Dirichlet's Boundary Value Problem. IV, 238 pages. 1972. DM 20,–

Vol. 269: Théorie des Topos et Cohomologie Etale des Schémas. (SGA 4). Dirigé par M. Artin, A. Grothendieck et J. L. Verdier. XIX, 525 pages. 1972. DM 50,–

Vol. 270: Théorie des Topos et Cohomologie Etale des Schémas. Tome 2. (SGA 4). Dirigé par M. Artin, A. Grothendieck et J. L. Verdier. V, 418 pages. 1972. DM 50,–

Vol. 271: J. P. May, The Geometry of Iterated Loop Spaces. IX, 175 pages. 1972. DM 18,–

Vol. 272: K. R. Parthasarathy and K. Schmidt, Positive Definite Kernels, Continuous Tensor Products, and Central Limit Theorems of Probability Theory. VI, 107 pages. 1972. DM 16,–

Vol. 273: U. Seip, Kompakt erzeugte Vektorräume und Analysis. IX, 119 Seiten. 1972. DM 16,–

Vol. 274: Toposes, Algebraic Geometry and Logic. Edited by. F. W. Lawvere. VI, 189 pages. 1972. DM 18,–

Vol. 275: Séminaire Pierre Lelong (Analyse) Année 1970–1971. VI, 181 pages. 1972. DM 18,–

Vol. 276: A. Borel, Représentations de Groupes Localement Compacts. V, 98 pages. 1972. DM 16,–

Vol. 277: Séminaire Banach. Edité par C. Houzel. VII, 229 pages. 1972. DM 20,–

Vol. 278: H. Jacquet, Automorphic Forms on GL(2). Part II. XIII, 142 pages. 1972. DM 16,–

Vol. 279: R. Bott, S. Gitler and I. M. James, Lectures on Algebraic and Differential Topology. V, 174 pages. 1972. DM 18,–

Vol. 280: Conference on the Theory of Ordinary and Partial Differential Equations. Edited by W. N. Everitt and B. D. Sleeman. XV, 367 pages. 1972. DM 26,–

Vol. 281: Coherence in Categories. Edited by S. Mac Lane. VII, 235 pages. 1972. DM 20,–

Vol. 282: W. Klingenberg und P. Flaschel, Riemannsche Hilbertmannigfaltigkeiten. Periodische Geodätische. VII, 211 Seiten. 1972. DM 20,–

Vol. 283: L. Illusie, Complexe Cotangent et Déformations II. VII, 304 pages. 1972. DM 24,–

Vol. 284: P. A. Meyer, Martingales and Stochastic Integrals I. VI, 89 pages. 1972. DM 16,–

Vol. 285: P. de la Harpe, Classical Banach-Lie Algebras and Banach-Lie Groups of Operators in Hilbert Space. III, 160 pages. 1972. DM 16,–

Vol. 286: S. Murakami, On Automorphisms of Siegel Domains. V, 95 pages. 1972. DM 16,–

Vol. 287: Hyperfunctions and Pseudo-Differential Equations. Edited by H. Komatsu. VII, 529 pages. 1973. DM 36,–

Vol. 288: Groupes de Monodromie en Géométrie Algébrique. (SGA 7 I). Dirigé par A. Grothendieck. IX, 523 pages. 1972. DM 50,–

Vol. 289: B. Fuglede, Finely Harmonic Functions. III, 188. 1972. DM 18,–

Vol. 290: D. B. Zagier, Equivariant Pontrjagin Classes and Applications to Orbit Spaces. IX, 130 pages. 1972. DM 16,–

Vol. 291: P. Orlik, Seifert Manifolds. VIII, 155 pages. 1972. DM 16,–

Vol. 292: W. D. Wallis, A. P. Street and J. S. Wallis, Combinatorics: Room Squares, Sum-Free Sets, Hadamard Matrices. V, 508 pages. 1972. DM 50,–

Vol. 293: R. A. DeVore, The Approximation of Continuous Functions by Positive Linear Operators. VIII, 289 pages. 1972. DM 24,–

Vol. 294: Stability of Stochastic Dynamical Systems. Edited by R. F. Curtain. IX, 332 pages. 1972. DM 26,–

Vol. 295: C. Dellacherie, Ensembles Analytiques, Capacités, Mesures de Hausdorff. XII, 123 pages. 1972. DM 16,–

Vol. 296: Probability and Information Theory II. Edited by M. Behara, K. Krickeberg and J. Wolfowitz. V, 223 pages. 1973. DM 20,–

Vol. 297: J. Garnett, Analytic Capacity and Measure. IV, 138 pages. 1972. DM 16,–

Vol. 298: Proceedings of the Second Conference on Compact Transformation Groups. Part 1. XIII, 453 pages. 1972. DM 32,–

Vol. 299: Proceedings of the Second Conference on Compact Transformation Groups. Part 2. XIV, 327 pages. 1972. DM 26,–

Vol. 300: P. Eymard, Moyennes Invariantes et Représentations Unitaires. II. 113 pages. 1972. DM 16,–

Vol. 301: F. Pittnauer, Vorlesungen über asymptotische Reihen. VI, 186 Seiten. 1972. DM 18,–

Vol. 302: M. Demazure, Lectures on p-Divisible Groups. V, 98 pages. 1972. DM 16,–

Vol. 303: Graph Theory and Applications. Edited by Y. Alavi, D. R. Lick and A. T. White. IX, 329 pages. 1972. DM 26,–

Vol. 304: A. K. Bousfield and D. M. Kan, Homotopy Limits, Completions and Localizations. V, 348 pages. 1972. DM 26,–

Vol. 305: Théorie des Topos et Cohomologie Etale des Schémas. Tome 3. (SGA 4). Dirigé par M. Artin, A. Grothendieck et J. L. Verdier. VI, 640 pages. 1973. DM 50,–

Vol. 306: H. Luckhardt, Extensional Gödel Functional Interpretation. VI, 161 pages. 1973. DM 18,–

Vol. 307: J. L. Bretagnolle, S. D. Chatterji et P.-A. Meyer, Ecole d'été de Probabilités: Processus Stochastiques. VI, 198 pages. 1973. DM 20,–

Vol. 308: D. Knutson, λ-Rings and the Representation Theory of the Symmetric Group. IV, 203 pages. 1973. DM 20,–

Vol. 309: D. H. Sattinger, Topics in Stability and Bifurcation Theory. VI, 190 pages. 1973. DM 18,–

Vol. 310: B. Iversen, Generic Local Structure of the Morphisms in Commutative Algebra. IV, 108 pages. 1973. DM 16,–

Vol. 311: Conference on Commutative Algebra. Edited by J. W. Brewer and E. A. Rutter. VII, 251 pages. 1973. DM 22,–

Vol. 312: Symposium on Ordinary Differential Equations. Edited by W. A. Harris, Jr. and Y. Sibuya. VIII, 204 pages. 1973. DM 22,–

Vol. 313: K. Jörgens and J. Weidmann, Spectral Properties of Hamiltonian Operators. III, 140 pages. 1973. DM 16,–

Vol. 314: M. Deuring, Lectures on the Theory of Algebraic Functions of One Variable. VI, 151 pages. 1973. DM 16,–

Vol. 315: K. Bichteler, Integration Theory (with Special Attention to Vector Measures). VI, 357 pages. 1973. DM 26,–

Vol. 316: Symposium on Non-Well-Posed Problems and Logarithmic Convexity. Edited by R. J. Knops. V, 176 pages. 1973. DM 18,–

Vol. 317: Séminaire Bourbaki – vol. 1971/72. Exposés 400–417. IV, 361 pages. 1973. DM 26,–

Vol. 318: Recent Advances in Topological Dynamics. Edited by A. Beck, VIII, 285 pages. 1973. DM 24,–

Vol. 319: Conference on Group Theory. Edited by R. W. Gatterdam and K. W. Weston. V, 188 pages. 1973. DM 18,–

Vol. 320: Modular Functions of One Variable I. Edited by W. Kuyk. V, 195 pages. 1973. DM 18,–

Vol. 321: Séminaire de Probabilités VII. Edité par P. A. Meyer. VI, 322 pages. 1973. DM 26,–

Vol. 322: Nonlinear Problems in the Physical Sciences and Biology. Edited by I. Stakgold, D. D. Joseph and D. H. Sattinger. VIII, 357 pages. 1973. DM 26,–

Vol. 323: J. L. Lions, Perturbations Singulières dans les Problèmes aux Limites et en Contrôle Optimal. XII, 645 pages. 1973. DM 42,–

Vol. 324: K. Kreith, Oscillation Theory. VI, 109 pages. 1973. DM 16,–

Vol. 325: Ch.-Ch. Chou, La Transformation de Fourier Complexe et L'Equation de Convolution. IX, 137 pages. 1973. DM 16,–

Vol. 326: A. Robert, Elliptic Curves. VIII, 264 pages. 1973. DM 22,–

Vol. 327: E. Matlis, 1-Dimensional Cohen-Macaulay Rings. XII, 157 pages. 1973. DM 18,–

Vol. 328: J. R. Büchi and D. Siefkes, The Monadic Second Order Theory of All Countable Ordinals. VI, 217 pages. 1973. DM 20,–

Vol. 329: W. Trebels, Multipliers for (C, α)-Bounded Fourier Expansions in Banach Spaces and Approximation Theory. VII, 103 pages. 1973. DM 16,–

Vol. 330: Proceedings of the Second Japan-USSR Symposium on Probability Theory. Edited by G. Maruyama and Yu. V. Prokhorov. VI, 550 pages. 1973. DM 36,–

Vol. 331: Summer School on Topological Vector Spaces. Edited by L. Waelbroeck. VI, 226 pages. 1973. DM 20,–

Vol. 332: Séminaire Pierre Lelong (Analyse) Année 1971-1972. V, 131 pages. 1973. DM 16,–

Vol. 333: Numerische, insbesondere approximationstheoretische Behandlung von Funktionalgleichungen. Herausgegeben von R. Ansorge und W. Tornig. VI, 296 Seiten. 1973. DM 24,–

Vol. 334: F. Schweiger, The Metrical Theory of Jacobi-Perron Algorithm. V, 111 pages. 1973. DM 16,–

Vol. 335: H. Huck, R. Roitzsch, U. Simon, W. Vortisch, R. Walden, B. Wegner und W. Wendland, Beweismethoden der Differentialgeometrie im Großen. IX, 159 Seiten. 1973. DM 18,–

Vol. 336: L'Analyse Harmonique dans le Domaine Complexe. Edité par E. J. Akutowicz. VIII, 169 pages. 1973. DM 18,–

Vol. 337: Cambridge Summer School in Mathematical Logic. Edited by A. R. D. Mathias and H. Rogers. IX, 660 pages. 1973. DM 42,–

Vol: 338: J. Lindenstrauss and L. Tzafriri, Classical Banach Spaces. IX, 243 pages. 1973. DM 22,–

Vol. 339: G. Kempf, F. Knudsen, D. Mumford and B. Saint-Donat, Toroidal Embeddings I. VIII, 209 pages. 1973. DM 20,–

Vol. 340: Groupes de Monodromie en Géométrie Algébrique. (SGA 7 II). Par P. Deligne et N. Katz. X, 438 pages. 1973. DM 40,–

Vol. 341: Algebraic K-Theory I, Higher K-Theories. Edited by H. Bass. XV, 335 pages. 1973. DM 26,–

Vol. 342: Algebraic K-Theory II, "Classical" Algebraic K-Theory, and Connections with Arithmetic. Edited by H. Bass. XV, 527 pages. 1973. DM 36,–

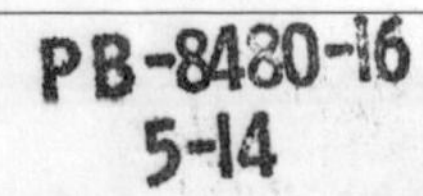